ADVANCES IN
HEAT PUMP–ASSISTED DRYING TECHNOLOGY

Advances in Drying Science & Technology

Series Editor: Arun S. Mujumdar

PUBLISHED TITLES

Advances in Heat Pump-Assisted Drying Technology
Vasile Minea

ADVANCES IN
HEAT PUMP-ASSISTED DRYING TECHNOLOGY

EDITED BY

Vasile Minea

CRC Press
Taylor & Francis Group
Boca Raton London New York

CRC Press is an imprint of the
Taylor & Francis Group, an **informa** business

CRC Press
Taylor & Francis Group
6000 Broken Sound Parkway NW, Suite 300
Boca Raton, FL 33487-2742

First issued in paperback 2019

© 2016 by Taylor & Francis Group, LLC
CRC Press is an imprint of Taylor & Francis Group, an Informa business

No claim to original U.S. Government works

ISBN-13: 978-1-4987-3499-8 (hbk)
ISBN-13: 978-0-367-87398-1 (pbk)

Visit the Taylor & Francis Web site at
http://www.taylorandfrancis.com

and the CRC Press Web site at
http://www.crcpress.com

Contents

Contents

Series Preface

It is well known that the unit operation of drying is a highly energy-intensive operation encountered in diverse industrial sectors such as agricultural processing, ceramics, chemicals, minerals processing, pulp and paper, pharmaceuticals, coal polymer, food, forest products industries, and waste management. Drying also determines the quality of the final dried products. The need to make drying technologies sustainable and cost-effective via application of modern scientific techniques is the goal of academic as well as industrial R&D activities around the world.

Drying is a truly multi- and interdisciplinary area. Over the past four decades, the scientific and technical literature on drying has seen an exponential growth. The continuously rising interest in this field is also evident from the success of numerous international conferences devoted to drying science and technology.

This new series entitled "Advances in Drying Science and Technology" is designed to provide authoritative and critical reviews and monographs focusing on current developments as well as future needs. It is expected that books in this series will be valuable to academic researchers as well as industry personnel involved in any aspect of drying and dewatering.

The series will also encompass themes and topics closely associated with drying operations, such as mechanical dewatering, energy savings in drying, environmental aspects, life cycle analysis, technoeconomics of drying, electrotechnologies, and control and safety aspects.

Arun S. Mujumdar
McGill University, Québec, Canada

Preface

This book, *Advances in Heat Pump-Assisted Drying Technology,* reviews recent innovations and system improvements proposed by academic and industry R&D communities. They focus on various aspects ranging from technological advancements in heat pumping in general to optimal dryer–heat pump coupling and control strategies, system modeling and simulation, and *in-field* long-term experiences.

Drying of solids is one of the most common, complex, and energy-intensive processes existing in many industrial sectors such as food, agricultural, pulp and paper, pharmaceutical, ceramic, and wood. It is a mix of fundamental sciences and technologies; it is also based on extensive experimental observations and operating experience.

The chapters in this book are written by recognized researchers throughout the world who are experts in various types of heat pump-assisted drying systems. It emphasizes several new design concepts and operating and control strategies that can be applied to improve the energy efficiency and environmental sustainability for drying processes.

This book is intended to serve both the practicing engineer involved in the selection and/or design of sustainable drying systems and the researcher as a reference work that covers the wide field of drying principles, various commonly used drying equipment, and aspects of drying in important industries.

The main aim of this book is to contribute to the increasing number of successful industrial implementations of heat pump-assisted dryers.

Vasile Minea
Hydro-Québec Research Institute
Québec, Canada

Acknowledgments

I gratefully acknowledge Dr. Arun S. Mujumdar for commissioning a new series of books entitled "Advances in Drying Science and Technology" as a valuable working tool for academic researchers and industry people involved in various drying technologies. Thanks to Dr. Arun S. Mujumdar, heat pump-assisted drying technique has been included in this series of books as one of the most energy-efficient and economical drying technologies.

I express many thanks for the support received from the contributing authors and for their constant efforts to enhance the scientific and technological content of different topics associated with heat pump-assisted drying of agro-food, lumber, and other valuable industrial products. The technical content of each chapter has been improved thanks to the valuable assistance, comments, and suggestions made by peer reviewers, who are internationally recognized experts in drying and industrial heat pump technologies.

I greatly appreciate the support of several publishers and companies that granted copyright permissions to use figures and tables. A special mention for Professor Dr. Hans-Jürgen Laue from IZW e. V. (Germany), who, as the operating agent of the International Energy Agency Heat Pump Technology's (IEA HPT) Annex 35/13 International Project on Industrial Heat Pumps, graciously permitted liberal use of material from the Annex's publications and final report. Also, journals such as *Drying Technology* and *International Journal of Refrigeration*, as well as publications of IEA HPT Centre (i.e., *Newsletters*), have been invaluable sources of information for this specialized book.

I also thank the publisher, CRC Press, for its sustained involvement, assistance, patience, and strong support throughout the process. Specifically, I would like to mention Allison Shatkin, Senior Editor, Books, Material Science & Chemical Engineering, and Laurie Oknowski, Project Coordinator, Editorial Project Development at Taylor & Francis Group for their courteous, professional, and helpful support, even when there were delays.

Finally, on behalf of the book's contributors, I warmly acknowledge the support and patience of their family members during the preparation and correction of the manuscripts, which sometimes continued during periods of special events such as childbirth, university exams, summer vacations, and holidays.

Series Editor

Arun S. Mujumdar is an internationally acclaimed expert in drying science and technologies. In 1978, he was the founding chair of the International Drying Symposium (IDS) series and has been the editor-in-chief of the *Drying Technology: An International Journal* since 1988. The fourth enlarged edition of his *Handbook of Industrial Drying*, published by the CRC Press, has just published. He is a recipient of numerous international awards including honorary doctorates from Lodz Technical University, Poland, and University of Lyon, France.

Please visit www.arunmujumdar.com for further details.

Editor

Vasile Minea is a PhD graduate of civil and mechanical building engineering from the Bucharest University, Romania. He worked as a professor at that university for more than 15 years, teaching various courses such as HVAC systems for civil, agricultural, and industrial buildings, as well as thermodynamics, heat transfer, and refrigeration. During this period, his R&D work focused on heat exchangers, heat pump and heat recovery systems, development and experimentation of advanced compression–absorption/resorption heat pump concepts, as well as on the usage of solar energy for comfort cooling processes and industrial cold and ice production. Since 1987, Dr. Minea has been working as a scientist researcher at the Hydro-Québec Research Institute, Canada. His research activity mainly focuses on commercial and industrial refrigeration, heat recovery and geothermal heat pump systems, low-enthalpy power generation cycles, and heat pump drying. During the past 15 years, he collaborated with the Canadian and American heat pump drying industry and R&D drying community in developing laboratory- and industrial-scale experimental prototypes. Drying of various products such as vegetables, agricultural and biological products, and wood has been theoretically and experimentally studied, and results have been published in several drying conference proceedings and in prestigious journals such as *Drying Technology* and the *International Journal of Refrigeration*.

Contributors

Lu Aye
Department of Infrastructure Engineering
Renewable Energy and Energy Efficiency
 Group
The University of Melbourne
Melbourne, Victoria, Australia

Y. Baradey
Department of Mechanical Engineering
Faculty of Engineering
International Islamic University Malaysia
Kuala Lumpur, Malaysia

Gerald Carrington
Department of Physics
University of Otago
Dunedin, New Zealand

Will Catton
Department of Physics
University of Otago
Dunedin, New Zealand

Ho Hsien Chen
Department of Food Science
National Pingtung University of Science and
 Technology
Neipu, Taiwan

M.N.A. Hawlader
Department of Mechanical Engineering
Faculty of Engineering
International Islamic University Malaysia
Kuala Lumpur, Malaysia

M. Hrairi
Department of Mechanical Engineering
Faculty of Engineering
International Islamic University Malaysia
Kuala Lumpur, Malaysia

A.F. Ismail
Department of Mechanical Engineering
Faculty of Engineering
International Islamic University Malaysia
Kuala Lumpur, Malaysia

Chung Lim Law
Department of Chemical and Environmental
 Engineering
University of Nottingham Malaysia Campus
Selangor, Malaysia

Vasile Minea
Laboratoire des technologies de l'énergie
Hydro-Québec Research Institute
Varennes, Québec, Canada

Sze Pheng Ong
Faculty of Engineering
University of Nottingham Malaysia Campus
Selangor, Malaysia

Conrad Oswald Perera
Food Science Programme
School of Chemical Sciences
The University of Auckland
Auckland, New Zealand

Zhifa Sun
Department of Physics
University of Otago
Dunedin, New Zealand

1 Advances in Industrial Heat Pump Technologies and Applications

Vasile Minea

CONTENTS

1.1 INTRODUCTION

Depletion of fossil fuels and increasing requirements for the environment protection have prompted academic and industrial R&D communities to develop and promote new, more efficient heating and cooling systems, as heat pumps recovering industrial waste heat (Srikhirin et al. 2001), combined or not with renewable energy sources, such as solar (Nguyen et al. 2001) and/or geothermal energies.

This chapter summarizes recent R&D advancements in heat pump technology and applications, including those reported for industrial drying. The review of R&D advancements refers to new components, such as compressors, working fluids, and heat exchangers, and advanced heat pumping cycles and control methods aiming at enhancing the system's overall energy performances, whereas new industrial applications mainly focus on heat pump integration with various energy sources, such as waste heat and solar energy, and industrial processes such as drying, evaporation, and distillation.

1.1.1 WORLD ENERGY CONTEXT

In 2008, the total world energy supply was 143 851 TWh (corresponding to about 15 TW of energy power), of which oil and coal combined represented over 60% (Figure 1.1a). Industrial users (agriculture, mining, manufacturing, and construction) consumed about 37%, personal and commercial transportation (20%), residential (heating, lighting, and appliances) (11%), and commercial buildings (lighting, heating, and cooling) (5%) of the total world energy supply. The rest, 32%, was lost in energy transmission and generation (IEA 2014) depending on the energy source itself, as well as the efficiency of end-use technologies.

Also in 2008, the world electricity generation was 20 181 TWh, of which more than 60% has been produced by using coal/peat and natural gas as primary energy sources (Figure 1.1b). Refrigeration, heat pump, and air conditioning industries consumed about 10%–15% of this total electric energy production (IEA 2014).

On the other hand, global CO_2 emissions came from electrical power generation (40%), industry (17%), buildings (14%), and transport (21%) energy consumptions (IEA 2014).

Such a world energy context opens up opportunities for developing alternative renewable and clean energy sources, such as solar, wind, hydrogen, water hydrokinetic, nuclear, ambient air, and geothermal. In addition, the considerable global energy use and CO_2 emissions could be reduced, especially in industry, if best available technologies were to be developed and applied worldwide (IEA 2015a).

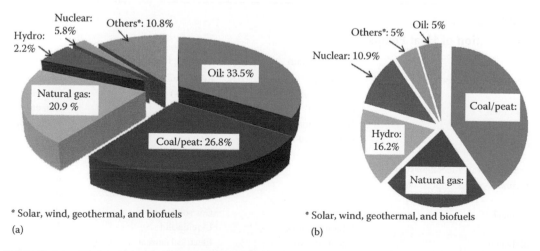

(a)

(b)

* Solar, wind, geothermal, and biofuels

FIGURE 1.1 World energy context in 2008: (a) Total energy supply and (b) total electricity generation by fuel. (From IEA, Key world energy statistics, 2014, http://www.iea.org/publications/freepublications; accessed May 15, 2015.)

Among these technologies, industrial heat pumps offer opportunities to recover and reuse waste and/or process heat for heating or preheating, or for building space and domestic hot water heating, and space cooling. They can significantly reduce primary electrical and fossil energy, and reduce power demand and greenhouse gas emissions in energy-intensive industrial processes such as drying, evaporation, distillation, and washing. However, despite such advantages, relatively few industrial heat pumps are currently installed in industry (IEA 2015a).

1.1.2 CLASSIFICATION OF HEAT PUMPS

Heat pumps are heat recovery devices that allow to recover free energy from ambient air or ground, or from industrial waste heat at relatively low temperatures and, simultaneously, to supply heat at higher temperature levels for domestic, commercial, or industrial usage.

This chapter refers only to the industrial heat pump systems listed in Table 1.1. Mechanical vapor compression heat pumps use electrical- or gas-driven compressors to compress synthetic, natural, or mixed refrigerants according to sub- or supercritical reverse Rankine-type thermodynamic cycles. Absorption, heat transformers, and compression–resorption heat pumps use two-component mixtures such as ammonia–water (NH_3/H_2O) and lithium–bromide (LiBr) as working fluids. They replace the traditional electrical- or gas-driven compressors by thermocompression processes. Mechanical vapor recompression heat pumps, which are among the most extensively applied systems in manufacturing industries, use a compressor or a high pressure blower to increase the pressure of the working fluid (generally, low pressure steam) in evaporation and/or distillation industrial processes. Thermal vapor recompression are mostly refrigeration machines without moving parts that recompress waste motive vapor from industrial boilers by using steam ejectors in order to provide cooling effects. Chemical heat pumps are systems that utilize organic or inorganic substances with relatively high thermochemical energy storage densities as well as reversible chemical reactions to upgrade the temperature of recovered thermal energy to higher temperature levels by absorbing (via endothermic reactions) and releasing heat (via exothermic reactions). Solid-state heat pumps such as thermoelectric, magnetocaloric, and thermoacoustic heat pumps are cooling or heating devices based on Peltier, magnetocaloric, and thermoacoustic effects. They eliminate conventional compressors and ozone-depleting or toxic working fluids, and generally include any moving components.

TABLE 1.1
Classification of Heat Pumps

Type	Variant	Working Fluids (Pairs)	Driven Energy
Mechanical vapor compression	Subcritical	Refrigerants (synthetic,	Electricity
	Supercritical	naturals)	Gas, oil
Absorption	Heat pump	NH_3-H_2O	Gas, oil
	Heat transformer	H_2O-LiBr	Waste heat
	Compression-resorption	New pairs	Solar
			Hybrid
Mechanical vapor recompression	Semi-open	Water vapor	Electricity
Thermal vapor recompression	Ejector	Water vapor	Steam
		Refrigerants	Waste heat
Chemical	Heat pump	Gas/solid	Waste heat
	Heat transformer	Liquid/solid	Solar
Solid-state	Thermoelectric	Electrical current	Waste heat
	Magnetic		Solar
	Thermo-acoustic		

1.1.3 INDUSTRIAL HEAT PUMPS' MARKET OUTLOOK

The number of industrial heat pumps implemented throughout the word in recent years is relatively low. This situation is attributed, among other factors, to lack of technology and process integration knowledge, low awareness of plants energy consumptions, relatively long payback periods (>5 years) and new requirements for high-pressure and temperature applications (compressors, lubricants, and refrigerants).

In Canada, among 339 questioned industrial plants (lumber drying; milk, cheese, and poultry processing; sugar refining; pulp production; and textile) in four Canadian provinces, only 7.7% use one or more industrial heat pumps for process and/or waste heat recovery. Specific barriers are related to low prices of natural gas and oil versus electricity costs and to the fact that, historically, many incentives were based on product quality and/or environmental concerns rather than energetic and/or economic (IEA 2015a). In Denmark, the industrial utilization of heat pumps is still limited today. The most important barriers are their rather low economic advantages, as well as lack of knowledge and in-field experiences (IEA 2015a). In France, even mechanical vapor closed-cycle compression (e.g., in breweries, meat processing, dairies, and lumber drying) and mechanical vapor recompression industrial heat pumps (e.g., in sugar plants) were largely used in the 1980s and 1990s; the actual market is far to be fully developed, eventhough the development potential is considered to be very high. As a specific barrier, the lack of specialized engineering companies is pointed out (IEA 2015a).

In Germany, machinery, automotive, food, and chemical industries show a high potential for low-temperature industrial heat pump applications up to 80°C. For high-temperature industrial heat pumps (i.e., up to 140°C), a huge potential has been found in food (pasteurization, sterilization, drying, and thickening), paper, textile, and chemical industries (polyethylene melting and rubber production). Natural refrigerants such as ammonia and CO_2 are frequently used as working fluids, and both electrically driven and gas-engine heat pumps are used. However, lack of documented successful applications of industrial heat pumps is noted as a specific barrier to persuade customers to implement heat pumps in Germany (IEA 2015a).

In Japan, industrial heat pumps are already adopted in greenhouse horticulture and hydroponic cultures, as well as in drying of agricultural, fishery, and lumber products, and food processing plants (washing). Heat pumps with ability of producing water at around 100°C, for coating and drying process at 120°C and steam, are under development. Mechanical recompression vapor heat pumps are increasingly used in beer factories for molt boiling and alcohol distilling processes. However, in Japan, there is still a need to develop higher efficient equipment (compressors and heat exchangers), especially for operating at temperatures over 100°C, so that heat pumps can become competitive in terms of lifetime cost with conventional heating systems (IEA 2015a). In Korea, the global heat pump market has grown rapidly in recent years, but the spread of industrial heat pump utilization still lags behind in market development. The main reason is the low price of natural gas that is one of the lowest in most OECD countries. For example, the price of electricity for domestic consumers is only 43% of that paid by UK domestic consumers (IEA 2015a). Finally, in the Netherlands, recent developments of heat pumps focused on higher heat delivery temperatures and lifts. Eight of the most representative applications still running in the Netherlands are mechanical vapor recompression, one thermal vapor recompression, and one large mechanical vapor compression heat pump. Promising markets for industrial heat pumps in this country include chemical (distillation), food (dairy), refrigeration, paper and pulp, and agriculture (drying) industries (IEA 2015a).

1.2 SUBCRITICAL MECHANICAL VAPOR COMPRESSION HEAT PUMPS

1.2.1 GENERALITIES

Most of industrial heat pumps work in the heating mode only according to the nonreversible subcritical Rankine reverse thermodynamic cycle (Figure 1.2a). The main components are compressor (reciprocating, scroll, screw, etc.), expansion device (thermostatic or electronic valve,

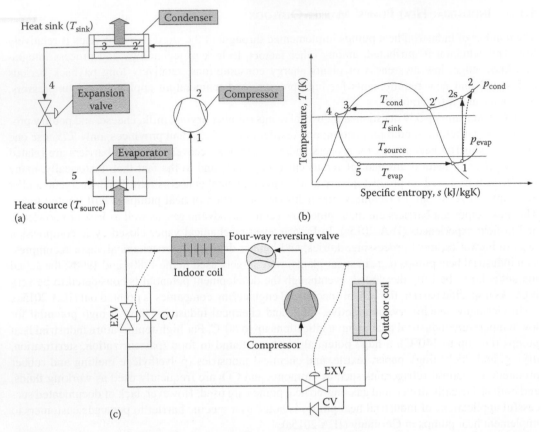

FIGURE 1.2 Schematic representation of a subcritical heat pump. (a) Nonreversible—operating in the heating mode only; (b) reversible—operating either in heating or cooling modes; and (c) thermodynamic cycle in T–s diagram. cond., condensing; CV, check valve; evap., evaporating; EXV, expansion valve; p, pressure; s, entropy; T, temperature.

orifice or capillary tube), and at least two heat exchangers (condenser and evaporator). Such a heat pump transfers heat from a low-temperature heat source to a higher temperature (warmer) heat (sink) source. To accomplish such a process, the heat pump uses the physical properties of a volatile working fluid (refrigerant), as well as some amount of external electrical (or fossil) primary energy to run the compressor and auxiliary equipment as blowers and/or fluid circulating pumps.

It can be seen that the entire subcritical thermodynamic process represented in the T–s diagram (Figure 1.2b) occurs below the critical point of the refrigerant being used. Heat absorption occurs by evaporation of the refrigerant at low pressure, and heat rejection takes place by condensing the refrigerant at a high pressure, but below that of the refrigerant critical point.

In the compressor, the refrigerant at state 1 (slightly superheated vapor) is adiabatically (theoretical process 1–2s) or polytropically (actual process 1–2) compressed up to the superheated states 2s (or 2), respectively. The electrical energy input is converted to shaft work to rise the pressure and temperature of the refrigerant. By increasing the vapor pressure, the condensing temperature is increased to a level higher than that of the heat source (T_{source}). In the condenser, the refrigerant is first desuperheated from superheated states 2s (or 2) to saturated vapor (state 2′) and then undergoes a two-phase condensation at constant temperature (T_{cond}) and pressure (p_{cond}) (process 2′–3). Before leaving the condenser, the saturated refrigerant is subcooled (process 3–4) in order to reduce the risks of flashing within the expansion valve (EXV). During all desuperheating, condensation, and subcooling processes, heat is rejected by the condenser to the heat sink medium (gas or liquid). After the condenser, the EXV expands the refrigerant at a constant enthalpy in order to reduce its pressure at a level corresponding to an evaporating temperature (T_{evap}) below the heat source temperature (T_{source}) (process 4–5). This device controls the refrigerant flow into the evaporator in order to ensure its complete evaporation and maintain a given superheat in order to avoid the liquid refrigerant to enter the compressor. However, excessive superheat may lead to overheating of the compressor. The refrigerant then enters the evaporator in a two-phase state (5), absorbs (recover) heat from the heat source thermal carrier and undergoes change from liquid-vapor to saturated vapor at constant pressure (p_{evap}) and temperature (T_{evap}). The saturated vapor is finally superheated slightly up to state 1 before entering the compressor. At this point, the cycle restarts.

By using a four-way reversible valve (Figure 1.2c), the subcritical heat pump may reverse the flow of refrigerant from the compressor through the outdoor or indoor coils in order to provide either heating or cooling, for example, to a building. In the heating mode, the outdoor coil acts as an evaporator, whereas the indoor is a condenser. The refrigerant flowing through the evaporator extracts thermal energy from outside air, water, or ground and changes its state from liquid to vapor. After compression, the refrigerant supplies heat to the indoor air or water to heat. In the cooling mode, the cycle is similar, but the outdoor coil is now the condenser and the indoor coil becomes the evaporator.

The most used subcritical nonreversible or reversible mechanical vapor compression heat pumps are the air- and ground-source heat pumps. Air-source heat pumps (see Figure 1.2a and b) extract heat from the ambient air or industrial waste gases and transfer it to building or industrial heating processes. Ground-source heat pumps extracts heat directly from the soil or from a water source (e.g., groundwater, river, lake, and sea) and transfer it to the building indoor air, to a water heating circuit (floor heating being the most efficient), or into a hot water tank for use as building and/or process hot water taps. These systems mainly use solar energy stored in shallow underground between 2 and about 200 m depth. To extract heat from the ground at very low temperatures (generally, between −5°C and 10°C during the heat pump normal operation) are used horizontal or vertical closed-loop ground heat exchangers (ASHRAE Handbook 2011). In the heating mode, an antifreeze mixture (brine) circulating, for example, through a vertical ground heat exchanger (Figure 1.3a) extracts heat from the ground (acting as a heat source), whereas the heat pump condenser, located inside the building, rejects it into the building's heating air acting as a heat sink medium. In the cooling mode (Figure 1.3b), the cycle is reversed, and the sensible and latent heat recovered from the building is rejected to the ground.

1.2.1.1 Design Outline

As could be seen from Figures 1.2 and 1.3, subcritical mechanical vapor compression heat pumps include two heat exchangers, that is, evaporator and condenser, and a compression device (compressor).

Optimum design of evaporator and condenser heat exchangers depends on their respective thermal capacities that are function of the operating temperature ranges, and on refrigerant flow rates. The theoretical thermal capacities of the evaporator and condenser (kW) can thus be calculated, for example, as functions of the refrigerant flow rate ($\dot{m}_R$, kg/s) and the refrigerant-side specific enthalpy (h, kJ/kg) changes (Figure 1.2b):

$$\dot{Q}_{evap} = \dot{m}_R \left(h_1 - h_5 \right) \tag{1.1}$$

$$\dot{Q}_{cond} = \dot{m}_R \left(h_2 - h_4 \right) \tag{1.2}$$

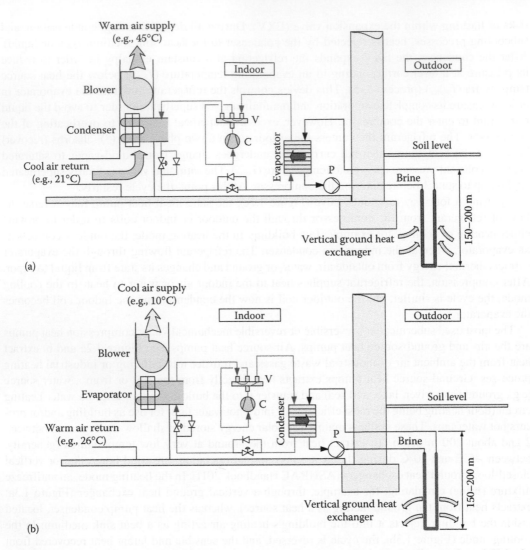

FIGURE 1.3 Schematic representation of a ground-source heat pump. (a) in the heating mode and (b) in the cooling mode. C, compressor; P, brine pump; V, 4-way reversible valve.

The isentropic efficiency of the actual compression process 1–2 versus the ideal (adiabatic) process (1–2s, where $s_1 = s_{2s}$), is defined as follows (Figure 1.2b):

$$\eta_s = \frac{h_{2s} - h_1}{h_2 - h_1} < 1 \tag{1.3}$$

Finally, the compressor theoretical electrical power input (kW) can be determined by the following energy conservation expression:

$$\dot{W}_{compr} = \dot{m}_R \left(h_2 - h_1 \right) = \dot{Q}_{cond} - \dot{Q}_{evap} \tag{1.4}$$

1.2.1.2 Performance Indicators

In the heating mode, a mechanical vapor compression heat pump based on the ideal Carnot cycle operates between two heat reservoirs having absolute temperatures T_{source} (K) (heat source) and T_{sink}

(K) (heat sink), respectively (Figure 1.2b). The heating coefficient of performance (COP) for such an ideal (Carnot) cycle is the maximum theoretical efficiency defined as follows:

$$\text{COP}_{\text{Carnot}}^{\text{heating}} = \frac{T_{\text{source}}}{T_{\text{source}} - T_{\text{sink}}} \approx \frac{T_{\text{cond}}}{T_{\text{cond}} - T_{\text{evap}}} \tag{1.5}$$

where:
T_{cond} is the condensing temperature (K)
T_{evap} is the evaporating temperature (K)

The heating COP of an actual (real) subcritical mechanical vapor compression heat pump is defined as the ratio between the condenser useful (supplied) thermal power output $(\dot{Q}_{\text{cond}})$ and the electrical power input at both compressor and blower $(\dot{W}_{\text{compr+blower}})$:

$$\text{COP}_{\text{subcritic}}^{\text{heating}} = \frac{\dot{Q}_{\text{cond}}}{\dot{W}_{\text{compr+blower}}} \approx 1 + \frac{\dot{Q}_{\text{evap}}}{\dot{W}_{\text{compr+blower}}} < \text{COP}_{\text{Carnot}}^{\text{heating}} \tag{1.6}$$

where $\dot{Q}_{\text{evap}}$ and $\dot{Q}_{\text{cond}}$ are defined by Equations 1.1 and 1.2, respectively.

In practice, the actual $\text{COP}_{\text{subcritic}}^{\text{heating}}$ of mechanical vapor compression heat pumps varies between 40% and 60% of the maximum $\text{COP}_{\text{Carnot}}^{\text{heating}}$ and depends on the difference (temperature lift) between the condensation (or heat sink) and evaporation (or heat source) temperatures: the smaller the difference, the higher the COPs. For example, air-to-air (or air-to-water) heat pumps operating in mild climates may achieve $\text{COP}_{\text{subcritic}}^{\text{heating}}$ up to 4.0, but at ambient temperatures below approximately $-8°C$, their $\text{COP}_{\text{subcritic}}^{\text{heating}}$ may drastically drop.

Similarly, in the cooling mode, the performance of a subcritical mechanical vapor compression heat pump can be described by the cooling COP:

$$\text{COP}_{\text{subcritic}}^{\text{cooling}} = \frac{\dot{Q}_{\text{evap}}}{\dot{W}_{\text{compr+blower}}} \approx \frac{\dot{Q}_{\text{evap}}}{\dot{Q}_{\text{cond}} - \dot{Q}_{\text{evap}}} < \text{COP}_{\text{Carnot}}^{\text{cooling}} \tag{1.7}$$

where

$$\text{COP}_{\text{Carnot}}^{\text{cooling}} = \frac{T_{\text{sink}}}{T_{\text{source}} - T_{\text{sink}}} \approx \frac{T_{\text{evap}}}{T_{\text{cond}} - T_{\text{evap}}} \tag{1.8}$$

represents the cooling COP of the equivalent ideal Carnot cycle.

In the United States and Canada, the heat pump's performance in the cooling mode is commonly described by instantaneous energy efficiency ratio (EER) or seasonal energy efficiency ratio (SEER), both expressed in Btu/Wh. The EER ratio is calculated by dividing the heat pump cooling capacity (expressed in Btu/h) by the compressor plus blower and circulating pumps power inputs (Watts) at a given set of rating conditions. The SEER ratio is defined as the cooling energy (Btu) provided by the heat pump during a given season (summer, year) divided by the total electrical energy consumed by the compressor plus blower and pumps during the same period of time, expressed in Wh) (ASHRAE 2013).

Note: $1\,\text{Btu/h} = 0.2931\,\text{W}$.

1.2.1.3 New Industrial Applications

An electrically driven (subcritical) mechanical vapor compression heat pump with HFC-134a as the refrigerant has been applied to recover waste heat from water chillers and air compressors of a meat processing plant in Austria (Figure 1.4). The temperature lifts varied between 30°C and 55°C in order to supply about $50\,\text{m}^3$/day of hot water for process cleaning and space heating (IEA 2015b).

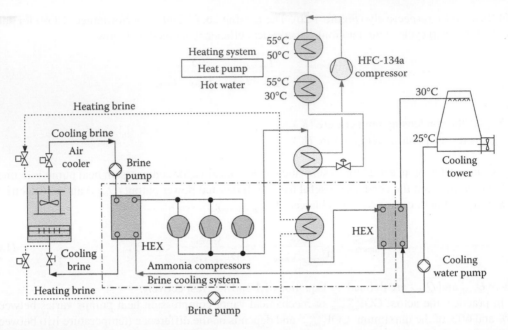

FIGURE 1.4 Process flow sheet of the subcritical mechanical vapor compression heat pump implemented in a meat processing plant in Austria. (From IEA, Industrial energy-related systems and technologies Annex 13, IEA Heat Pump Program Annex 35, Application of Industrial Heat Pumps, Final Report, Part 2, 2015b; redrawn and reprinted with permission from IEA HPT Annex 13/35 operating agent.)

Another 260-kW (thermal power supplied) (subcritical) mechanical vapor compression heat pump system has been installed in 2011 in a German metal forming and surface treatment plant (Figure 1.5). The heat pump recovers 180 kW of thermal power by cooling down from 27°C to 22°C, the cooling water required by five CO_2 laser cutting machines that run continuously. At the same time, the heat pump supplies process hot water at 65°C via one or several stratified hot water storage tanks (IEA 2015b).

Three 1.25 MW (thermal power) subcritical mechanical vapor compression heat pumps with twin-screw compressors and HFC-134a as the refrigerant have been implemented (2003) in a (tomatoes) greenhouse in the Netherlands (Figure 1.6). In the winter heating mode, the heat pump

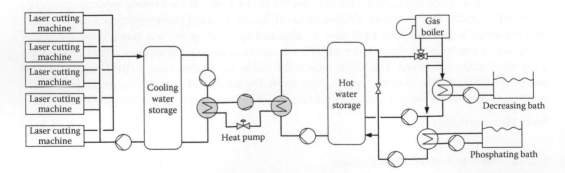

FIGURE 1.5 Subcritical mechanical vapor compression heating and cooling heat pump integrated in a metal processing plant in Germany. (From IEA, Industrial energy-related systems and technologies Annex 13, IEA Heat Pump Program Annex 35, Application of Industrial Heat Pumps, Final Report, Part 2, 2015b; redrawn and reprinted with permission from IEA HPT Annex 13/35 operating agent.)

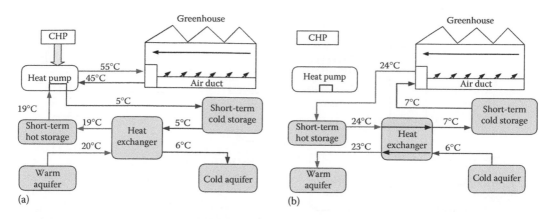

FIGURE 1.6 Mechanical vapor compression heat pump implemented in a commercial greenhouse in the Netherlands. (a) Winter heating mode and (b) summer cooling mode. CHP, combined heating and power plant. (From Raaphorst, M., Optimale teelt in de gesloten kas—Teeltkundig verslag van de gesloten kas bij Themato in 2004. http://www.hdc.org.uk/sites/default/files/research_papers/PC%20256%20final%20report%202007. pdf, 2005; IEA, Industrial energy-related systems and technologies Annex 13, IEA Heat Pump Program Annex 35, Application of Industrial Heat Pumps, Final Report, Part 2, 2015b; redrawn and reprinted with permission from IEA HPT Annex 13/35 operating agent.)

recovers heat from groundwater (240 m³/h at 20°C) and delivers hot air to the greenhouse at temperatures up to 55°C (Figure 1.6a). Backup heating is provided by combined heat and power (CHP) gas-fired engines. In the summer cooling mode, the heat pump does not work and the excess heat is rejected to the groundwater (Figure 1.6b). The reported simple payback period was relatively high (14.9 years), but the energy cost savings were estimated at 29% and the reduction of CO_2 emissions, at between 40% and 60%. In addition, the system provided better temperature and humidity interior conditions, higher CO_2 concentrations inside the greenhouses, higher crop production (about 17% of annual increase) and quality, and reduced by 80% pesticide use (IEA 2015b).

Two subcritical mechanical vapor compression heat pumps with HFC-134a as the refrigerant have been installed in Japan for heat recovery within industrial cutting and washing processes (Figure 1.7). In conventional systems, the cutting water is cooled by a chiller up to 20°C, while the washing liquid is heated by electric or hot steam boilers to around 60°C. The implemented heat pumps eliminate both water chiller and electric heaters by recovering heat (30 kW) from the cutting

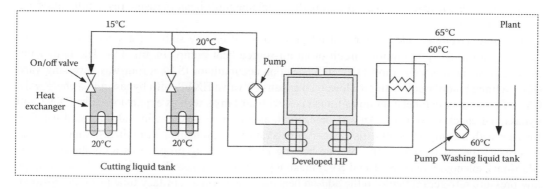

FIGURE 1.7 Subcritical mechanical vapor compression heat pump installed for industrial cutting and washing processes in Japan. (From IEA, Industrial energy-related systems and technologies Annex 13, IEA Heat Pump Program Annex 35, Application of Industrial Heat Pumps, Final Report, Part 2, 2015b; redrawn and reprinted with permission from IEA HPT Annex 13/35 operating agent.)

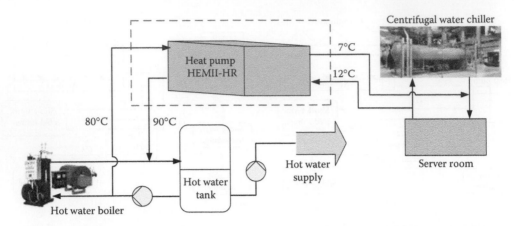

FIGURE 1.8 Subcritical mechanical vapor compression heat pump implemented in the server room of a Korean Internet Data Centre. (From IEA, Industrial energy-related systems and technologies Annex 13, IEA Heat Pump Program Annex 35, Application of Industrial Heat Pumps, Final Report, Part 2, 2015b; redrawn and reprinted with permission from IEA HPT Annex 13/35 operating agent.)

liquid to keep it at 20°C and, simultaneously, delivering heat (21.8 kW) at 60°C for heating the washing liquid. As a result, the system global COP, accounting for both simultaneous cooling and heating effects, was of about 5 (IEA 2015b).

An energy saving system using a subcritical mechanical vapor compression heat pump has been implemented to recover waste heat from water coming at 12°C from the cooling system of a server room in a Korean Internet data center (Figure 1.8). The heat pump supplies hot water at 90°C for industrial processes and cleaning (IEA 2015b).

A dual-energy source subcritical mechanical vapor compression heat pump has been developed for cold and very cold climates (Figure 1.9) (Minea 2011). In such climates, the heating performance of conventional air-source heat pumps sharply drops at outdoor temperatures below −8°C. As a result, conventional heat pumps use add-on electrical heaters installed in the duct work of forced hot air systems. They operate whenever additional heat is required and the heat pump is out of operation for significant periods of time during the winter. Moreover, when outdoor temperatures fall below −12°C, most electrical grids may achieve peak power demand loads at relatively high costs. Using fossil fuels as backup heating energy sources may raise issues such as insufficient space for incorporating the supplementary heaters inside residential buildings (Guilbeault 1987).

The new developed add-on concept integrates a fossil energy source (propane, oil, and natural gas) within the refrigeration circuit of an air-source heat pump (Figure 1.9). The indoor unit includes a variable speed compressor (C), a suction accumulator (SA), a four-way reversing valve (RV), a finned indoor coil with air blower, an expansion valve EXV2 with bypass and a check valve CV2. The outdoor cabinet contains a finned coil with air fan as well as expansion valve EXV1 with bypass and check valve CV1. An add-on cabinet, also located outdoor, includes a small gas-fired furnace with a compact combustion gas-to-refrigerant heat exchanger. It preheats, vaporizes, and superheats the refrigerant in the backup heating mode by using propane as a combustible. Two additional solenoid valves (SV1 and SV2) make it possible to bypass the outdoor coil and to supply low-pressure refrigerant liquid to the add-on heat exchanger via a capillary tube installed upstream of SV2. Finally, the check valve CV3 allows the refrigerant vapor leaving the add-on coil to bypass the four-way reversible valve and to flow, via the SA, to the compressor suction line in the backup heating mode.

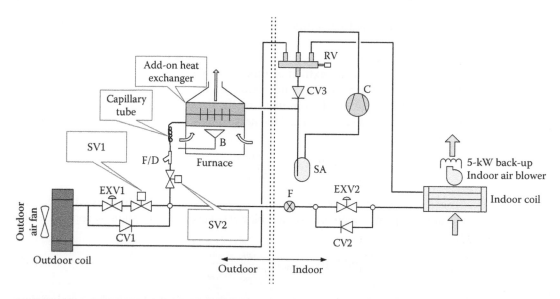

FIGURE 1.9 Schematic of the dual-energy source heat pump. B, burner; C, variable speed compressor; EXV, expansion valve; F, refrigerant flow meter; F/D, filter-drier; RV, 4-way reversible valve; SA, suction accumulator; SV, Solenoid valve. (From Minea, V., Dual-energy source heat pump, in *10th IEA Heat Pump Conference*, May 16–19, Tokyo, Japan, 2011; reprinted with permission from IEA HPP Centre.)

1.2.2 SUBCRITICAL MECHANICAL VAPOR COMPRESSION HEAT PUMP-ASSISTED DRYERS

Drying is an energy-intensive operation consuming between 9% and 25% of national energy of the developed countries (Mujumdar 1995). Such a process may consume up to 70% of the total energy in wood drying industry, 50% in the manufacturing of finished textile fabrics and over 60% from farm corn production (Mujumdar 1987). Consequently, improving the energy efficiency of drying equipment in order to reduce and/or recover a part of energy losses, mostly because of the moist air venting (representing about 85% of existing industrial dryers), is one of the most relevant objective of R&D activities throughout the world (Chua et al. 2002).

Among other methods aiming at reducing global energy consumption (electrical and fossil) in the drying industry there are heat pump-assisted dryers. These systems generally integrate air-to-air subcritical mechanical vapor compression heat pumps, acting as simultaneous dehumidifiers and heating devices, and slightly modified conventional drying enclosures (chambers). They recover sensible and latent heat by condensing moisture from a part of the hot and humid drying air and supply it back to the dryer by heating the same air stream. Such a process may accelerate the drying cycles, preserve the quality of dried products, and reduce the overall energy consumption.

1.2.2.1 Energy Efficiency

Energy performances of heat pumps used as dehumidification and heating devices in drying processes can be characterized by their COPs ($COP_{drying}^{heating}$) defined as the total heat supplied to the dryer by the heat pump's condenser (Q_{cond}) (kWh) divided by the compressor and blower electrical energy consumption ($E_{compr+blower}$) (kWh) during each drying cycle:

$$COP_{drying}^{heating} = \frac{Q_{cond}}{E_{compr+blower}} = \frac{Q_{evap} + E_{compr+blower}}{\dot{W}_{compr+blower} * \tau} \qquad (1.9)$$

where:

$\dot{W}_{compr+blower}$ is the electrical power input of the heat pump compressor and blower (kW)

τ is the drying time

Generally, to be energetically and economically acceptable, the average $\text{COP}_{\text{drying}}^{\text{heating}}$ must be higher than 4.

On the other hand, the dehumidification performance of a heat pump-assisted dryer can be described by the dryer's specific moisture extraction ratio (SMER) (expressed in $\text{kg}_{\text{water}}/\text{kWh}_{\text{input, total}}$) defined as the ratio between the amount of water extracted from the dried material (m_{water}, kg) and the total energy (electrical and fossil) consumed by the dryer ($E_{\text{input, total}}^{\text{dryer}}$, kWh), including the dryer (fans, supplementary heat, and controls) and the heat pump (compressor, blower, and controls) energy consumptions:

$$\text{SMER}_{\text{dryer}} = \frac{m_{\text{water}}}{E_{\text{input, total}}^{\text{dryer}}} \tag{1.10}$$

As in many published studies, the dehumidification performances of heat pump-assisted dryers can be related only to the heat pump's (compressor and blower) energy consumption, according to the following equation:

$$\text{SMER}_{\text{heat pump}} = \frac{m_{\text{water}}}{E_{\text{compr + blower}}^{\text{heat pump}}} \tag{1.11}$$

Such a performance parameter depends, among other parameters, on the heat pump running time and ranges between 1.5 and 4. This performance indicator must not be compared with the $\text{SMER}_{\text{dryer}}$ of conventional air convective driers that, generally, ranges between 0.12 and 0.8 $\text{kg}_{\text{water}}/\text{kWh}$).

Specific energy consumptions of heat pump-assisted dryers ($\text{SEC}_{\text{dryer}}$, $\text{kWh/kg}_{\text{water}}$) and heat pumps ($\text{SEC}_{\text{heat pump}}$, $\text{kWh/kg}_{\text{water}}$), defined as reciprocals of $\text{SMER}_{\text{dryer}}$ and $\text{SMER}_{\text{heat pump}}$, respectively, are also used in practice.

1.2.2.2 Simple Payback Period

The *simple payback period* (SPB) of a heat pump-assisted dryer can be defined as the total initial capital investment cost (TIC), (expressed, e.g., in US $) divided by the annual net energy savings (ANES) (in US $/year):

$$\text{SPB}_{\text{investment}} = \frac{\text{TIC}}{\text{ANES}} \tag{1.12}$$

The initial investment cost, very site-specific but decreasing with the dryer size, generally includes the costs of system design, procurement, drying chamber fabrication, installation and commissioning, compressor, refrigerant/lubricant, blower and fans, heat exchangers, ductwork, air filters, and controls. To increase the heat pump-assisted dryer's SPB, the total investment cost must be reduced as much as possible, namely the cost of the heat pump itself that may represent more than 25% of the system total cost. Higher $\text{SMER}_{\text{dryer}}$ or $\text{SMER}_{\text{heat pump}}$, both strongly influenced by the electricity price and the heat pump COPs, provide lower operating costs and, consequently, shorter simple payback periods. For the drying industry, acceptable $\text{SPB}_{\text{investment}}$ values are normally between 2 and 3 years. If it is assumed that the useful heat generated by the heat pump partially replaces heat from an existing boiler with thermal efficiency η_{boiler}, the simple payback period can be expressed using the following equation (Mujumdar 2008):

$$\text{SPB}_{\text{boiler}} = \frac{1}{\left[(\text{FEP}/\eta_{\text{boiler}}) - (\text{EEP}/\text{COP}_{\text{heat pump}})\right] * 8760 - \text{AMC}} \tag{1.13}$$

where:
   FEP is the fuel energy price (US $/kWh)
   EEP is the electrical energy price (US $/kWh)
   $\text{COP}_{\text{heat pump}}$ is the heat pump's coefficient of performance
   AMC is the heat pump's annual operating and maintenance costs (US $/kWh/year).

It can be observed that SPB_{boiler} is sensitive to fuel prices and heat pump's operating and maintenance costs. It may decrease at higher fuel prices and lower electricity costs.

1.2.2.3 Advantages and Limitations

The main advantage of using heat pump technology in drying operations are the energy-saving potential by recovering sensible and latent heat of evaporation from the drying air (Kiang and Jon 2006) and the ability to control drying temperature and air humidity (Chou and Chua 2006; Mujumdar 2006; Claussen et al. 2007).

Other advantages of heat pump-assisted drying are (1) a wide range of drying temperatures, typically from 20°C to 110°C; (2) excellent control of the drying environment for high-value products and reduced energy consumption; (3) drying parameters practically unaffected by ambient thermal conditions products, because drying is achieved in almost closed systems (Dincer 2003); (4) improved quality of heat-sensitive products compared to conventional hot air dryers (Prasertsan et al. 1997; Soponronnarit et al. 1998; Strommen et al. 1999; Islam and Mujumdar 2008); and (5) environmental-friendly technology because of lower energy requirement and no release of gases and fumes into the atmosphere (note that management of condensate environmental impacts seems easier) (Perera and Rahman 1997). The following are the current limitations of heat pump-assisted dryers (Islam and Mujumdar 2008): (1) higher initial capital and maintenance costs compared to conventional dryers due to additional refrigeration components; (2) more complex operation compared to simple air convective dryers; (3) requirement for auxiliary (backup) heating, especially for high-temperature drying systems operating in moderate and cold climates; (4) requirement for regular maintenance of heat pumps and air duct components (compressors, blowers, fans, and air filters); (5) risk of refrigerant leakage; and (6) additional floor space requirement.

1.2.2.4 Technological Advances

Since 1980s, under the rise of fossil energy prices and new legislation on environmental pollution, several technological advances in drying technology, as well as new opportunities for R&D and demonstration activities have been reported and/or suggested (Mujumdar 1996, 2002, 2006, 2008, 2009; Hawlader et al. 1998; Mujumdar and Wu 2008).

However, several of proposed technological improvements were rather complex dryer-heat pump integrated concepts leading to higher initial costs and more sophisticated control strategies (see Minea 2014a for more details). This section highlights some simple integrated drying concepts, as well as a number of design and operation improvements.

1.2.2.4.1 Optimization of System Integration

As represented in Figure 1.10, a heat pump-assisted dryer can operate in both *batch* and *continuous* (*tray*) (Jolly et al. 1990). However, they are more suitable for *batch* drying operations equipped with compact- (Figure 1.10a) or split-type (Figure 1.10b) heat pumps. That is because *batch* drying systems allow partial recirculation of drying air with very low leakage rates, thus achieving higher thermal efficiencies (Perera and Rahman 1997). In such dryers, the material to be dried is placed on a tray inside the drying chamber and removed once the desired moisture content (MC) is reached. The drying air can flow parallel or perpendicular to the surfaces of the dried material.

In the case of batch heat pump-assisted dryers with *compact-type* heat pumps (Figure 1.10a), the heat pump's main components (compressor, evaporator, condenser, EXV, refrigerant piping, fans, and controls) are installed outside the drying chamber in a closed cabinet. The hot and humid air from the drying enclosure enters both the evaporator EV and the condenser CD. A mixing process using motorized air dampers and constant (or variable speed) blowers is achieved prior reheating the air through the condenser CD. Backup heating coils may supplement the heat that the drying air absorbed from the condenser. The dryer fan circulates the air through the stack of dried product and its rotation direction periodically changes during the dehumidification process. The air vents open when the dryer fan changes rotation direction in

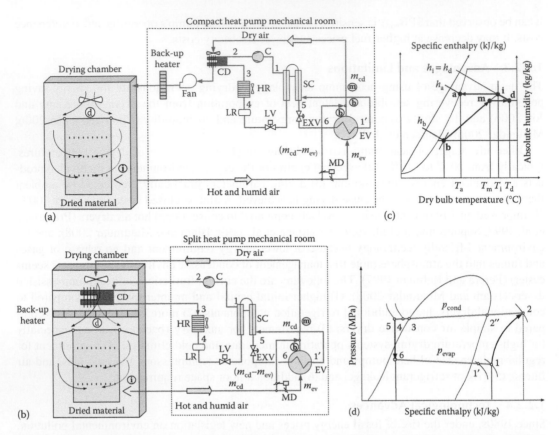

FIGURE 1.10 (a) Heat pump dryer with compact-type heat pump; (b) heat pump dryer with split-type heat pump; (c) air thermodynamic cycle in a North American-type Mollier diagram; and (d) refrigerant thermodynamic cycle in p–h diagram. C, compressor; CD, condenser; HR, air-cooled heat rejection heat exchanger; LR, liquid receiver; LV, liquid valve; EV, evaporator; EXV, expansion valve; $\dot{m}$, drying airflow rate; CD, condenser.

order to avoid air implosion hazards and, also, when the dryer dry-bulb temperature exceeds its setting point in order to avoid excessive superheating.

In the case of heat pump-assisted batch dryers with *split-type* heat pumps (Figure 1.10b), there are also two chambers thermally isolated from one another, that is, a drying enclosure where the heat pump condenser is placed, and a mechanical room where all temperature-sensitive components (compressor, evaporator, variable speed blower, EXV, and controls), as well as air ducts, are installed and, thus, thermally separated from the product drying area.

In an optimized heat pump-assisted batch dryer, a part (i.e., a variable mass flow rate $\dot{m}_{ev} < \dot{m}_{cd}$) of the hot and humid air leaving the dryer at state i and mass flow rate $\dot{m}_{cd}$ passes through the heat pump evaporator where it is first cooled (process i–a) and, then, dehumidified when condensation of moisture occurs as the air temperature goes below its dew point (process a–b) (Figure 1.10c). Within the heat pump, the working fluid (refrigerant) at low pressure and temperature is vaporized and superheated in the evaporator (process 6–1′) by heat drawn from the dryer exhaust air with mass flow rate $\dot{m}_{ev}$ (Figure 1.10d). After the mixing process around the evaporator, the air at state m and flow rate $\dot{m}_{cd}$ enters the condenser where the refrigerant condenses (process 2–3) and, thus, returns the sensible and latent heat recovered from the drying air at the evaporator, plus the equivalent heat corresponding to the compressor electrical energy input. In other words, heat is removed from the condensing refrigerant and is transferred to the drying air. At the exit of the heat pump condenser, the hot and dry air at state d is allowed to pass through an auxiliary (backup) heater and, then, through the drying chamber where it gains moisture from the product to be dried through an isenthalpic process d–i. After passing through a liquid subcooler (SC) (process $1'$–1), the refrigerant enters the compressor, which quasi-adiabatically raises the enthalpy and pressure of the refrigerant (process 1–2) and discharges it as superheated vapor (state 2). The liquid valve LV allows controlling the refrigerant migration during the compressor ON/OFF cycling and standby periods. The refrigerant liquid leaving the heat pump condenser at state 3 may be further air-subcooled (process 3–4) in a heat rejection heat exchanger as well as within the SC (process 4–5), throttled to a low pressure through the EXV (process 5–6), prior entering the evaporator to complete the cycle (see Figure 1.10d).

In continuous heat pump-assisted dryers, the product is placed on a tray positioned on a conveyor belt system of which speed can be varied. Such drying systems may involve faster loading and unloading of dried products could be less labor intensive and can potentially be a better option for drying specialty crops. However, very few studies and applications have been reported on heat pump-assisted continuous (tray) dryers compared to heat pump-assisted batch dryers (Adapa and Schoenau 2005). Although *batch* and *continuous* heat pump-assisted dryers are the most commonly used today, other types have been proposed, as heat pump-assisted fluidized beds (Strommen and Jonassen 1996), and rotary heat pump-assisted dryers.

For parallel drying chambers operating simultaneously at different drying temperature and humidity conditions, an advanced and useful concept includes two air streams coming from two separate drying chambers, each passing through two separate evaporators of the same heat pump (Figure 1.11). In this system, the subcooled refrigerant liquid is split into two streams at the exit of the heat pump condenser. One stream enters the expansion valve EXV1 of the high-temperature evaporator EV1 and the second enters the expansion valve EXV2 of the low-temperature evaporator EV2. A two-temperature valve in the suction line keeps the low-side pressure of the refrigerant in evaporator EV2 at a higher pressure than in the evaporator EV1. A check valve, located in the suction line coming from the colder evaporator EV1, prevents the warmer, higher pressure low-side vapor from entering the colder evaporator EV1 during the off cycles. The vaporized refrigerant is returned to the compressor where it is compressed and becomes high-pressure, high-temperature superheated vapor (Minea 2010b).

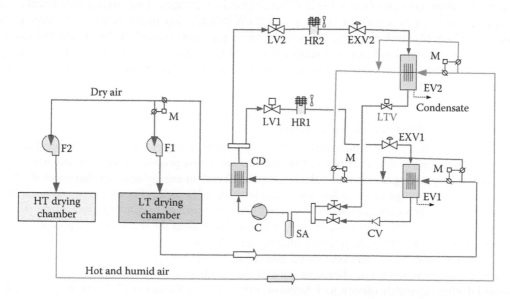

FIGURE 1.11 Batch heat pump-assisted dryer with parallel drying chambers. C, compressor; CD, condenser; CV, check valve; F, fan; LV, liquid valve; EV, evaporator; EXV, expansion valve; HR, heat rejection air-cooled heat exchanger; M, motor; SA, suction accumulator.

1.2.2.4.2 Multiple Heat Source (Hybrid) Drying Systems

Heat pump-assisted dryers may include multiple (hybrid) heating sources, such as radio frequency (i.e., electromagnetic radiation with frequencies between 300 kHz and 300 MHz) or microwave (i.e., electromagnetic radiation with frequencies from 300 MHz to 300 GHz) (Lawton 1978; Metaxas and Meredith 1983; Jia et al. 1993), infrared (i.e., electromagnetic radiation with frequencies from 300 GHz to 430 THz), and solar energy.

Materials such as ceramics, bricks, and glass fiber, which are difficult to dry with air convection heating dryers because of poor heat transfer characteristics particularly during the falling period, can be candidates for radio frequency-assisted (hybrid) heat pump-assisted dryers (Patel and Kar 2012). Such hybrid dryers comprise a mechanical vapor compression heat pump coupled with a pulsed, high-frequency radio-frequency device that generates heat within the wet material to heat it and simultaneously evaporate the water at relatively low temperatures, usually not exceeding 82°C (Thomas 1996; Marshall and Metaxas 1998). In the drying chamber, an alternating field is created between two metal-plate electrodes connected to the generator, and the product is dried by placing or passing it between these electrodes. This drying process is rapid because the heat required to dry the wet product is generated within the product itself by applying an alternating polarity of the electric field causing the rotation of dipole water molecules. This rotation causes molecular friction and, at radio-frequency drying, enough heat to produce temperatures exceeding the boiling temperature of water. Among the reported advantages of radio frequency-assisted heat pump dryers can be mentioned: (1) improved SMER$_{dryer}$ because of reduced global energy consumption (Marshall and Metaxas 1999) and (2) reduction of discoloring of products that are highly susceptible to surface color change and elimination of cracking and other defects caused by stress due to uneven shrinkage during drying (Kudra and Mujumdar 2001).

Infrared-assisted air convective heat pump dryers, where heat for drying is generated by radiation from infrared generators, can also be used for faster removal of surface moisture during the initial stages of drying, followed by intermittent irradiation coupled with convective air drying over the rest of the drying cycle. In the case of food heat-sensitive products, such a hybrid drying system can reduce the drying time (Zbicinski et al. 1992) and may ensure faster initial drying rate because of the direct transfer of energy from the infrared heating element to the product surface without heating the surrounding air (Jones 1992). Other reported advantages of infrared-assisted heat pump dryers were (1) high heat transfer rates (up to 100 kW/m^2); (2) easy to direct the heat source to the product surface; (3) quick response times allowing easy process control of the drying process; (4) system compactness, simplicity and relatively low cost; and (5) possibility of significant energy savings and enhanced product quality because of reduced residence time in the dryer chamber (Paakkonen et al. 1999).

1.2.2.4.3 Solar-Assisted Heat Pump Dryers

In most of the tropical and subtropical countries, sun-only drying is the common method used to preserve agricultural, such as fruits and vegetables, and forest products (Basunia and Abe 2001). Even if sun-only drying uses a renewable, clean, and abundant energy source (Sharma et al. 2009) and involves any energy costs; it has some disadvantages: (1) variability of the intensity incident radiation; (2) product degradation, spoilage, and losses because of rain, wind, dust, birds, insects, and animals; and (3) labor-intensive process requiring large areas for spreading the products to dry (Diamante and Munro 1993; Imre 2006; Sharma et al. 2009).

Combining solar energy and heat pump-assisted dryer technology is generally considered as an attractive concept to reduce or eliminate some of the disadvantages of using solar and heat pump-assisted drying separately (Sporn and Ambrose 1955: cited by Fadhel et al. 2011). A typical solar-assisted heat pump dryer mainly comprises a mechanical vapor compression heat pump (evaporator, condenser, compressor, and EXV) and a solar collector (Imre et al. 1982; Chaturvedi and Shen 1984; Morrison 1994; Kuang et al. 2003; Imre 2006). The solar collector can act as an additional

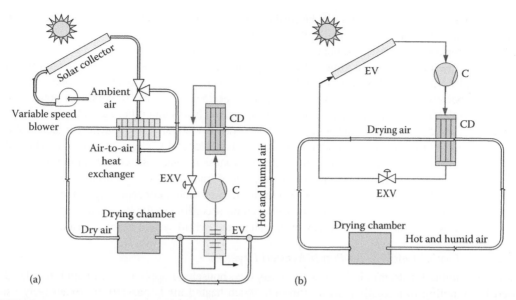

FIGURE 1.12 Schematics of conventional solar-assisted heat pump dryers. (a) With indirect solar collector and (b) with direct solar collector (evaporator). C, compressor; CD, condenser; EV, evaporator; EXV, expansion valve.

air heater allowing further increase in the air drying temperature prior to entering the drying chamber and, thus improving the overall energy efficiency of the drying system (Figure 1.12a) or as the heat pump's direct expansion evaporator (Figure 1.12b). In the solar-assisted heat pump dryer with indirect solar collector (Figure 1.12a), the solar radiation heats the air passing through the solar collector. This heated air then serves to further heat, via an air-to-air heat exchanger, the drying air leaving the heat pump condenser, prior to being rejected outside. On the other hand, the hot and humid air leaving the drying chamber passes over the heat pump evaporator where it is cooled and dehumidified. Both sensible and latent heats are absorbed by the evaporating refrigerant and the resulted vapor is compressed by the compressor.

The high-pressure superheated refrigerant vapor condenses inside the condenser by transferring the recovered heat plus the compressor equivalent electrical energy consumed to the drying air. After being further heated in this heat exchanger, the hot and dry air passes through the drying chamber and picks up moisture from the dried product. Such a combined system offers the flexibility of operating with the solar collector only, with the heat pump only or, simultaneously, with both systems. Such indirect solar-assisted heat pump dryers may provide advantages as energy savings and higher operating drying temperatures compared to stand-alone heat pump drying systems (Kiang and Jon 2006). However, these systems have higher capital costs required for additional solar panels, blowers, heat exchangers, and controls, whereas the amount of available solar energy varies significantly throughout the day and/or the year (Chou and Chua 2006).

Experimental research works on solar-assisted heat pump dryer with indirect solar collectors have been conducted in different climate regions (Cervantes and Torres-Reyes 2002; Kuang et al. 2003) in order to dry peanuts (Auer 1980; Baker 1995), rice (Best et al. 1994), and other similar products (Daghigh et al. 2010). The solar-assisted dryer system used to dry peanuts was equipped with a mechanical vapor compression heat pump and a solar collector, installed on top of the drying chamber and connected to a closed water loop with storage tank (Auer 1980; Baker 1995); part of the hot and moist air leaving the dryer flows through the heat pump evaporator where it is cooled and dehumidified. The sensible and latent heat recovered plus the equivalent heat of the compressor power input were used to heat the cold water intake within the condenser. The hot water produced is stored in a storage tank where it can be further heated with solar energy or used as a heat source to preheat the inlet ambient air (Auer 1980; Imre et al. 1982; Imre 2006).

Direct expansion solar-assisted heat pump dryers (see Figure 1.12b) consist of a thermal solar or photovoltaic/thermal collector acting as the heat pump's evaporator, and a heat pump (compressor, condenser, EXV, etc.). In this case, the refrigerant is directly vaporized inside the solar collector–evaporator because of the solar energy input. Such an experimental 1.5 kW (compressor power input) prototype with HFC-134a as the refrigerant for drying green beans and other agriculture products at 45°C, 50°C, and 55°C has been developed and extensively investigated under different meteorological conditions of Singapore by Hawlader et al. (2003), Hawlader and Jahangeer (2006), and Hawlader et al. (2008). According to Kara et al. (2008), direct expansion solar-assisted heat pump dryers may have advantages such as the elimination of evaporators of traditional heat pump-assisted dryers and higher thermal efficiency of solar collectors. However, even such a concept may reduce the system initial costs, the choice of the working fluid (refrigerant) according to the local ambient temperature conditions, as well as the control strategy of the integrated hybrid system are very critical issues.

1.2.2.4.4   Cloth/Laundry Heat Pump-Assisted Dryers

Domestic tumble (or drum) cloth/laundry dryers (i.e., household appliances in which textiles are dried by tumbling in a rotating drum, through which heated air is passed) are extensively used throughout the world (Werle et al. 2011).

There are two conventional types of dryers aiming at removing the water from dried clothes/laundry: (1) air-vented and (2) condensing dryers (Nipkow and Bush 2009, 2010; Werle et al. 2011). Air-vented dryers are open systems where fresh air is drawn from ambient, heated up by electric resistances, passed through a drum where it absorbs water evaporated from the wet clothes and, finally, exhausted as hot and moist air into the room or, through a vent, to the outdoor. Ventless (condensing) cloth/laundry dryers, more suited for cold or moderate climate, are closed systems that remove moisture from the drying air by the aid of a heat exchanger where the colder room air helps cooling down and condensing a part of the moisture removed from the dried clothes. According to Bansal et al. (2001), the energy efficiency of condensing clothes dryers is about 7% higher compared to that of conventional air-vented drum dryers. However, at room temperatures above 30°C, the condensing efficiency of the heat exchanger declines and, because these dryers require very high water consumption, they are no longer sold on the market (Nipkow and Bush 2009). Finally, cloth/laundry heat pump-assisted dryers combine condensing dryers with sub- or supercritical mechanical vapor compression heat pumps, making them a highly efficient alternative to conventional dryers (Nipkow and Bush 2009, 2010; Werle et al. 2011). In such systems, warm and humid airflows out from the laundry drum through the heat pump evaporator, where it is cooled and then dehumidified, prior being reheated into the condenser and returned back to the drum. However, in steady-state operation, the heat pump may deliver more heat than recovered, so a part of heat must be rejected outside (Meyers et al. 2010). Some manufacturers in Europe and Japan produce such devices, but none of them is available on the North American market yet (Meyers et al. 2010).

Reported advantages of heat pump-assisted clothes dryers (Nipkow and Bush 2009, 2010; Werle et al. 2011) are (1) the best heat pump-assisted dryers consume about 0.23 kWh/kg of laundry, whereas conventional clothes dryers consume at least 0.4 kWh/kg of laundry, both at 60% initial moisture; (2) as a consequence, up to 50% less electrical energy may be consumed compared to conventional electric resistance air-vented clothes dryer (Meyers et al. 2010; Werle et al. 2011); (3) rejected heat in the operation room is about 50% lower, which is advantageous in the summer at temperatures above 30°C; (4) no smelling and steaming exhaust air is rejected as with conventional air-vented dryers; and (5) in the winter, there is no cooling down of the operating room as caused by compensation air for vented dryers. The main market obstacle of heat pump-assisted clothes dryers is that they contain additional, more expensive components compared to conventional dryers, which causes higher manufacturing costs and, therefore, higher purchase prices (Nipkow and Bush 2009, 2010).

In the future, in order to increase its overall efficiency, additional R&D efforts must be provided to optimize the efficiency of heat pump's evaporator (which may claim more space than usually available models), use more efficient compressor, fan, and drums. An alternative technology would be to use an air cycle heat pump based on the reversed Brayton cycle. Such a cycle uses recirculated air as the working fluid and removes sensible and latent heat from the air leaving the dryer drum. The recovered heat plus the equivalent heat of the compression electrical energy heats the air stream reentering the dryer drum. The air cycle heat pump-assisted dryer prototypes developed and tested by Takushima et al. (2000) provided moisture removal rates of up to about six times higher compared to those of conventional dryers. A conventional air cycle tumbler clothes dryer has been coupled with a heat pump based on the reversed Brayton cycle with expander as expansion work recovery device (Braun et al. 2002). According to this author, such a system should operate at atmospheric pressure and temperatures higher than 0°C, and the temperature of the air entering and leaving the drum should be less than 130°C and 80°C, respectively. In addition, for acceptable energy performances with a drum power input of 270 W, the rate of steady-state moisture evaporation from clothes should be 3.5 kg/h, the process airflow rate entering the drum should be lower than 200 kg/h, and the design room air conditions must vary around 20°C at 60% relative humidity. The reported results showed performance improvements up to a 40% over conventional air-vented electric dryers (Braun et al. 2002).

1.2.2.4.5   Heat Pump Sizing Rules

Dryers equipped with medium- or high-temperature split-type heat pump, as represented in Figure 1.10b, are subject to particular design and control requirements because of the following factors: (1) as with any other heat pump-assisted dryer, the heat pump dehumidification capacity has to be matched with the product thermal dewatering rate; (2) the refrigerant thermophysical properties may require specific flow control strategies, as for the evaporator superheating amount; (3) remote condensers are installed inside the drying chamber and, consequently, potentially exposed to sudden thermal shocks if not adequately protected; and (4) relatively high temperature differences exist between the dryer chamber and the heat pump mechanical room.

If all or even a part of such design and control requirements are not met in practice, frequent compressor cycling can occur, as for example, when the thermal dewatering rate of the dried product and the heat pump dehumidification capacity, are unbalanced (Minea 2010a, b) for more details.

Sizing the heat pump's nominal capacity for drying purposes implies a good comprehension of the typical material drying curve, that is, the profile of MC variation during the entire drying cycle. Over- or undersized heat pumps may drastically penalize both the energy and drying performance of heat pump-assisted dryers. To achieve as high as possible $COP_{heat\,pump}$ and $SMER_{heat\,pump}$, the dried material must give off enough moisture to be able to provide sufficiently high absolute humidity to the drying air at the heat pump evaporator inlet. This will ensure enough moisture condensing enthalpy (latent heat) to be removed. Such a challenge could be overcome if the nominal heat pump's dehumidification capacity is sized as low as possible in order to match the quantity of moisture that can be removed from the product stack. Such an approach may help keep the heat pump compressor in continuous running mode while avoiding excess heat rejection outdoor. Figure 1.13 suggests the optimum location (not shown to scale) of the heat pump nominal design point for batch-type dryers. This point must be determined for each dried materials (agro-food, wood, etc.), and, among the parameters previously mentioned, according to the material initial and final MCs (Minea 2015).

More precisely, in the case of wood drying, for example, the heat pump nominal design point must be at, or just above the wood fiber saturation point (FSP), defined as "the point in dried softwood at which all free moisture has been removed from the cells themselves while the cell walls remain saturated with absorbed moisture" (Cech and Pfaff 2000) (Figure 1.13a). At MCs above FSP,

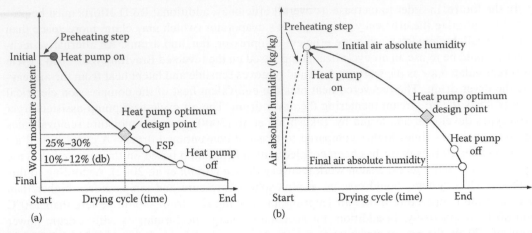

FIGURE 1.13 Relative position of heat pump nominal sizing (design) point versus the drying curves as a function of (a) wood's actual MC and (b) air drying actual absolute humidity. FSP, fiber saturation point. (Note: point positions are not to scale.) (From Minea, V., High-temperature heat pump-assisted softwood dryer: Sizing and control requirements & energy performances, in *24th International Congress of Refrigeration (ICR2015)*, August 16–22, Yokohama, Japan, 2015; reprinted with permission from International Refrigeration Institute, IIR/IIF, Paris, France.)

the heat pump $COP_{heat pump}$ and $SMER_{heat pump}$ could be relatively high, because the absolute humidity of the drying air can be kept high enough, and there is practically no change in volume (shrinkage or swelling), while MC still decreases. However, as the FSP approaches, surface moisture diffusion becomes the limiting factor of the moisture removal rate. As a consequence, the $COP_{heat pump}$ and $SMER_{heat pump}$ decrease and, at the FSP or at a certain absolute humidity of the drying air, the heat pump must be shut down in order to save energy and preserve wood quality. To prevent frequent heat pump shut downs, the dehumidification rate must be progressively decreased using different methods in order to match the material thermal dewatering rate. On the other hand, Figure 1.13b suggests the relative location of the heat pump optimum design point (also, not shown to scale) versus the actual absolute humidity of the drying air. When the absolute humidity of the drying air drops below a certain value, generally corresponding to wood's FSP, the evaporator recovers less heat and the heat pump $COP_{heat pump}$ and $SMER_{heat pump}$ decrease, even if the compressor consumes less electrical energy. It can be concluded that if sized at an optimum nominal capacity, the heat pump may operate within a certain dehumidifying rate range (max, min) corresponding to the product MCs and/or the drying air absolute humidity, each varying, like the heat pump dehumidification rate, between their maximum and minimum values.

1.2.2.4.6 Matching Thermal Capacities

During drying cycles, the heat pump dehumidification capacity has to be continuously adapted to the material's actual thermal dewatering process. Methods such as evaporator air bypass and variable velocity are common in refrigeration industry, but often they are forgotten in some heat pump-assisted drying studies.

The heat pump's main purpose is to recover as much sensible and latent heat as possible from the drying air. This recovered heat partially helps to overcome the water retention forces inside the dried material and the dryer heat losses and, also, to remove moisture from the product surfaces. In order to achieve thermally balanced, high-performance heat pump-assisted drying systems, without excessively cycling the compressor, the moisture removal rate and the heat pump dehumidification capacity must be, as far as possible, continuously matched. This difficult but fundamental task may be achieved by using the following design and operating measures: (1) size the heat pump nominal dehumidification rate according to the dryer capacity (volume) and the product initial and final MCs, as suggested in Figure 1.13; an insufficient quantity of material to be dried means small quantities of moisture and heat to be extracted, excessively low evaporating and compressor suction pressures and temperatures with associated risks of humidity freezing on the evaporator finned tubes, and short compressor running cycles; conversely, if the heat pump dehumidification capacity is oversized, the equivalent heat corresponding to the electrical energy consumed (compressor and blower) will progressively build up inside the adiabatic drying enclosure and thus must be periodically rejected outdoor by means of external parallel air- or water-cooled condensers, or refrigerant SCs; such heat rejection must be avoided in order to optimize the heat pump-assisted dryer energy performance and (2) continuously match the heat pump dehumidification capacity with the product thermal dewatering rate, for example, by using evaporator air bypassing, variable speed blowers and/or compressors, multiple-rack compressors (Minea 2010a, b), or intermittent drying (Chua et al. 2002).

The use of a variable airflow rate (velocity) through the evaporator by evaporator bypassing (with motorized air dampers) or by using variable speed air blowers (as in the case of split-type heat pumps) can be justified by the fact that at the beginning of each drying cycle, the material to be dried is able to supply a lot of moisture providing high absolute humidity to the drying air but, toward the end of the drying cycle the amount of removed moisture decreases. In the case of bypass dampers, a temperature sensor controls both primary and motorized bypass dampers, so that the temperature of the air leaving the evaporator is (approximately) the same as the temperature of the refrigerant leaving the evaporator. It is however a relatively imprecise method for industrial applications because of both low temperature reading accuracy and low damper reliability. According to one of expressions shown in Equation 1.14, if, at the beginning of each drying cycle, the airflow rate (i.e., velocity) through the evaporator finned tubes is low, it will remove a higher quantity of moisture providing to the evaporator a dehumidification (cooling) capacity close to its maximum designed rate ($\dot{Q}_{evap}^{max}$):

$$\Delta\omega_{air} \approx \frac{K}{\dot{m}_{air}}; \quad \Delta\omega_{air} \approx \frac{K}{\dot{v}_{air}} \tag{1.14}$$

where:

$\Delta\omega_{air}$ is the variation of air absolute humidity through the evaporator ($kg_{water}/kg_{dry\,air}$)
$\dot{m}_{air}$ is the mass flow rate of the drying air through the evaporator (kg/s)
K is a constant function of the evaporator thermal efficiency and heat transfer area, air density (considered constant), and water-specific condensing enthalpy

However, toward the end of the drying cycle, the MC of the dried product, as well as the absolute humidity of the air at the evaporator inlet and $\Delta\omega_{air}$ through the evaporator heat exchange surface, decrease progressively. Consequently, the drying airflow rate (i.e., velocity) through the evaporator has to be progressively increased in order to lower the heat pump dehumidification (cooling) capacity and, thus, match it as much as possible with the product thermal dewatering rate. In this way, the heat pump dehumidification (cooling) capacity will remain within its designed range, and the compressor will quasi-continuously run:

$$\dot{Q}_{evap} \in \left(\dot{Q}_{evap}^{max}, \dot{Q}_{evap}^{min}\right) \tag{1.15}$$

Intermittent drying can also help match the heat pump dehumidification capacity with the material thermal dewatering rate and, thus, avoid any shrinkage eventually caused by drying stresses. This technique allowing matching the heat pump dehumidification capacity with the material dewatering rate is well adapted to capillary porous materials such as woods and ceramics, because their drying processes are often accompanied by a cracking phenomenon caused by the drying induced stresses (Kowalski and Pawlovski 2008). The intermittency ratio is defined using the following equation:

$$\alpha = \frac{\tau_{ON}}{\tau_{ON} + \tau_{OFF}} \qquad (1.16)$$

where τ_{ON} and τ_{OFF} are the *on* and *off* durations of the heat pump running periods, respectively.

During tempering periods, the moisture migrates from the inner core of the dried product and is redistributed on the surface. Consequently, improved drying rates are obtained when the heat pump restarts. Moreover, thermal energy consumption and operating cost savings are apparently possible due to shorter effective drying times and the absence of energy consumption during the tempering periods. However, the *intermittent* heat supply time period must be carefully determined as, sometimes, long tempering periods may increase the product overall drying time (Chua et al. 2002).

1.2.2.4.7 Evaporator Superheating Temperature Range

The amount of refrigerant superheating required for preventing the liquid from entering the compressor suction line is controlled by the EXV based on the saturated evaporating pressure and temperature, and the compressor actual suction temperature. For most common refrigerants (Figure 1.14a), suction superheating of about 5°C is sufficient to prevent moisture from entering the compressor. However, with some high-temperature refrigerants, the isentropic curves can be more vertically oriented compared to those of common refrigerants (e.g., HFC-245fa) (Figure 1.14b), and thus higher amounts of suction superheating are required to prevent the isentropic compression curve from crossing the vapor saturation line. For the HFC-245fa high-temperature refrigerant, it was found that the optimum superheating range is from 15°C to 20°C at any saturated suction pressure (Minea 2010a, b).

1.2.2.4.8 Wet-Bulb Temperature Setting

For a more efficient drying process, the set point of wet-bulb temperature (WBT) of the drying air can be automatically modified during the drying cycle, for example, according to the following equation:

$$WBT_{set}^{i} = WBT_{set}^{i-1} - 2.5°C \qquad (1.17)$$

where i is the current WBT set value. The control strategy could be described as follows: the heat pump compressor starts at the end of the preheating step when the WBT of the air inside the drying chamber reaches the first set value. Because the heat pump's compressor is now running, the difference between the actual air WBT and the first preset WBT gradually decreases over time. When this

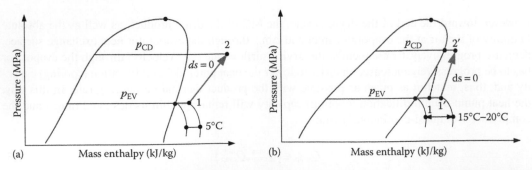

FIGURE 1.14 Comparison of saturated curves and adiabatic compression processes in p–h diagrams of (a) most conventional refrigerants and (b) as for HFC-245fa (for legend, see Figure 1.10). (From Minea, V., High-temperature heat pump-assisted softwood dryer: Sizing and control requirements & energy performances, in *24th International Congress of Refrigeration (ICR2015)*, August 16–22, Yokohama, Japan, 2015; reprinted with permission from International Refrigeration Institute, IIR/IIF, Paris, France.)

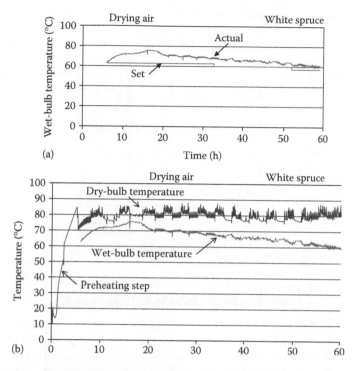

FIGURE 1.15 Example of an optimized wood drying process. (a) Profiles of actual and set wet-bulb temperatures and (b) profile of dry- and wet-bulb temperatures. (Note: the difference between these two temperatures is known as the process temperature depression.) (From Minea, V., High-temperature heat pump-assisted softwood dryer: Sizing and control requirements & energy performances, in *24th International Congress of Refrigeration (ICR2015)*, August 16–22, Yokohama, Japan, 2015; reprinted with permission from International Refrigeration Institute, IIR/IIF, Paris, France.)

difference reaches, for example, 5.5°C, the system process controller may automatically lower the first preset air WBT, for instance, by 2.5°C. This operation will continue up to the end of the drying cycle when the compressors stop (Minea 2010a, b, 2015). Such a strategy aims at setting the drying air WBT according to the actual measured values (as can be seen in Figure 1.15a for wood drying), coordinated with the drying air temperature depression, defined as the difference between the air drying dry-bulb temperature and WBT (Figure 1.15b). In this way, the heat pump will run quasi-continuously, thus improving its long-term reliability and performances.

1.2.2.4.9 Control of Refrigerant Migration

In the case of medium- and high-temperature heat pump-assisted driers, significant temperature differences (up to 100°C) may exist between the drying enclosure and the heat pumps mechanical room. This particular situation requires appropriate management of refrigeration migration, particularly during each compressor starting sequence in large-scale industrial dryers. It is well known that during the preheating step, the heat pump is not running and the refrigerant naturally migrates toward the *colder* compressors and SAs. When the compressor starts, it will first try to pump the stored incompressible liquid refrigerant and, therefore, automatically will shut down. Several such on/off cycles may occur and contribute to reducing the useful life of reciprocating compressors and/or rapidly damaging them. In the case of relatively small-scale heat pump dryers, to manage this process, a known method consists in installing a liquid valve LV between the liquid receiver LR and the EXV (Figure 1.16a) in order to achieve simple pump-down sequences. However, in the case of large-scale high-temperature split-type heat pumps, such pump-down sequences may not be entirely safe. In order to reinforce the control of refrigerant migration, a modified SA (MSA) may

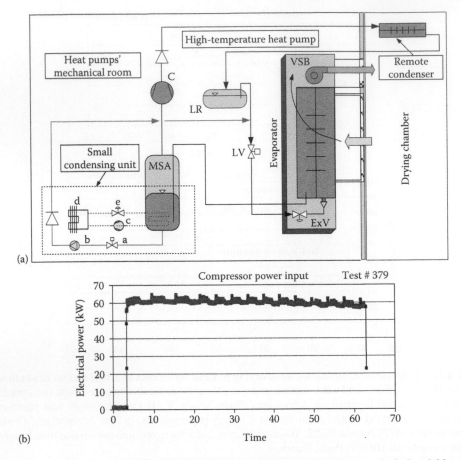

(a)

(b)

FIGURE 1.16 (a) Means of managing liquid migration in large- and very large-scale industrial heat pump-assisted dryers. (From Minea, V., High-temperature heat pump-assisted softwood dryer: Sizing and control requirements & energy performances, in *24th International Congress of Refrigeration*, August 16–22, Yokohama, Japan, 2015; reprinted with permission from International Refrigeration Institute, IIR/IIF, Paris, France.) (b) Example of an optimized running profile of the compressor of an industrial large-scale heat pump-assisted dryer. (From Minea, V., Improvements of high-temperature drying heat pumps, *International Journal of Refrigeration*, 33(1), 180–195, 2010a; Minea, V., Industrial drying heat pumps, *Refrigeration: Theory, Technology and Applications*, Nova Science Publishers, New York, pp. 1–70, 2010b.) a, Solenoid valve; b, refrigerant liquid pump; c, small compressor; d, small air-cooled condenser; e, expansion valve; C, hea pump compressor; LR, heat pump liquid receiver; LV, heat pump liquid valve; VSB, heat pump variable speed blower.

be coupled to a small air-cooled condensing unit, as shown in Figure 1.16a. A few minutes before the first compressor start-up, the condensing unit starts and cools the liquid stored inside MSA. As a consequence, the liquid eventually stored inside the compressor crankcase naturally migrates toward the colder MSA. Then, the SV (a) opens, the liquid pump (b) starts-up and the liquid is pumped toward the heat pump liquid receiver LR. After having pumped the whole quantity of liquid, the SV (a) closes, the liquid pump (b) shuts down, and the heat pump compressor C starts-up safely. Figure 1.16b represents an improved, continuous compressor running profile achieved by preventing the refrigerant liquid from entering the compression suction port (Minea 2015).

1.2.2.4.10 Avoiding Thermal Shocks
Dehumidification drying systems using heat pumps need smaller quantities of supplementary (backup) energy compared to most conventional systems. The reason is that the heat pumps recover

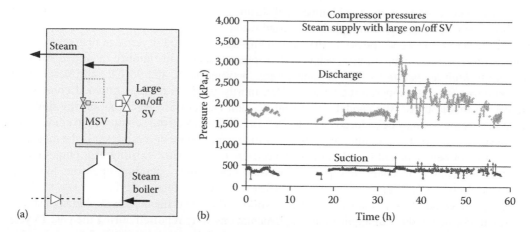

FIGURE 1.17 (a) Backup steam boiler and distribution header with large size solenoid valve (SV) for steam supply during the preheating step and small size modulating SV (MSV) for continuous steam supply during the heat pump running. (b) Abnormal compressor running profile with intermittent backup steam supply using the large on/off steam valve SV. (From Minea, V., High-temperature heat pump-assisted softwood dryer: Sizing and control requirements & energy performances, in *24th International Congress of Refrigeration (ICR2015)*, August 16–22, Yokohama, Japan, 2015; reprinted with permission from International Refrigeration Institute, IIR/IIF, Paris, France.)

sensible and latent heat from the dryer warm, humid air, and returns it to the dryer chamber, along with the equivalent electrical energy consumed by their compressors and blowers. Some additional heat is however needed to offset dryer heat losses and air leakages, and other uncontrollable energy losses, especially in cold climates.

During the preheating step, the steam-to-air heating coils installed inside the drying enclosure are supplied with high-pressure steam via a large on/off SV that fully (100%) opens (Figure 1.17a). After the heat pump start-up, the quantity of additional heat required is much lower compared to that of the preheating step. Therefore, a small, high precision modulating valve, activated by a pressure sensor, must be installed in parallel with the main SV. This valve will supply small but constant steam flow rates in order to prevent sudden thermal shocks at the remote condenser air inlet and, thus, ensure discharge pressure and temperature within normal, safe operating ranges. As can be seen in Figure 1.17b (Note: compressor discharge pressure profile is represented in light gray), if the large on/off SV is used to intermittently supply additional (backup) heating steam, excessively high compressor discharge pressures can be achieved (Minea 2015).

1.2.2.5 Industrial Applications of Heat Pump-Assisted Drying

Wood (timber and lumber), as solid, porous and hygroscopic material, is an interesting material for drying with heat pumps. In many countries, such as New Zealand, Canada, Poland, northern European countries, and the United States, wood drying, a complex and energy-intensive process, is one of the largest markets for heat pump dryers. The fundamental reason for drying wood (such as softwoods and hardwoods) is to provide useful products by enhancing their properties in order to minimize quality losses, conserve natural resources and, at the same time, make economic profits. Among other advantages of dried wood (lumber), the following can be noted: (1) efficient utilization of recovered sensible and latent heat; (2) controlled drying rates result in drying low energy costs; (3) equivalent or even reduced drying time; (4) enhanced sawmill productivity; (5) improved wood quality; (6) dried lumber at less than 20% MCs has no risk of developing stain, decay, or mold as a result of fungal activity; (7) dried lumber is typically more than twice as strong and nearly twice as stiff as wet wood; (8) dried lumber weighs 40 to 50% less than wet, undried lumber; (9) products in service made from properly dried lumber will shrink very little; and (10) gluing, machining, and finishing are much easier to accomplish with dried wood.

Depending on the wood species to dry, heat pump-assisted wood dryers can operate at (1) *low temperatures* (i.e., at temperatures below 54°C), often loaded with lumber of mixed MCs; (2) *medium to high-temperatures* (i.e., at temperatures between 82°C and 93°C); and (3) *very high-temperatures* (i.e., at temperatures up to 110°C), mostly used for drying softwood construction lumber; in this last case, the capital investment per unit capacity is higher, but drying times and energy consumption may be reduced by up to 25% and 50%, respectively, compared to conventional air convective dryers. For medium, high, and very high temperature dryers, steam is the most common source of heat supplied by oil-burned or electrical boilers. Most existing heat pump-assisted wood dryers contain mechanical vapor compression heat pumps operating upon the subcritical cycle (see Section 1.2). Even they can provide efficient and cost-effective drying of low-, medium-, or high-grade wood species, relatively few R&D works and/or experimental research/demonstration studies and/or new industrial applications/implementations were reported during the past 15 years (Minea 2014a, b). On the other hand, in the past, the reliability and energy performance of heat pump-assisted wood dryers using fossil fuels as backup energy (e.g., bark, natural gas, propane, or oil) were often disappointing, because of, among other factors, relatively inadequate heat pump sizing and control strategies.

Carrington et al. (1995) analyzed the operating characteristics of a medium-temperature heat pump-assisted drying system using HFC-134a as a refrigerant with *passive evaporator-economized and staged liquid subcooling*. The authors reported that the heat pump maximum dehumidification performance (SMER) could reach $5.11 \, kg_{water}/kg$ at 50°C dry-bulb temperature and 90% relative humidity.

Bannister et al. (1998, 1999) concluded that the energy efficiency of a dehumidifier kiln should be greater than an equivalent air vented kiln drier by a factor of the order of two. Heat losses of the drying chamber and performances of a heat pump dehumidifier with HFC-134a as the refrigerant and a 5-kW compressor have been evaluated for a timber dryer working at dry-bulb temperatures up to 60°C (Bannister et al. 1998). The reported results showed that heat losses due to heat conduction and uncontrolled air leaks accounted for approximately 90% of the total energy input to the kiln. Because of these heat losses, the drying process mean temperature (around 43°C) was much lower compared to the designed operating temperature (60°C), adversely affecting both the productivity and running cost of the kiln. In their review, Chua et al. (2002) reproduced the schematic diagram of the heat pump-assisted dryer used by Bannister et al. (1999) for drying timber. Even though it did not show any means of controlling the heat pump dehumidification rate, as well as the temperature and relative humidity of the supplied drying air, it included an auxiliary heater, which is absent in many other heat pump-assisted drying schematics.

Relatively few new applications of heat pump-assisted drying systems have been reported during the past 10 years. Among them, there are some laboratory- and industrial-scale heat pump-assisted kiln (timber) drying systems. In a $13 \, m^3$ laboratory-scale hardwood heat pump-assisted drying system, the drying process of deciduous trees, such as hard maple, yellow and white birch, oak (hardwoods), with a 5.6 kW low-temperature electrically driven subcritical mechanical vapor compression heat pump has been extensively studied (Figure 1.18a) (Minea 2015). Electricity and natural gas (steam) have been alternatively used as auxiliary and backup energy sources. The heat pump (compressor and blower) electrical energy consumption varied between 25% and 30% of the total equivalent energy consumption of each drying cycle. The dryer central fan represented 8%–9% and the electrical (or fossil) auxiliary/backup energy, between 62% and 66% of the total equivalent drying energy consumption. For initial MCs above 41%, the total water quantities extracted above the fiber saturation point (FSP = 25%) were up to 2.9 times higher than those removed below FSP. Consequently, in these cases, the heat pump dehumidification efficiency ($SMER_{heat\ pump}$) was up to three times higher above the FSP than below this value. The heat pump-assisted hybrid drying cycles reduced natural gas consumption by 56% and the equivalent energy costs by 21.5%, compared to conventional drying cycles with natural gas as unique heating energy source (Minea 2012,

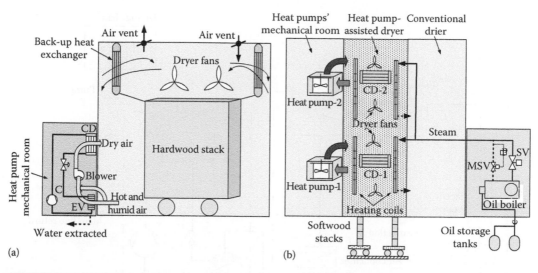

FIGURE 1.18 (a) Experimental setup of a low-temperature heat pump-assisted hardwood dryer with compact-type heat pump. (b) Configuration of a heat pump-assisted high-temperature softwood dryer with split-type heat pump. CD, heat pump remote condenser; MSV, modulating steam valve; SV, on/off steam valve. (From Minea, V., High-temperature heat pump-assisted softwood dryer: Sizing and control requirements & energy performances, in *24th International Congress of Refrigeration*, August 16–22, Yokohama, Japan, 2015; reproduced with permission from International Refrigeration Institute, IIR/IIF, Paris, France.)

2014a, b, 2015). Another 335 m³ batch industrial-scale, air-forced convective, high-temperature dryer for drying coniferous (resinous) lumber (e.g., white spruce, balsam fir, also known as *softwoods*) (Figure 1.18b) was modified and coupled with two 65 kW (compressor nominal shaft power input) split-type electrically driven subcritical mechanical vapor compression heat pumps using HCF-245fa as the refrigerant (Minea 2004, 2012, 2015). The compressors, evaporators, variable speed blowers, and controls were installed inside an adjacent mechanical room, whereas the remote condensers are located inside the kiln along with the dryer fans. The dryer fan rotation changed direction every 3 h, at the beginning, and every 2 h toward the end of the drying cycles in order to achieve uniform drying conditions through the softwood stacks and efficient moisture removal. Three air vents opened when the dryer fan rotation changes direction and when the actual air dry-bulb temperature exceeded the set point. An oil-burned boiler supplied high-pressure saturated steam to the dryer's heating coils in order to preheat the softwood stacks and, when required, to provide supplemental (backup) heat. The average SMER$_{\text{heat pump}}$ of this hybrid system varied from 2.35 kg$_{\text{water}}$/kWh (for white spruce) to 1.5 kg$_{\text{water}}$/kWh (for balsam fir), whereas the average COP$_{\text{heating}}$ varied from 3.0 (at the end) up to 4.6 (at the beginning of drying cycles). The cycle duration ranged from 2.5 days (for white spruce) to 6.3 days (for balsam fir), including the initial preheating steps.

The feasibility of a mechanical vapor compression heat pump has been studied for a timber drying process of a prefabricated house manufacture in Germany (Figure 1.19). In this facility, the timber used for prefabricated houses has to be dried up to final average MCs of 15% (dry basis) to avoid cracks. The main heat source for the facility is provided by 8.2 MW biomass power plant using residual wood. About 5 MW of the wasted thermal power by the extraction–condensation turbine are used to cover the factory's heating demand (e.g., wood presses, drying chambers, and space heating) via a heat extraction heat exchanger. An air-cooled condenser condenses the remaining steam of the biomass power plant. Instead, installing one additional oil-fired boiler to cover the increasing heat demand for wood drying, especially in the winter, a more environmental-friendly and cost-effective solution, that is, a heat pump system, was proposed. In this concept, the heat pump recovers heat from the power plant condensate at around 55°C and supplies 180 kW of thermal power (i.e., about 73%

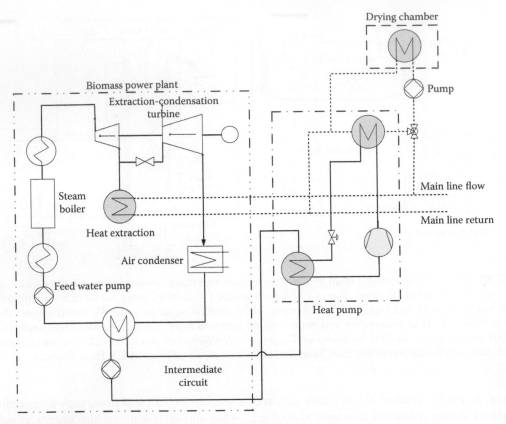

FIGURE 1.19 Integration of heat pump-assisted drying system in a German prefabricated house manufacture. (From IEA, Industrial energy-related systems and technologies Annex 13, IEA Heat Pump Program Annex 35, Application of Industrial Heat Pumps, Final Report, Part 2, 2015b; redrawn and reprinted with permission from IEA HPT Annex 13/35 operating agent.)

of the total heating power demand) to the drying chambers at temperatures varying between 65°C and 90°C. Higher temperatures are needed only during the preheating drying steps. Payback periods of 4–5.5 years have been estimated. During the drying process, the air temperatures are increased stepwise from 50°C to 90°C, and the temperature is held constant several days (IEA 2015b).

1.3 SUPERCRITICAL MECHANICAL VAPOR COMPRESSION HEAT PUMPS

As shown in Section 1.2, subcritical mechanical vapor compression heat pumps recover and release heat by refrigerant evaporation and condensation, respectively, at constant pressures and temperatures, lower than those of the refrigerant critical point.

On the other hand, a typical supercritical mechanical vapor compression heat pump includes similar components with those of subcritical cycle, without (Figure 1.20a) or with (Figure 1.20b) internal heat exchanger (IHEX). Compared to subcritical heat pump cycles (see Figure 1.2a and b), the most significant difference consists in replacing the two-phase heat transfer condenser by a gas cooler where a single-phase, sensible cooling process takes place with large temperature glide (defined as the difference between the refrigerant temperatures at the gas cooler inlet and outlet ports) to transfer heat to the heat sink fluid in the refrigerant supercritical region, where it is neither gas nor liquid, and its temperatures and pressures are thus independent properties. However, the evaporator still absorbs heat from the heat source by a two-phase convective vaporization process at subcritical constant pressure and temperature.

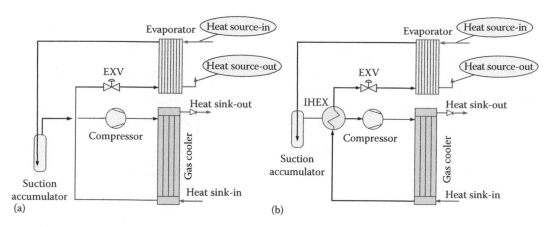

FIGURE 1.20 Basic configurations of mechanical vapor compression supercritical heat pumps: (a) without and (b) with internal heat exchanger. EXV, expansion valve; IHEX, internal heat exchanger.

This section will refer only to supercritical mechanical vapor compression heat pumps using CO_2 (carbon dioxide) as a working fluid.

Identified as a natural refrigerant in 1850, CO_2 has been widely used since 1900, mainly in marine refrigeration and air conditioning. It more or less disappeared when artificial refrigerants (CFC) were launched in the 1930s (Neksa et al. 1998, 1999, 2010). Among other fluids, CO_2 was chosen as alternative to conventional refrigerants for use in supercritical heat pumps due to its particular thermo-physic properties (Miyara et al. 2012). CO_2 is a pure (99.9%) natural fluid, relatively low cost, readily available, nonflammable, with high volumetric capacity and, at concentrations below 5% by volume in air, a nontoxic fluid. It has high specific heat and thermal conductivity, and low viscosity. Its global warming potential (over a 100 year integration period) is negligible (GWP = 1) when used as a confined refrigerant. Compared to conventional refrigerants, its critical temperature is very low (31.1°C), a remarkable property allowing CO_2 heat pumps to work above the critical pressure (7.3773 MPa) and reject heat over large temperature glides in special heat exchangers (known as *gas coolers*) (Kim et al. 2004). To achieve a heat pump supercritical thermodynamic process, the refrigerant vapor pressure is increased via a quasi-adiabatic compression process (1–2) into the supercritical region, that is, at pressures and temperatures higher than the refrigerant critical point (Figure 1.21a and b). Temperature changes of heat source and heat sink thermal carriers in the evaporator and gas cooler, respectively, are represented in light gray in Figure 1.21b. In the case of supercritical CO_2 heat pump hot water heaters, heat source temperatures typically vary from −5°C to maximum 25°C (at the evaporator inlet, due to the low critical temperature of CO_2) and heat sink temperatures up to 90°C (at the gas cooler outlet).

After compression, a CO_2 vapor isobaric cooling process (2–3) takes place within the gas cooler with a significant temperature glide. Compared to subcritical condensation process, the gliding temperature profile of refrigerant matches more closely the temperature profile of the heat sink thermal carrier fluid (see Figure 1.21b), which improves the gas cooler effectiveness. Moreover, the system energy performance is enhanced when the gas cooler temperature glide increases (Cecchinato et al. 2005) and the heat sink inlet temperature decreases (Laipradit et al. 2007). The isenthalpic expansion process (3–4) occurs in the expansion device EXV and the refrigerant isobaric evaporation process (4–1), within the evaporator. Contrary to subcritical cycles, supercritical cycles use high-pressure EXVs from which the modulation control comes from the high side of the system, rather than controlling refrigerant flow from the low-pressure side of the cycle. In other words, the expansion valves control refrigerant injection into the evaporator by opening and closing based on the increase or decrease in the gas cooler pressure. Thus, the EXV does not directly control the evaporator superheat, it being indirectly regulated by the system charge and the heat

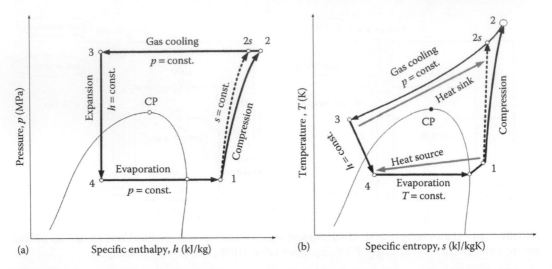

FIGURE 1.21   Heat pump supercritical cycle. (a) Pressure–enthalpy diagram; and (b) temperature-entropy diagram. CP, critical point; h, specific enthalpy; s, specific entropy.

load while avoiding the liquid refrigerant to return to the compressor. As a consequence, compared to other refrigerants, the performance of supercritical CO_2 heat pumps is much more sensitive to under- or over-charging conditions, and thus an optimum (critical) refrigerant charge exists (Cho et al. 2005).

Compared to conventional subcritical cycles, the CO_2 pressures within the gas cooler and the evaporator are much greater, that is, up to 130 bars (thus, not limited by the critical temperature of the refrigerant) and 65 bars, respectively. Even if these cycles operate at pressures 5–10 times higher compared to those of subcritical heat pumps, the development of new, more compact components resistant at such high pressures (such as compressors, evaporators, gas coolers, IHEXs, and pipes) allowed eliminating associated safety risks.

Supercritical CO_2 heat pumps are competitive with subcritical heat pumps, especially in domestic and process hot water heating applications (Neksa et al. 1998; Kim et al. 2004), where they take advantage of the high compressor discharge temperatures and large gliding temperatures in gas coolers.

1.3.1 DESIGN OUTLINE

Design of CO_2 supercritical heat pumps is based on transport characteristics of the refrigerant (CO_2) and secondary thermal carrier fluids. It usually involves simplifying assumptions as: (i) steady-state operation; (ii) adiabatic compression; (iii) isenthalpic expansion process; and (iv) negligible changes in kinetic and potential energies, and negligible heat losses in piping.

The evaporator thermal power can be expressed for both the refrigerant (CO_2) (Equation 1.18) and the heat source (e.g., water) (Equation 1.19) sides:

$$\dot{Q}_{evap} = \dot{m}_{CO_2} \left(h_1 - h_4 \right) \tag{1.18}$$

$$\dot{Q}_{evap} = \dot{m}_{source} * c_{p,\,source} \left(T_{source,\,in} - T_{source,\,out} \right) \tag{1.19}$$

where:
   $\dot{m}_{CO_2}$ is the refrigerant mass flow rate (kg/s)
   h_1 and h_4 are the refrigerant specific enthalpy at the evaporator outlet and inlet, respectively (kJ/kg) (see Figure 1.21a)

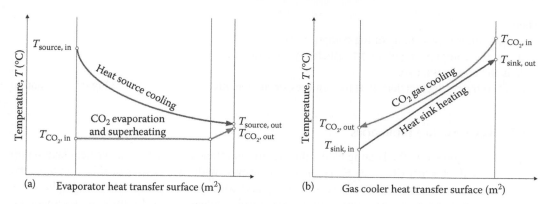

FIGURE 1.22   Temperature–heat transfer surface diagrams: (a) evaporator and (b) gas cooler.

$\dot{m}_{\text{source}}$ is the heat source (e.g., water) mass flow rate (kg/s)

$c_{p,\text{source}}$ is the heat source specific heat (kJ/KgK)

$T_{\text{source, in}}$ and $T_{\text{source, out}}$ are the heat source temperatures at the evaporator inlet and outlet, respectively (°C) (see Figure 1.22a)

Similarly, the gas cooler thermal power can be expressed for both superheated CO_2 vapor and heat sink (e.g., water) sides:

$$\dot{Q}_{\text{gc}} = \dot{m}_{CO_2}\left(h_2 - h_3\right) \tag{1.20}$$

$$\dot{Q}_{\text{gc}} = \dot{m}_{\text{sink}} * c_{p,\text{sink}}\left(T_{\text{sink, out}} - T_{\text{sink, in}}\right) \tag{1.21}$$

where:

$\dot{m}_{CO_2}$ is the CO_2 mass flow rate (kg/s)

h_2 and h_3 are the refrigerant specific enthalpy at the gas cooler inlet and outlet, respectively (kJ/kg) (see Figure 1.21a)

$\dot{m}_{\text{sink}}$ is the heat sink (e.g., water) mass flow rate (kg/s)

$c_{p,\text{sink}}$ is the heat sink specific heat (kJ/KgK)

$T_{\text{sink,out}}$ and $T_{\text{sink,in}}$ are the heat sink temperatures at the gas cooler outlet and inlet, respectively (°C) (see Figure 1.22b)

For the real (actual) compression process, the compressor input power (kW) can be calculated as follows:

$$\dot{W}_{\text{comp}} = \frac{\dot{m}_{CO_2}}{\eta_{\text{isentropic}}}\left(h_2 - h_1\right) \tag{1.22}$$

where:

$\dot{m}_{CO_2}$ is the refrigerant mass flow rate (kg/s)

$\eta_{\text{isentropic}}$ is the compression isentropic efficiency

h_2 and h_1 are the CO_2 specific entropy at the compressor outlet and inlet, respectively (kJ/KgK) (see Figure 1.21a)

The refrigerant mass flow rate can be determined with the following equation:

$$\dot{m}_{CO_2} = \dot{V}_{\text{swept}} * \eta_{\text{vol}} * N * \rho_{CO_2}^{\text{suction}} \tag{1.23}$$

where:

$\dot{V}_{swept}$ is the swept volume of the compressor (m^3)

η_{vol} is the compressor volumetric efficiency (function of the compressor pressure ratio)

N is the compressor speed

$\rho_{CO_2}^{suction}$ is the refrigerant density at the compressor suction inlet (m^3/kg) (state 1 in Figure 1.21a and b)

1.3.2 PERFORMANCE INDICATORS

As previously shown, in the heating mode, the maximum efficiency of a CO_2 subcritical heat pump is defined by the Carnot COP (see Equation 1.5). However, with CO_2 supercritical heat pumps, the $COP_{Carnot}^{heating}$ can not be used, because there is no two-phase condensation process, but single-phase gas cooling heat transfer with large temperature glides inside the gas cooler. In this case, the theoretic (maximum) heating COP is defined by the modified Lorentz cycle, more suitable as a theoretical reference cycle, as follows (see Figure 1.21a and b) (Stene 2005, 2007, 2008):

$$COP_{Lorentz}^{heating} = \frac{T_{gc}^{mean}}{T_{gc}^{mean} - T_0} = \frac{T_2 - T_3}{(T_2 - T_3) - T_0 \ln(T_2/T_3)} \tag{1.24}$$

where:

$$T_{gc}^{mean} = \frac{T_2 - T_3}{\ln(T_2/T_3)} \tag{1.25}$$

is the thermodynamic average temperature during heat rejection and T_0—the CO_2 evaporation temperature (K).

Similar to $COP_{Carnot}^{heating}$, the $COP_{Lorentz}^{heating}$ cannot be reached in practice due to the system thermodynamic losses, according to the following equation:

$$COP_{supercritic}^{heating} = \eta_{Lorentz} COP_{Lorentz}^{heating} \tag{1.26}$$

where $\eta_{Lorentz} < 1$ is the Lorentz efficiency factor.

1.3.3   R&D AND TECHNOLOGICAL ADVANCES

Researches on supercritical CO_2 heat pumps were initiated in the late 1980s (Lorentzen 1989, 1993) in order to replace environmentally harmful chemical refrigerants (CFCs, HCFCs). Since the CO_2 revival, CO_2 heat pump and refrigeration technologies have undergone tremendous development (Neksa et al. 2010). During the last ten years, heating performances of supercritical CO_2 heat pump water heaters, in particular, have been increased mostly by using IHEXs, by improving the design and control operation of gas coolers, evaporators and compression processes, by recovering the expansion work with refrigerant expanders and ejectors, as well as by using multi-stage compression cycles with or without vapor injection.

1.3.3.1   Internal Heat Exchanger

The IHEX (see Figure 1.20b) serves to further subcool the CO_2 vapor prior entering the EXV and, simultaneously, further superheat the refrigerant vapor leaving the gas cooler before entering the compressor suction port. The use of IHEX is controversial, because it increases the refrigerant vapor superheating and specific volume at the compressor inlet resulting in lower mass flow rate. The reduction of refrigerant mass flow rate would typically lead to reduced capacities, but, on the other hand, the relatively constant compressor work input increases the discharge temperature and enthalpy which leads to an increase in the driving temperature difference through the gas cooler that can offset the penalty because of the reduced mass flow rate (Kim et al. 2004). According to several R&D studies (Goodman et al. 2010), the IHEX may increase the $COP_{supercritic}^{heating}$ by 4% up to a certain gas cooler pressure and heat transfer surface (Kim et al. 2005) and 7% (Robinson and Groll 1998), or by 7.1%–9.1% at compressor frequencies varying from 40 to 60 Hz (Cho et al. 2007).

1.3.3.2 Compression Process

Modifications of the compression process of CO_2 supercritical heat pumps by using, for example, two-stage compression with intercooling, may further improve the $COP_{supercritic}^{heating}$ up to 9%. In this case, the addition of an IHEX may provide an additional 7.6% of $COP_{supercritic}^{heating}$ improvement (Cavallini et al. 2005). By using two-stage compression with intercooling, each compressor will achieve a lower pressure ratio, which in turn improves the isentropic efficiency and, at the same time, lower the discharge temperatures of both CO_2 compressors (Cecchinato et al. 2009).

Another indirect modification of the evaporation process consists in the addition of a flash gas bypass where the refrigerant expands through a single throttling device to the evaporator pressure and, then, enters a flash tank. From the flash tank, only saturated liquid enters the evaporator, while the vapor bypasses it. The evaporation process is thus improved because the vapor quality at the evaporator inlet is significantly reduced. Such a flash gas bypass system may also improve the refrigerant distribution inside the evaporator and increase the evaporator's heat transfer capacity up to 9%, and the $COP_{supercritic}^{heating}$ up to 7%. These improved performances were mainly attributed to reduced pressure drop and increased heat transfer coefficient inside the evaporator (Elbel and Hrnjak 2004).

Another technological advance consists in using flash intercooling for cooling the CO_2 between the two compression stages. The inter-stage CO_2 temperature is lowered by mixing it with expanded CO_2 vapor within a flash tank. However, such a solution may decrease the heat pump COP compared to equivalent single stage compression cycle, mainly because the CO_2 mass flow rate through the second-stage compressor increases significantly, as well as the actual compression work.

A two-stage compression system with additional high-pressure IHEX installed prior to the low-pressure compressor was experimentally tested for water heating. Compared to a system using only two-stage compression with intercooler, this concept improved the heat pump $COP_{supercritic}^{heating}$ by 7.5%. Cecchinato et al. (2009) theoretically analyzed a similar system, which resulted in an increase of $COP_{supercritic}^{heating}$ by 16.8%–28.7% compared to a single-stage compression system with intercooler only.

1.3.3.3 Gas Cooler

The design of gas cooler is strongly dependant on the type and flow characteristics of the heated (heat sink) fluid. Its heat transfer improvements can significantly impact the performance of CO_2 supercritical heat pumps. In the case of supercritical CO_2 water heaters, the ratio between (internal) vapor CO_2 (h_{CO_2}) and (external) water (h_{H2O}) convective heat transfer coefficients are of major importance. Because $h_{CO_2} < h_{H2O}$, h_{CO_2} is the primary factor that determines the magnitude of the overall heat transfer coefficient (U). Contrarily, in air-coupled gas coolers, the overall heat transfer coefficient will be more sensitive to the (external) air convective heat transfer coefficient because $h_{air} < h_{CO_2}$ (Fronk and Garimella 2011). Because up to 90% of the gas cooler irreversibility is due to the temperature difference between the cooled CO_2 vapor and the heated fluid (water or air), while the irreversibility due to pressure drop is negligible in comparison (Sarkar et al. 2004), many R&D works have been conducted in order to reduce this temperature difference and, thus, improve the gas cooler thermal performances.

The most significant technological advancement consisted of using microchannels (see Section 1.2.3.8) in order to reduce, in addition, the overall size and weight of the gas cooler (Sarkar et al. 2009; Fronk and Garimella 2011). CO_2 is a refrigerant well suited for use with microchannel heat exchangers (such as gas coolers, evaporators, and condensers) because of its high working pressures and vapor density that reduce the problem of fluid maldistribution experienced by some common refrigerants, and may enhance the overall heat transfer coefficients up to 33% compared to conventional heat exchangers (Yun et al. 2007).

1.3.3.4 Expansion Work Recovery

Expansion devices (valves, capillary tubes, or orifices) provide a pressure difference between the gas cooler and the evaporator, and maintain proper refrigerant distribution through the evaporator tubes. According to the second law of thermodynamic analysis, using such devices results in cycle irreversibility and energy losses, because any useful work is done during the expansion process. In fact, during the expansion process, a significant portion of the kinetic energy due to the passage from high pressure to the low pressure is dissipated in the fluid. The process is then not entirely isenthalpic, and these losses reduce the system heating and cooling efficiency.

1.3.3.4.1 Expanders

With CO_2 as a refrigerant in supercritical heat pumps, the greater pressure difference between compressor discharge and suction pressures results in greater expansion losses, thus making work recovery more feasible and more beneficial. In such systems, the expansion power recovered can be used to reduce the power consumption of the CO_2 compressor and, thus, to improve the system energy efficiency. One of methods developed to recover the work losses during the expansion process consists in replacing the EXV by an expander (Robinson and Groll 1998; Nickl et al. 2005; Subiantoro and Oil 2013), as shown in Figure 1.23 for a single-stage supercritical CO_2 heat pump.

An expander, a standard device in cryogenics (e.g., in air liquefaction) (Quack 1999, 2000), is a kind of reverse compressor that increases the heat pump heating capacity through performing a

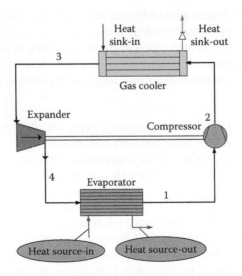

FIGURE 1.23 Typical expansion power recovery with expander in a single-stage supercritical CO_2 heat pump.

near-isentropic expansion, hence reducing the enthalpy of the refrigerant at the evaporator inlet, or by recovering the expansion energy, hence reducing the externally electrical power requirement of the compressor (Nickl et al. 2005). The expansion work recovered with expanders provides two advantages: (1) system specific heating and cooling capacities, as well as associated $COP^{heating}_{supercritic}$ are improved and (2) discharge pressures and compression works of CO_2 compressors are lowered in both single- and two-stage supercritical CO_2 heat pumps. For example, such a device may reduce the discharge pressure of the CO_2 compressor from 100 bars to less than 80 bars (Nickl et al. 2003).

Existing expanders must be designed for the special case of CO_2 supercritical heat pumps because of the following features of these thermodynamic systems: (1) very high pressure differences at relatively small volumetric flow rates; (2) expansion into the two-phase region, mainly on the left side of the critical point; and (3) amount of efficient work recovery, which is even more important for $COP^{heating}_{supercritic}$ improvement than supplying additional electrical power to CO_2 compressor. According to Robinson and Groll (1998), by recovering a part of the expansion work with expanders in supercritical CO_2 heat pumps with intermediate heat exchangers, the compressor discharge pressures and pressure ratios could be reduced, the evaporator performances—improved, and the system $COP^{heating}_{supercritic}$—increased. However, the IHEX may become effective at increasing $COP^{heating}_{supercritic}$ only if less than 30% of the work recovered by the expander is used to reduce compressor work. According to Yang et al. (2005), using expanders in place of conventional expansion devices may reduce up to 50% the exergy losses, resulting in 33% higher $COP^{heating}_{supercritic}$. However, although replacing the EXV with an expander can significantly improve the performance of supercritical CO_2 cycles, such additional devices may not be economically feasible for many practical applications, especially for small capacity supercritical CO_2 heat pump water heaters. The question is where the expanders should be integrated into the heat pump systems, and how the liquid must be distributed to the evaporators. In the case of two-stage supercritical CO_2 heat pumps, turbine-type expanders can transfer the expansion work recovered to directly drive, via gear boxes, the low-pressure, high-pressure compressor, or the low-pressure compressor provided with intermediate intercooler (Yang et al. 2007). According to Yang et al. (2007), compared to single-stage compression heat pumps with EXVs, the best performance is achieved by the expander directly driving the high-pressure compressor, that is, not via a gear box.

A three-stage reciprocating free piston expander directly powering the second stage compressor in a supercritical CO_2 refrigeration system (with liquid–vapor separator between the second and third stage expanders) with isentropic efficiency varying between 65% and 70%, increased the $COP_{cooling}$ by 40% compared to a similar cycle using a conventional throttling valve (Nickl et al. 2005). When the shaft of a scroll-type expander is directly coupled to the first-stage compressor of a two-stage compression supercritical CO_2 system with intercooling, the compressor efficiency decreases as inlet pressure increases. The expander reduces the required external compressor work by about 12%, which results in an increase in the cooling capacity and $COP_{cooling}$ by 8.6% and 23.5%, respectively (Kim et al. 2008). An experimental two-cylinder reciprocating piston expander for work recovery in a supercritical CO_2 cooling system, with an isentropic efficiency of 10.27%, improved the $COP_{cooling}$ by 6.6% compared to the system using an EXV (Baek et al. 2005). A rolling piston expander in a supercritical CO_2 cycle experimentally recovered 14.5% of expansion work and showed that there is an optimal rotational speed of the expander which maximizes its efficiency, the amount of work recovered and the system $COP_{cooling}$ (Hua et al. 2010). A swing piston expander experimentally achieved isentropic efficiency between 28% and 44% depending upon operating conditions. The expander efficiency increased as expander inlet temperature (gas cooler outlet temperature) increased, but the system heating capacity decreased (Haiqing et al. 2006).

She et al. (2014) proposed a hybrid heat pump system where the output power of the expander was employed to drive the compressor of an auxiliary subcooling cycle, whereas the refrigerant at the outlet of condenser was subcooled by an evaporative cooler. The main cycle included an evaporator, a compressor, a water-cooled condenser, an evaporative cooler, and an expander, whereas the auxiliary subcooling cycle consisted of an evaporative cooler, a compressor, a water-cooled condenser, and a throttle valve. In this system, the expansion power recovered from the main refrigeration system is employed to supply for the compressor in the auxiliary subcooling cycle, which provided the cooling capacity of evaporative cooler to further subcool the CO_2 in the main cycle. According to She et al. (2014), such a system provided a significantly higher COPs, which indicated that the subcooling method using expansion power recovery (expander efficiency of 0.35) can become an efficient way to improve the performance of mechanical vapor compression heat pumps.

1.3.3.4.2 Ejectors

It is well known that air-to-air and air-to-water heat pumps are attractive technologies for either domestic hot water or space heating applications (Hepbasli and Kalinci 2009; Bourke and Bansal 2010), but their heating capacity and COPs degrade at low or very low ambient temperature conditions (Bertsch and Groll 2008). In order to partially remediate this issue, a lot of methods have been proposed, such as using ejectors and refrigerant injection (Chua et al. 2010; Heo et al. 2011), as well as hybrid systems as dual-source and compression-absorption heat pumps (Lazzarin 2012) (see Section 1.4.7). Heat recovery ejector refrigeration systems work similarly with conventional vapor compression cycles, but instead of mechanical compression devices, they use ejectors to provide cooling effect and then compress the refrigerant vapor (Hsu 1984). In addition, the use of ejectors as expansion devices replacing the throttling valves in the mechanical vapor compression heat pumps and refrigeration systems seems to be one of the efficient ways of reducing the expansion losses (irreversibility).

The ejector-expansion devices have advantages such as low cost, no moving parts, and the ability to handle two-phase flow, making it attractive for the development of high-performance CO_2 refrigeration system (Yari 2009). The ejector-expansion reduces the compressor work by raising the suction pressure to a level higher than that in the evaporator, leading to the improvement of system overall COPs (Kornhauser 1990). In such cycles, the high-pressure CO_2 from the gas cooler enters the nozzle of the ejector where its velocity is increased and pressure is decreased. This lower pressure draws CO_2 vapor from the evaporator into the ejector mixing chamber where the pressure increases. A diffuser is utilized to increase CO_2 pressure while also lowering the velocity. Compressed CO_2 then enters a liquid–vapor separator from which vapor is drawn into the compressor via the SA, while the separated liquid re enters the evaporator (Sarkar 2008). In other words, inside the ejector, the refrigerant from the high pressure side of the system (gas cooler) expands in the motive nozzle and the high-speed two-phase flow at the motive nozzle outlet mixes with the evaporator flow (Sarkar 2009).

Kornhauser (1990) studied the thermodynamic performance of the ejector-expansion refrigeration cycle using HCFC-12 as a refrigerant based on a constant mixing pressure model. The author found a $COP_{cooling}$ improvement of up to 21% over the standard cycle under standard operating conditions (i.e., evaporator temperature of −15°C and a condenser temperature of 30°C). However, improvements in the order of 3.9% to 7.6% have been achieved in the subsequent experiments with HFC-134a (Harrell and Kornhauser 1995). Menegay and Kornhauser (1996) tested an air-to-air refrigeration cycle using a two-phase ejector as an expansion device and found that the $COP_{cooling}$ improvement was poor because of nonequilibrium effect in the motive nozzle. To find a remedy for this problem, they developed a bubbly flow tube installed upstream of the motive nozzle to reduce the thermodynamic nonequilibrium. Such an ejector improved up to 3.8% of the $COP_{cooling}$ over the conventional cycle under standard conditions with CFC-12 as the working fluid. According to Elbel and Hrnjak (2008), compared to conventional systems with EXVs, a supercritical CO_2 ejector prototype improved the $COP_{cooling}$ and the cooling capacity by 7% and 8%, respectively. Dokandari et al. (2014) evaluated, based on the first and second laws of the thermodynamics, the ejector impacts on the performance of a cascade cycle using CO_2 and NH_3 as refrigerants According to Dokandari et al. (2014), their novel CO_2-NH_3 ejector-expansion cascade cycle achieved maximum COPs approximately 7% higher as compared to the conventional cycle. In addition, the exergy destruction rate decreased by about 8% because of lower exergy losses through EXVs and the

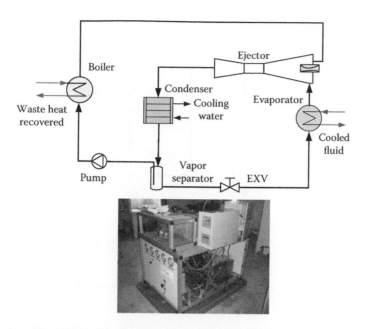

FIGURE 1.24 Schematic of a CO_2 heat recovery prototype with ejector for cooling purposes. EXV, expansion valve; F, flowmeter; P, pressure sensor. (From IEA, Industrial energy-related systems and technologies Annex 13, IEA Heat Pump Program Annex 35, Application of Industrial Heat Pumps, Final Report, Part 1, 2015a; redrawn and reprinted—or reproduced—with permission from IEA HPT Annex 13/35 operating agent.)

system second law efficiency increased by 5% as compared to the equivalent conventional system. An ejector refrigeration system (Figure 1.24) may recover waste heat at temperatures >65°C and provide cooling at temperatures <5°C (Minea 2013). Huang et al. (2014) studied a solar-assisted ejector cooling system that integrated an ejector refrigeration system with a mechanical vapor compression air conditioning heat pump. The ejector is driven by the refrigerant vapor evaporated by a solar collector. The intercooler heat exchanger acts as the evaporator of the ejector cooling system and subcools the liquid refrigerant leaving the condenser of an air conditioner with inverter-type compressor. According to Huang et al. (2014), the use of the ejector cooling concept in the first stage of such a solar-assisted ejector cooling and heating system reduced the condensing temperature of the air conditioner by 7.3°C–12.6°C and the compressor input power by 34.5%–81.2%. In addition, reported results showed that the overall system COPs, including auxiliary energy consumptions, increased by between 33% and 43%.

1.3.3.4.3 Vapor Injection

As previously noted, subcritical mechanical vapor compression air-to-air and air-to-water heat pumps are widely used in the residential houses and institutional/commercial buildings for both cooling and heating applications (see Section 1.2.1). However, such heat pumps work very well in climates with moderate ambient temperatures, but in colder climates, the degradation of the heat pump COP becomes significant when the outdoor temperature decreases under −8°C in the heating mode, or increases over 35°C in the cooling mode. Effectively, in the winter heating mode, low ambient temperatures result in low refrigerant density and mass flow rate at the compressor suction inlet port, reduced volumetric efficiency of the compressor, low isentropic efficiency of the compression process and, thus, excessively high compressor power input. Therefore, the heat pump heating capacity and COP are reduced compared to those provided at ambient temperatures lower than about −8°C. Contrarily, in the summer cooling mode, the refrigerant is compressed at higher pressure and temperature, which excessively increases the compressor discharge temperature and power input, which result in lower cooling capacity and cooling COPs. High compressor discharge

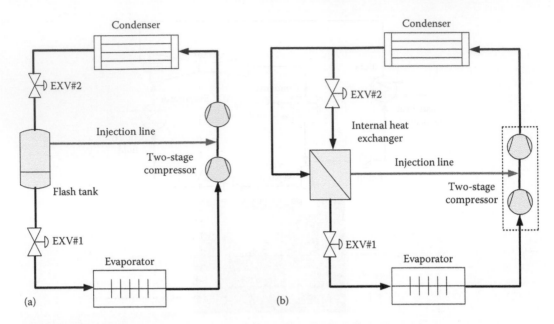

FIGURE 1.25 Heat pump concepts with refrigerant injection. (a) Via a flash tank (phase separator) and (b) via an internal heat exchanger (economizer). EXV, expansion valve.

temperatures might also degrade the lubricating oil and thus reduce the reliability of the system. To partially fix these problems, refrigerant vapor injection techniques, where a part of the vapor after the EXV is extracted and injected to the compressor, have been patented, proposed, and/or applied in practice.

There are two vapor injection concepts: (1) with phase separator (flash tank) (Figure 1.25a) (see Elbel and Hrnjak 2004 for more information) and (2) with IHEX (economizer) (Figure 1.25b) (Aikins et al. 2013).

In the flash tank (phase separator) vapor injection cycle (Figure 1.25a), the flash tank receives partially expanded refrigerant diverted from the upper-stage EXV and injects it into the two-stage compressor at an intermediate pressure. Refrigerant has to be injected into the compressor in a superheated state in order to avoid injecting wet or liquid into the two-stage compressor. It can be seen that, due to the two-phase separation in the flash tank, the liquid entering the evaporator has lower enthalpy compared to that of a single-stage cycle. Thus, the enthalpy difference across the evaporator is greater than that of a single-stage cycle, and the vapor injection reduces the refriger-ant mass flow rate through the evaporator. However, the increased enthalpy difference increases the two-phase heat transfer area in the evaporator. Therefore, the overall effect is that the system capacity is increased, as well as the system COP. The saturated vapor from the flash tank also has lower temperature than that of the vapor in the compressor, which helps to reduce the com-pressor discharge temperature. Contrary to flash tank injection cycle, IHEX vapor injection cycle may employ the evaporator outlet superheating amount as the control parameter to regulate the openings of both EXVs by means of sensing bulbs and pressure balancing lines placed to the evaporator outlet.

In the IHEX vapor injection cycle (Figure 1.25b), the injected vapor is generated by the econo-mizer (SC) resulting from the temperature difference along the upper-stage EXV. After the con-denser, the refrigerant liquid is separated into two paths: one of them flows through the upper-stage EXV and enters the IHEX, where it provides subcooling to the refrigerant coming from the other circuit. The two-phase refrigerant absorbs heat within the IHEX and becomes vapor, which is then injected to the compressor. At the same time, the subcooled liquid is further subcooled by the two-phase refrigerant and enters the lower-stage EXV, the evaporator and, finally, the compressor. Because of the (total) liquid subcooling process (3–5), after the lower-stage EXV, the enthalpy difference across the evaporator becomes higher than that of the single-stage cycle.

Compared to liquid injection strategy, which mainly aims at reducing the compressor discharge temperature, the vapor injection technique provides some benefits as (1) evaporator cooling capacity and COPs improvement because of reduced quality of the inlet vapor (Ma et al. 2003; Hwang et al. 2004; Ma and Chai 2004; Tian and Liang 2006; Tian et al. 2006; Bertsch and Groll 2008; Cho et al. 2009); (2) condenser heating capacity improvement in severe climates (i.e., heat pumping at ambient temperature lower than −8°C and air conditioning at ambient temperatures higher than 35°C) because of the added refrigerant from the injection line; (3) system capacity can be varied by controlling the injected refrigerant mass flow rate, which allows some energy savings by avoiding intermittent operation of the compressor; (4) lower compressor discharge temperature than that of conventional cycles, because the injected vapor temperature is lower than that of the vapor at the compressor inlet; therefore, the compression process is closer to isentropic behavior (Xu 2012); and (5) although the vapor injection reduces the refrigerant mass flow rate through the evaporator, the increased enthalpy difference increases the two-phase heat transfer area in this heat exchanger and, therefore, the system capacity increases. From a thermodynamic point of view, the performance of flash tank and IHEX cycles should have similar performance (Xu 2012). Their working principle is to decrease the evaporator inlet enthalpy by two-stage expansion. The only difference is to achieve it by subcooling through the additional heat exchanger or by two-phase separation in the flash tank. However, the actual performance of the flash tank cycle is superior to that of the IHEX cycle.

In the case of air-source supercritical CO_2 heat pump water heaters, by applying the vapor injection techniques, due to the increase of refrigerant mass flow rate through the condenser, leads to greater heat transfer through, the heat pump capacity can also be improved (Xu 2012). Ma and Zhao (2008) showed that the heating capacity and COP of the flash tank cycle are 10.5% and 4.3% higher than those of the IHEX cycle, respectively, mainly because the IHEX introduces additional pressure drops to the injected vapor, contrary to the flash tank, which leads to a saturated state of refrigerant and, therefore, there is no additional pressure drop in the vapor injection line. On the other hand, the IHEX cycle has a much wider operating range than that of the flash tank cycle (Ma and Zhao 2008).

Control strategy of vapor injection is still a challenging task today. This is because improper control of both refrigerant EXVs would result in undesirable amount of vapor injected to the compressor or poor overall performances of the heat pump. Even the flash tank cycle typically shows better performance than the cycle with IHEX, its control is not yet clearly defined today and is still a challenging task (Xu 2012). This is because, the vapor refrigerant separated in the flash tank is in a saturated state, and, therefore, a conventional thermostatic EXV does not operate properly, because it causes valve hunting. As a result, electronic EXV is more suitable for the flash tank cycle, but a liquid level sensor is needed to measure the liquid refrigerant level in the flash tank, which increases the system cost. The development of effective control strategy for flash tank vapor injection cycle is thus critical for successful commercial applications. Leimbach and Heffner (1992) proposed a refrigerant injection valve based on temperature responsive sensor that controls the opening of injection port by measuring the compressor discharge temperature. Ma et al. (2003) reported that an air source heat pump with vapor injection via an IHEX operated reliably at ambient temperatures as low as −15°C with improved heating capacity and COPs. He et al. (2006) conducted field tests with a vapor injection heat pump with IHEX using HCFC-22 as the refrigerant. Their results showed that, compared to conventional air-to-air heat pumps, the heating capacity and COP have been improved by 34% and 6%, respectively, at −20°C (outdoor) and 20°C (indoor) temperatures. Tian et al. (2006) also studied an air-source heat pump employing vapor injection coupled with an IHEX and shown that the heating COP was over 2.0 at −25°C evaporating and 50°C condensing temperatures. Lifson et al. (2006) proposed a fast-acting control valve placed on the vapor injection line near the compressor vapor injection port to control the duration of each injection cycle. Bertsch and Groll (2008) tested a two-stage heat pump system at low ambient temperature. The heating COP was found to be 2.1 at the ambient temperature of −30°C. Wang (2008) conducted a series of testing using the vapor injection technique, and reported a maximum capacity and COP improvement of 33% and 23%, respectively, when the ambient temperature was −18°C. Fan et al. (2008) tested a heat pump water heater using HCFC-22 as the refrigerant at ambient temperatures of −30°C to 12°C. The hot water temperature was between 55°C to 60°C. With vapor injection, the hot water heating capacity increased from 2.0 to 2.8 kW, together with a COP increase from 1.5 to 1.8.

Baek et al. (2014) reported that at low ambient temperatures, the air-to-water supercritical CO_2 heat pumps using variable speed twin rotary compressors and SC vapor injection or flash tank vapor injection with suction line heat exchanger, improve their performances compared to those of conventional heat pumps using HFC-410A as the refrigerant, mainly because the thermophysical properties of CO_2 in the supercritical region strongly affected (i.e., improved) the injection mass flow rate and the intermediate pressure in the twin rotary compressor. Theoretical results obtained shown that at standard heating conditions, the flash tank vapor injection cycle is slightly more effective in improving heating performance than the other vapor injection cycles. In addition, the inter-cooling effect of the vapor injection CO_2 cycles ensures the reliability of the compressor by decreasing the high discharge temperature to the optimum level. The injection cycles also reduce the performance degradation under low outdoor temperatures by increasing the total mass flow rate through the gas cooler in the particular case of CO_2 supercritical heat pumps. At the outdoor temperature of −15°C and the compressor frequency of 55 Hz, the heating capacity and COP of the flash tank vapor injection cycle at the optimum injection ratio were 18.3% and 9.4% higher than those of the noninjection cycle, respectively. In addition, based on the performance comparison of the optimized vapor injection CO_2 cycles at the outdoor temperature of −15°C and the compressor frequency of 55 Hz, the heating capacity and COP of the flash tank vapor injection cycle were 12.1% and 12.7% higher than those of the flash tank vapor injection cycle, respectively, because the total mass flow rate in the SC vapor injection cycle was higher than that in the flash tank vapor injection cycle because of the large temperature and pressure differences in the SC of the SC vapor injection cycle. The optimum injection ratio of the SC vapor injection cycle yielding the maximum COP decreased from 0.59 to 0.50 with the increase in the compressor frequency from 45 to 55 Hz at the outdoor temperature of −15°C, while that of the flash tank vapor injection cycle increased from 0.29 to 0.34 because of increase in the optimum injection opening of the EXV. Ma and Chai (2004) tested a heat pump cycle with hot water supply. The prototype demonstrated high temperature and capacity water supply even at the ambient temperature of −10°C to −15°C.

During the past two decades, a number of R&D works have been conducted in the field of heat pump applications at low ambient temperatures (i.e., in cold climates) (Xu 2012). Today, several manufacturers provide air-source heat pumps designed specifically for cold climates with vapor injection able to achieve higher heating capacities by maintaining proper compression ratios at higher two-stage compressor operating frequencies. For example, Mitsubishi Electric launched, split-ductless air-source heat pumps with inverter-driven compressors, vapor flash injection and HFC-410A as the refrigerant for northern regions, which can extract useful heat from ambient air as cold as −37°C with heating COPs of 0.9 without auxiliary (backup) heaters (http://www.mitsubishielectric.ca/en/hvac/residential.html, accessed May 15, 2015). Such a product consists of a compressor and an injection loop. The refrigerant leaving the phase separator is divided into two lines: one is the main line entering the second expansion device and the other is the injection line entering to connection pipe of a two-stage compressor (Figure 1.26).

Such a ductless mini-split air-to-air heat pump has been tested under the Canadian cold climate (Sager 2013). Even the author did not explicitly state whether backup heat sources have been

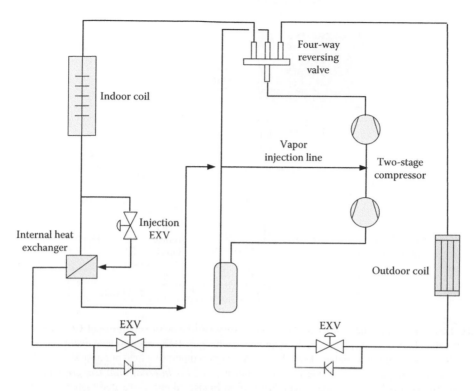

FIGURE 1.26 Schematic drawn of a cold climate air-to-air heat pump with inverter-driven two-stage compressor and vapor injection with internal heat exchanger in the heating mode only. (From http://www .mitsubishielectric.ca/en/hvac/residential.html, accessed May 15, 2015.)

used for outdoor temperatures below −37°C; it was claimed that the cold climate air-source heat pump provided 60% energy (not energy cost) savings compared to natural gas heating, without considering the efficiency of electricity generation. At high ambient temperatures, vapor injection techniques can also enhance the cooling capacity of heat pumps operating in the cooling (e.g., air conditioning) mode in very hot and humid climates (Xu 2012). Bertsch and Groll (2008) tested a vapor injection heat pump cycle at the ambient temperature of 50°C and shown that due to the vapor injection technique, the compressor discharge temperatures remained below 105°C. Cho et al. (2009) reported that a two-stage CO_2 heat pump cycle with vapor injection operating in the cooling mode enhanced the cooling COP by 16.5% over that of the two-stage noninjection cycle, while the compressor discharge temperature decreased by 5°C to 7°C, because of the inter-cooling effect from the vapor injection. Wang et al. (2009) reported that both the IHEX cycle and the flash tank cycle tested at severe climate conditions showed comparable performance improvement compared to the baseline without injection. They found that the cooling COP and capacity improvements at ambient temperature of 46°C were 2% and 15%, respectively.

1.3.4 Applications of Supercritical CO_2 Heat Pumps

1.3.4.1 Industrial Waste Heat Recovery

Supercritical CO_2 heat pumps have many application opportunities such in medium-sized industries (such as food processing) and institutional buildings (such as hospitals), where process and/or domestic hot water is required at temperatures above 60°C for process, cleaning, washing or

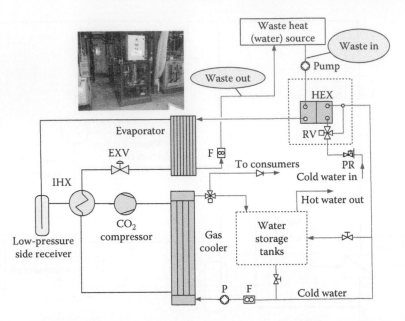

FIGURE 1.27 Two-stage laboratory-scale heat recovery system with supercritical CO_2 heat pump. EXV, expansion valve; F, flow meter; HEX, preheating heat exchanger; IHE, internal heat exchanger; PR, pressure regulator; RV, 3-way regulating valve. (From Minea, V., High-temperature heat pump-assisted softwood dryer: Sizing and control requirements & energy performances, in *24th International Congress of Refrigeration (ICR2015)*, August 16–22, Yokohama, Japan, 2015; IEA, Industrial energy-related systems and technologies Annex 13, IEA Heat Pump Program Annex 35, Application of Industrial Heat Pumps, final report, Part 1, 2015a; reprinted with permission from IEA HPT Annex 13/35 operating agent.)

domestic usages, and residential sector. In industry, these devices can simultaneously provide heating (i.e., hot water or air) and cooling (i.e., cold water or air), thus maximizing the systems' overall efficiency and consuming up to 27% less electricity than two separate systems operating under identical thermal conditions (Byrne et al. 2009). In many industrial facilities, low-enthalpy waste heat in liquid or air forms is often available at temperatures of below 45°C, whereas the cold water to be heated comes from municipal water networks at relatively low temperatures. To efficiently recover of low-temperature waste heat from industrial processes effluents, a two-stage heat recovery systems, including a preheating heat exchanger and a CO_2 supercritical heat pump as the first and second stages, respectively, can be successfully implemented (Figure 1.27) (IEA 2015a; Minea 2015). In such a system, if the waste heat fluid enters the heat recovery system at temperatures between 15°C and 45°C, it can be cooled down to between 10°C and 38°C, depending on the actual inlet temperature of the cold water. In cold climates, it generally varies from a minimum of 7°C (in the winter) to a maximum of 17°C (in the summer). By bypassing (or not) the preheating heat exchanger, the cold city water can be heated up to 38°C before entering the supercritical CO_2 heat pump. The process is controlled by the regulating valve RV according to the relative magnitude of the inlet temperatures of the heat source and heat sink thermal carriers.

An industrial-scale supercritical CO_2 heat pump has been implemented in a Canadian dairy plant (Minea 2014c). It includes a 25 kW compressor (nominal power input) and is integrated between two of the plant's industrial processes (Figure 1.28a). It recovers energy from one process at temperatures below 0°C and provides heat to another process at temperatures above 85°C. Because of the low critical temperature of CO_2, these boundary thermal conditions allow the heat pump to operate according to a supercritical cycle with evaporation at subcritical pressures and heat rejection at pressures above the CO_2 critical pressure (7.377 MPa-a). So, unlike conventional subcritical heat pump cycles, heat is not supplied by means of refrigerant condensation, but by

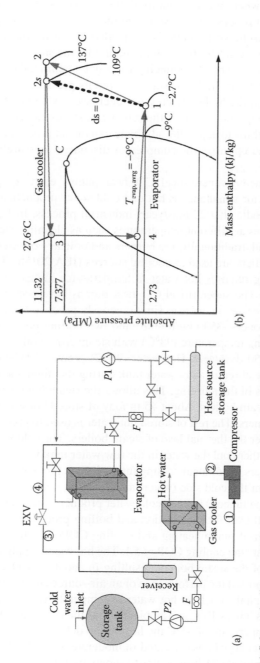

FIGURE 1.28 (a) Schematic diagram of the industrial-scale supercritical CO$_2$ heat pump implemented in a Canadian dairy plant. (From Minea, V., Efficient process integration and cooling & heating energy performance of supercritical CO$_2$ heat pumps, in *11th IEA Heat Pump Conference 2014*, May 12–16, Montréal, Québec, Canada, 2014c; reprinted with permission from IEA HPP Centre.) (b) Example of a typical (average) thermodynamic cycle. C, critical point; P1, hot brine pump; P2, cold water pump; EXV, expansion valve; F, flow meter.

cooling compressed high-pressure CO_2 inside a special heat exchanger called *gas cooler*. As can be seen in Figure 1.28a, IHEX is included in the heat pump thermodynamic cycle. Because both the cooling and heating thermal effects of the unit can be used by the industrial facility, the overall system energy efficiency is further improved. In this system, the superheated refrigerant (state 1) enters the compressor where it is brought to higher pressures and temperatures (state 2) by a nonisentropic compression process (1–2) (Note: the theoretical isentropic process is shown as 1-2s) (Figure 1.28b). The discharge pressure is controlled based on the varying amount of refrigerant inside the high-pressure side of the system and not by the saturation pressure as is the case with conventional HFC refrigerants. By reducing the valve opening, the CO_2 mass flow rate decreases, the CO_2 accumulates in the high-pressure side and its pressure thus rises. Conversely, by increasing the valve opening, the high-side pressure decreases, whereas the excess charge is stored as a liquid inside the low-pressure side of the refrigeration circuit. This particular control strategy requires special algorithms to adjust the adiabatic expansion process in order to keep the evaporation almost constant and optimize the compressor discharge pressure and the gas cooler outlet temperature.

Two 79.9 kW (heating capacity) supercritical CO_2 heat pumps each equipped with 24-kW CO_2 reciprocating compressors for simultaneous hot and cold water production have been implemented in Japan in a frozen noodle (10,000 tons/year) industrial process. In Japan, food processing consists of multiple processes at different temperatures such as cleaning, sterilization, boiling, cooling, freezing, and drying and, traditionally, gas burners and/or heavy oil-fired steam boilers are used as heating sources and chillers are used as cooling sources (IEA 2015b). The boiling process of the frozen noodle processing requires hot water at temperatures higher than 80°C, traditionally supplied by steam boilers. The supercritical CO_2 heat pumps have been introduced between the boiling and the cooling processes in order to simultaneously produce hot and cold water. The process includes three steam boilers (3500 kg/h total) and two boiling pools (3000 L) where the water is reheated near the boiling temperature (98°C) with steam from boilers. The plant operates about 16 h/day during about 250 days/year. Hot water (90°C) produced by the heat pump flows through a heat exchanger and is stored in a hot water tank during the night and delivered at higher temperature to the boiling pools in the morning. This allows the steam boilers to start up later and thus reduce the peak load of steam boilers. About 45 m³/day of stored hot water is daily delivered from the storage tank to consumers. The rest of stored hot water is used to preheat the boilers' feed water, which contributes to lower the thermal load of steam boilers. The cold water supplied by the heat pump at 7°C is used to further cool the water in the raw water tank for the cooling pools, thus reducing the load of water chillers. Additional cooling (up to 3°C) is provided by the existing water chillers. By absorbing heat from the cold side (water) and rejecting heat to the hot side (water) of the industrial process, these simultaneous hot and cold water producing heat pumps are able to heat water from 17°C to 85°C (3000 L/h) for sterilization and boiling processes, and cool water from 10°C to 5°C (5000 L/h), with heat pump heating and cooling COPs of 3.0, and 2.1, respectively, and overall simultaneous heating and cooling COP of 5.1. The hot water supply to the boiling pools also reduced the heating load of the steam boilers, resulting in about 4% CO_2 emissions reduction (IEA 2015b). Anstett (2006) reported the performance of an air-source supercritical CO_2 heat pump water heater installed in a hospital to deliver hot water at temperatures between 60°C and 80°C with water inlet temperatures as low as 10°C and ambient air (as the heat source) temperatures from –20°C to 40°C. At ambient temperature of –5°C the heat pump delivered hot water at 70°C with a COP of 2.5. In the case of low-energy houses located in moderate and cold climates, Stene (2005, 2007) proposed to apply supercritical CO_2 heat pumps to supply multiple, simultaneous or alternate, space and hot water heating loads. In this concept, the gas cooler was partitioned to separately supply domestic water preheating, space heating, and domestic water reheating. Hihara (2006) also reported that supercritical CO_2 heat pumps can provide combined domestic hot water and hydronic floor heating with seasonal COPs of 2.7, value that includes motor efficiency and transient thermal losses from the hot water storage tank.

1.3.4.2 Industrial Drying

A supercritical CO_2 heat pump has been implemented in Japan as an air heater for a drying process (Figure 1.29) (IEA 2015b). In order to reduce the fuel consumption for drying as well as the drying time, hot air at around 120°C is frequently used for drying in industrial casing painting of electrical transformers (about 35,000 units per year). To generate such a drying hot air, conventional boilers, burners, or electric heaters are usually used. In traditional painting processes, air circulated in the drying ovens, after the electrodeposition process, is heated up to about 170°C by liquid propane gas burners. After the top coating process, the air circulated in the ovens is heated up to about 155°C. Partial ventilation is necessary to prevent the circulated air from contamination, which causes a decrease in the thermal efficiency of the facilities. In addition, exclusive chillers are used for keeping the temperature of the electrocoating baths at 29°C. Because heat pumps using conventional HFC refrigerants can heat air up to maximum 70°C, supercritical CO_2 heat pumps have been employed to generate hot air at temperatures higher than 80°C. Hence, a 110 kW (heating capacity) supercritical CO_2 heat pump equipped with a semi-hermetic, reciprocating, inverter controlled capacity (30–65 Hz) CO_2 compressor, has been installed in order to preheat the fresh air and, thus, simultaneously cooling (35 kW cooling capacity) the electrocoating baths (Figure 1.29). In this system, outdoor fresh air is first heated up to between 80°C and 120°C by the supercritical CO_2 heat pump by recovering waste heat at temperatures varying from −9°C to 35°C. Then, the air is heated further up to required temperatures by the liquid propane gas burners prior entering the air drying ovens. The preheating operation with

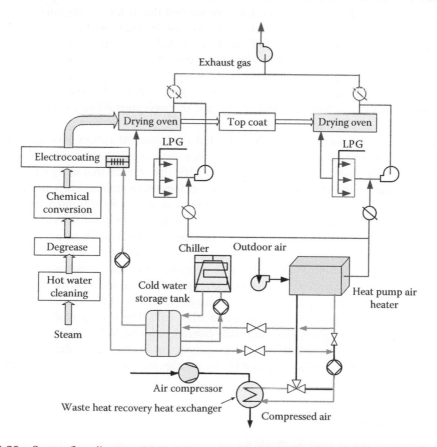

FIGURE 1.29 System flow diagram of drying process in painting application in Japan. (From IEA, Industrial energy-related systems and technologies Annex 13, IEA Heat Pump Program Annex 35, Application of Industrial Heat Pumps, Final Report, Part 2, 2015b; redrawn and reprinted with permission from IEA HPT Annex 13/35 operating agent.)

supercritical CO_2 heat pump reduces the liquid propane gas burners' heating load. CO_2 heat pump may also recover heat from the cold water stored at 15°C. After being cooled in the evaporator, cold water at 10°C returns to the cold water storage tank. This heat recovery process reduces the thermal load of the water chiller. If the heat recovered from the cold water is not sufficient, the heat pump will recover more heat from the air compressor discharge. This system has reduced the process average natural gas consumption by 24.1%, the running costs by 12% and the CO_2 greenhouse gas emissions by 13.1%.

Drum dryers have been used for a long time in homes for drying of laundries (see Section 1.2.2.2). They are household appliances used to remove moisture from clothes and other textiles. Because traditional refrigerants, as HFC-134a (GWP factor of 1300), have negative environmental impacts, the academia and industry started looking at supercritical CO_2 heat pumps as drum dryers about 15 years ago. Schmidt et al. (1998) compared the thermodynamic behavior of two dehumidification heat pump cycles: a subcritical process with R-134a and a supercritical process with CO_2. Their simulation results showed that both cycles are equivalent in terms of energy efficiency. However, better compression efficiency was expected with CO_2, as well as improved energy efficiency. Klöcker et al. (2001) studied the feasibility of using CO_2 as a working fluid for laundry heat pump dryers and compared it with R-134a systems. They found that the former did not need more energy than the latter, and that the CO_2 supercritical cycle was suitable for heat pump dryers due to the good environmental characteristics and thermal properties of CO_2. These authors compared the energy performance of two optimized CO_2 heat pumps with those of a conventional electrically heated dryer and concluded that if the drying time is not the key requirement, energy savings of about 65% can be achieved. In addition, if a high water extraction rate is sought, energy savings of about 53% can be provided, including fan energy consumption, but with much shorter drying times than that of the conventional system. Klöcker et al. (2002) built a laboratory prototype laundry dryer using CO_2 as a working fluid. Their experiments showed that the heat pumps used in laundry dryers at 50°C–60°C exhibited a significant energy savings potential (between 53% and 65%) compared to conventional high temperature (130°C) drying methods. However, their drying cycles with supercritical CO_2 heat pumps in batch mode were relatively short (about 54 min). The reported experimental results showed that the end of the drying cycles was dependent on the maximum air temperature entering the evaporator (40°C) rather than the material final MC. These authors also showed that up to 35% of each drying cycle was a transient process with air temperatures leaving the gas cooler (entering the drum) being below the desired set points (50°C–60°C). Honma et al. (2008) presented an experimental study on a compact heat pump dryer using CO_2 as a refrigerant for domestic cloth washers/dryers. The authors mentioned the requirement to balance the amount of heat provided by the air with the heat supplied by the gas cooler and to adapt the refrigerant cycle based on the progress of the drying cycle. They controlled the EXV to keep the superheat constant in order to maintain the targeted heating capacity at a constant level of 2.7 kW and reduce the drying time. The gas cooler had a heat transfer surface about 40% higher than that of the evaporator for the same airflow rate, and the heat pump COP was estimated at 4.07. The tests performed revealed that the prototype was able to reduce electric power consumption by 59.2% and the drying time by 52.5% in comparison with electrical heater drying systems. Furthermore, the authors estimated that the drying time can be further reduced by 3% by keeping the refrigerant superheating temperature at 6°C–10°C.

1.3.5 REFRIGERANTS

Selection of refrigerants as working fluids for mechanical vapor compression heat pumps depends on their thermodynamic properties and environmental impacts. The environmental impacts are due to leakages in the atmosphere and greenhouse gas emissions associated with energy use.

Refrigerants are classified as follows (UNEP Ozone Secretariat):

- Chlorofluorocarbons (CFCs), as CFC-11 (trichlorofluoromethane) and CFC-12 (dichloro-difluoromethane), with strong destructive impact on the ozone layer if released into the atmosphere and important influence upon the global warming; Molina and Rowland (1974) have first suggested that CFCs might be responsible for the destruction of the stratospheric ozone layer; the manufacture of these refrigerants was discontinued in January 1996.
- Hydrochlorofluorocarbons (HCFCs), as HCFC-22, with reduced impact on the ozone layer and on the greenhouse effect (i.e., with relatively low ODPs and GWPs), have been used as short-term alternatives to CFCs.
- Hydrofluorocarbons (HFCs) such as HFC-134a (1,1,1,2-tetrafluoroethane) and HFC-410A are harmless to the ozone layer and with less than 1% of the global warming effect of all greenhouse gases and are used as long-term alternatives to CFCs and HCFCs; although HFCs do not contain chlorine atoms and are very stable fluids, their GWP values are relatively high (see Table 1.1); HFCs are continuously subjected to progressive phase-down in order to eliminate all adverse environmental effects.
- Inorganic natural refrigerants as carbon dioxide (CO_2, R744) and ammonia (NH_3, R717), and flammable hydrocarbons (HCs) (R-600-propane, R-600A-isobutane, ethylene, and propylene); all these fluids are harmless to the ozone layer, and have less or no influence upon global warming.

In ANSI/ASHRAE Standard 34 (2013), refrigerants are also classified, according to their toxicity and flammability in six safety groups coming from the least to the most hazardous.

During the past few decades, CFCs, HCFCs, and HFCs have been extensively used in refrigeration, heat pump, and air-conditioning fields due to their excellent thermodynamic characteristics. However, a disadvantage of these synthetic refrigerants is their strong contribution to the greenhouse effect in case of massive leakage. Consequently, in 1987, Montreal Protocol has regulated the production and trade of ozone depleting substances. As a result, CFCs were completely phased out in the developed countries and extensive R&D activities have been initiated to find new, more environmental-friendly refrigerants.

The impact of refrigerants on ozone destruction is estimated with the ozone depletion potential (ODP), which is a measure of the destructive potential of a particular fluid because of a complex reaction with chlorine compounds present in refrigerants, relative to depletion caused by an equal amount of a reference substance. CFC-11 is typically defined as the reference compound, and is assigned an ODP of 1.0. On the other hand, GWP quantifies the capability of refrigerants to absorb infrared radiation relative to carbon dioxide. The GWPs of synthetic refrigerants are, for example, 1300–2100 times higher compared to that of the natural refrigerant CO_2 taken as a reference.

The total equivalent warming impact (TEWI), also used as an indicator for environmental impact of heat pumps over their entire lifetime, is the sum of the direct refrigerant emissions, expressed in terms of equivalent CO_2 (direct effect) and the indirect emissions of CO_2 from the system's energy use over its service life (indirect effect) (Fischer et al. 1991; Sand et al. 1999).

Generally, direct effects of common refrigerants are negligible, but indirect effects are significant because of energy consumption. The TEWI index can be calculated by using the following equation (ASHRAE 2013):

$$\text{TEWI} = n * l * \text{GWP} + m_R \left(1 - \alpha_{\text{rec}}\right) \text{GWP} + n * \beta * E$$

where:

n is the system technical life

l is the annual leakage rate in the system (3%–8%)

GWP is the global warming potential of refrigerant for 100 years

m_R is the refrigerant charge (kg)

α_{rec} is the refrigerant recycling factor (0.70%–0.85%)

β is the CO_2 emission factor

$E = \dot{W}_{compr} * t$ is the annual energy consumption (kWh)

$\dot{W}_{compr}$ is the compressor shaft input power (kW)

t is the system running time (hours/year)

Finally, the life cycle climate performance (LCCP) of heat pumps includes the TEWI index and adds the effects of direct and indirect emissions associated with refrigerant manufacturing processes.

1.3.5.1 Low-Temperature Synthetic Refrigerants

Table 1.2 lists the most common low-temperature synthetic refrigerants (i.e., fluids with relatively low critical temperature, normally allowing the industrial heat pumps working in the subcritical regime), as well as their most promising alternatives with low GWP indexes. For small-, medium-, and large-size industrial heat pumps, HFC-134a and HFC-410A are traditionally used as low-temperature refrigerants, but they have relatively high GWPs over a 100-year time span, while, for example, the European Union has approved the phase-out of mobile air conditioning systems using refrigerants with GWP greater than 150 (Brian and Sumathy 2011).

This is the reason for what new, zero-ODP, and very low-GWP working fluids are intended to replace the traditional low-temperature HFC refrigerants in industrial heat pumps (Higashi 2010). Among them, there are hydrofluoroolefins (HFOs) refrigerants as HFO-1234yf (2,3,3,3-tetrafluoropropene) and R-1234ze (E) (Calm 2008; Brown 2009) and HFO-1234ze (E) (trans 1,3,3,3-tetrafluoropropene) (Ueda et al. 2012) as promising low GWP substitutes to HFC-134a. The GWPs of HFO-1234yf and HFO-1234ze (E) are as low as 4 and 6, respectively, but these new refrigerants are slightly flammable and relatively not stable.

HFO-1234yf should replace HFC-134a, which has a high contribution to the greenhouse effect (GWP = 1430), especially in the automotive air conditioning systems. Besides, HFO-1234yf has good thermophysical properties very similar to HFC-134a can be directly charged into conventional refrigeration systems with only minor modifications, has a lower pressure and higher cooling capacity and energy efficiency. As an alternative to HFC-134a in mobile air conditioning systems, HFO-1234yf shown cooling capacity and COPs between 4% and 8% lower than those of HFC-134a (Minor and Spatz 2008; Zilio et al. 2011; Ozgur et al. 2014), but, at the same operating conditions, lower compression ratios and discharge temperatures. Navarro-Esbri et al. (2013a, b) shown that the cooling capacity and COPs obtained with HFO-1234yf are about 9% and between 5% and 30% lower,

TABLE 1.2
Some Properties of Low-Temperature Refrigerants

Refrigerant	ODP (CFC-11 = 1)	GWP (CO_2 = 1; 100-Year ITH)	Critical Temperature	Critical Pressure
–	–	–	°C	MPa
Present Refrigerants				
HFC-134a	0	1430	101.08	4.0603
HFC-410A	0	1890	72.13	4.9261
Replacements				
HFO-1234yf	0	4	124.69	3.382
HFO-1234ze(E)	0	6	109.35	3.632

respectively, than those obtained with HFC-134a, and that these performances further decrease when the condensing temperature increases and if internal heat exchanges are used.

On the other hand, HFO-1234ze (E) is considered as the best medium pressure, low GWP refrigerant available on the market. It is an energy-efficient alternative to traditional refrigerants (as HFC-134a) in air-cooled and water-cooled chillers for supermarkets and commercial buildings, as well as in other medium temperature applications such as heat pumps, refrigerators, vending, beverage machines, air dryers, and CO_2 cascade systems in commercial refrigeration. Field tests with air-cooled chillers using HFO-1234ze (E) instead propane (R-290) showed significantly lower energy consumption.

1.3.5.2 Natural Refrigerants

The worldwide efforts for fossil energy savings are also accompanied by the replacement of synthetic refrigerants (CFCs, HCFCs, and HFCs) by natural refrigerants such as carbon dioxide, ammonia and water, and HCs (Table 1.3). Because of its negligible impact on climate change, carbon dioxide (CO_2, R-744) is a promising natural refrigerant for usage in direct or indirect refrigeration systems, as well as in sub- (see Section 1.2) or supercritical heat pumps (see Section 1.3), often in combination with ammonia in cascade-type installations. CO_2, actually recovered from other industrial processes, is not toxic, nonflammable, noncorrosive, relatively inexpensive, and readily available. However, the high working pressures required and the (too) low critical temperature (31.1°C at a critical pressure of 73.7 bars) limit the usage of CO_2 as a working fluid in certain refrigeration and heat pump applications. For example, in conventional (subcritical) heat pumps, the low critical temperature limits the operating temperature range, that is, heat cannot be delivered at temperatures greater than this value. In addition, saturation pressure is generally between 3 and 5 MPa higher than that of most HFC refrigerants. CO_2 has however some interesting environmental features, such as low GWP, zero ODP, and nonflammability.

Ammonia (NH_3, R-717) is suitable for usage in industrial low- and high-temperature heat pumps, because it can easily provide temperatures as high as 80°C, and even 90°C, with relatively high energy efficiency, but with specific safety measures.

Water (H_2O, R-718) can be used as a very high-temperature (above 100°C) refrigerant because at such temperatures, the working pressures are much lower compared to other refrigerants. A disadvantage of the use of water is its low density of the gaseous phase, which requires relatively high compressor capacities. For heat pumps working at temperatures above 100°C, additional R&D works are required (IEA 2015a).

HC refrigerants, such as R-600a (isobutane) and R-290 (propane), have been already widely used in small domestic refrigerators and drink coolers, which require a reduced refrigerant charge

---

TABLE 1.3

Properties of Today's Most Promising Natural Refrigerants for Heat Pumps

Refrigerant	ODP[a]	GWP[b]	Critical Point	
			Temperature	Pressure
–	–	–	°C	MPa
Carbon dioxide (CO_2, R-744)	0	1	30.98	7.474
Ammonia (NH_3, R-717)	0	0	132.25	11.483
Water (H_2O, R-718)	0	0	373.946	22.06
Propane (R-290)	0	3	96.7	4.30
Isobutane (R-600a)	0	3	134.7	3.677

[a] Based on ODP of CFC-11 = 1.
[b] Based on GWP of CFC-11 = 1 over 100 year ITH.

and are probably the most suitable low GWP substitute for HFC-134a. Because compared to other refrigerants, the increase of pressure with temperature of these natural refrigerants, they are suitable for use in heat pumps providing temperatures higher than 80°C.

1.3.5.3   High-Temperature Refrigerants

The most common available refrigerants, in particular HFCs, are limited to heat distribution temperatures of maximum 80°C (IEA 2015a). To provide heat at higher temperatures (i.e., at 90°C–120°C) by recovering industrial waste heat at temperatures up to around 50°C–60°C and, thus, reduce the specific energy consumption (kWh/product unit), several synthetic and/or natural fluids are available, or will be developed, for industrial heat pumps (Table 1.4). Natural refrigerants as *carbon dioxide* (CO_2, R-744), ammonia (NH_3, R-717), and water (R-718) (see Section 1.2.3.2) are promising environmental-friendly high-temperature refrigerants for industrial heat pumps. Other high-temperature synthetic refrigerants, such as R-236fa (Table 1.4), are also employed since 2000.

HFC-245fa (1,1,1,3,3-pentafluoropropane) is a hydrofluorocarbon that has a relatively high GWP value (950) (i.e., 950 times the global warming effect of CO_2), no ODP, is nearly nontoxic and practically nonbiodegradable with a lifetime of 7.2 years when it eventually does escape into the atmosphere. This refrigerant provides comparable and, in some cases, improved performance in high-temperature heat pumps developed, for example, for industrial drying, to reheat circulating water and to generate steam from waste heat available at 50°C–60°C (Zyhowski et al. 2002).

HFO-1234ze(E), with a relatively high critical temperature (108.91°C) could be considered a moderate high-temperature refrigerant (see Section 1.2.3.1) and the isomer HFO-1234ze(Z) are today attractive candidates substitute for HFC-236fa and HFC-245fa in high temperature heat pumps because of their low GWP (Brown at al. 2009; Fukuda et al. 2014). In particular, HFO-1234ze (Z) exhibits a very high critical temperature (around 150°C) that allows operating sub-critical cycles at the high temperatures required by industrial heat pumps. Therefore, it seems to be the most promising low-GWP refrigerant for high- and very high-temperature heat pumps. During condensation, HFO-1234ze (Z) showed heat transfer coefficients and frictional pressure drop higher than HFC-236fa, as well as weak sensitivity to saturation temperature and great sensitivity to refrigerant mass rate. For example, at mass flux higher than 15 kg/m²s, forced convection condensation occurs and the heat transfer coefficients increase by 30%. By doubling the refrigerant mass flux, HFO1234ze(Z) exhibits higher heat transfer coefficients (between 48% and 82%) and frictional pressure drop (73%–82%) than those of HFC-236fa under the same operating conditions (Longo et al. 2014).

TABLE 1.4
Some Properties of Available High-Temperature Refrigerants for Heat Pumps

Refrigerant	ODP	GWP	Critical Temperature	Critical Pressure
–	–	–	°C	MPa, a
Carbon dioxide (CO_2, R-744)	0	1	30.98	7.474
Ammonia (NH_3, R-717)	0	0	132.25	11.483
Water (R-718)	0	0	373.946	22.06
HFC-236fa	0	6300	124.9	3.12
HFC-245fa	0	950	154.01	3.651
HFO-1234ze(E)	0	6	109.35	3.632
HFO-1234ze(Z)	0	<5	153.7	3.97
HFC-365mfc	0	950[a]	189.85	3.266

[a] At 9.9 years atmospheric lifetime.

Finally, R-365mfc, a new hydrofluorocarbon (1,1,1,3,3-pentafluorobutane), is considered to be suitable as a refrigerant of high-temperature heat pumps aiming at generating steam using waste heat because of its very high critical temperature (189.85°C), but its GWP value is relatively high (i.e., 950 at 9.9 years atmospheric lifetime). Therefore, further R&D works are required to develop substitutes for R-365mfc (IEA 2015a).

1.3.5.4 Refrigerant Mixtures

Refrigerant mixtures can be zeotropic or nonzeotropic (or a-zeotropic). Zeotropic refrigerants (e.g., HFC-245fa/HCF-134a used as the working fluid for heat pumps generating steam at >120°C) are chemical mixtures (blends) of two (or more) fluids that have different boiling points, and undergo phase changes at varying temperatures during constant-pressure evaporation/boiling and condensation process. In other words, these processes achieve temperature *glides*, contrary to pure single-component refrigerants that evaporate and condense at constant temperatures. Their boiling and condensation processes are thus nonisothermal, and dew and bubble-point curves do not touch each other over the entire composition range. If the temperature *glide* fits with the temperature change of the source/sink heat, it will contribute to higher heat pump COPs (Pan and Li 2006). A drawback of zeotropic mixtures is the preferential leakage of more volatile component(s) leading to change in the mixture composition over time.

Azeotropic refrigerants, mostly used in low-temperature refrigeration applications, are mixtures in which the dew and bubble-point curves are touching each other at least at one point, indicating the same composition in the vapor and liquid phase. They evaporate and condense at constant temperatures acting as a single substance in both liquid and vapor phase, and their components cannot be separated by simple distillation.

Refrigerant mixtures can be

1. HC-based (e.g., R-1270/R-290/R-152a) as alternatives to R-502 in low-temperature refrigeration applications.
2. HFC based-mixtures (e.g., R-404A/R-407C/R-410A) as potential alternatives to HCFC-22 in refrigeration and heat pump systems.
3. Carbon dioxide-based (e.g., R-744/R-41, R-744/R-32, R-744/R-23, and R744/R125) as working fluids in low-temperature applications.
4. Ammonia-based mixtures (e.g., R-717/R-170) having lower compressor discharge temperature compared to R-717 and good miscibility with mineral oil, thereby reducing the usage of highly hygroscopic synthetic oils (Cox et al. 2009).

1.3.5.5 Nanorefrigerants

Nanorefrigerants are mixtures of base (host) fluids and nanoparticles. The base fluid is a conventional heat-transfer fluid, such as water (the most commonly used, having the greatest aptitude to suspend noncoated nanoparticles, in comparison with other base fluids such as ammonia, HCs, HFCs, and HCFCs), light oils, ethylene glycol, or refrigerants, whereas nanoparticles are colloidal suspensions, such as metallic (Cu and Au) or metal oxides (Al_2O_3, TiO_2, and ZrO_2), with very small dimensions (1–100 nm), allowing limiting, fouling, sedimentation, erosion, and high-pressure drops (Henderson et al. 2010).

Such mixtures provide much higher thermal conductivity and boiling heat transfer coefficients due to the nature of base fluid (e.g., highly wetting), particles' concentration, surface properties, and higher heat transfer areas, as well as better dispersion stability with predominant Brownian motion that prevents gravity settling and agglomeration of particles, reduced particle clogging and pumping power as compared to the base fluid (Choi 1995; Barber et al. 2011).

In mechanical vapor compression heat pumps, nanoparticles are added to lubricants in order to enhance the solubility of mineral oils and, at the same time, improve the convective boiling heat transfer in evaporators up to 30%, depending on the type of refrigerant and nanoparticles used (Peng et al. 2009). Such heat transfer improvement is normally achieved by varying the particle concentration in the base fluid in order to increase its thermal conductivity and surface wettability, while changing the fluid viscosity and density (Barber et al. 2011). Most researches also confirmed an enhancement in the critical heat flux during nanofluid convective boiling (by up to 53%), which would allow for more compact and effective industrial heat pumps. However, this enhancement of the critical heat flux could be related to the nanoparticles (as alumina—Al_2O_3, zinc oxide, and diamond) deposition and cluster on the heat transfer surfaces, especially in microchannel heat exchangers (Lee and Mudawar 2007; Kim et al. 2010; Barber et al. 2011).

Adding nanoparticles to refrigerants may also lead to increase of pressure drops. However, if the penalty in pressure drops is not considerable in comparison with heat transfer enhancement, the use of nanorefrigerant can be helpful in industrial heat pumps.

In the case of a domestic refrigerator using TiO_2/CFC-12 as a nanorefrigerant working fluid, the compression work has been reduced by 11% and the COP increased by 17% (Sabareesh et al. 2012). A refrigeration system using the Al_2O_3/R-600a as nanorefrigerant reduced the compressor power input by about 11.5% compared to the conventional POE oil, whereas the system COP was enhanced by about 19.6% (Kumar et al. 2013). With TiO_2-R600a as nanorefrigerant working fluid, domestic refrigerators reduced the energy consumption by 9.6%, whereas the COP increased by 19.6% (Bi et al. 2008). However, it can be noted that in heat pump and refrigeration applications, where the main heat transfer processes are convective vaporization (boiling) and condensation, increasing the internal forced convective heat transfer coefficients will have little or any impact on the heat exchangers' overall heat transfer coefficients. As a consequence, this may not improve the heat pumps' overall performance, because the actual limits for the global heat transfer are the airside and/or water-side heat transfer coefficients that are too small compared to internal convective vaporization/boiling and condensing heat transfer coefficients.

1.3.6 COMPRESSORS

Four different types of compressors are traditionally used in mechanical vapor compression heat pumps: scroll (in small and medium heat pumps up to 100 kW heat output), reciprocating (up to approximately 500 kW), screw (up to around 5 MW), and turbo compressors in large systems above 2 MW, as well as oil-free turbo compressors above 250 kW.

For industrial heat pumps, the majority of compressors is of the positive displacement (or reciprocating, most of them being limited to compression ratios on the order of 8:1), screw or rotary vane type. The vast majority of such compressors are available with variable-speed drives to run the drive motor at speeds that match the heat pump load heating at any given time and, thus, reduce energy costs by as much as 30% compared with traditional fixed-speed compressors. However, historically their technological advances are relatively slow, mostly being depended economically dependent on the market size. Despite the improvements already achieved, there is a continuing need to make heat pump compressors operate more quietly, reliably, and with higher efficiencies over a long service life.

During the past two decades, novel single- and two-stage reciprocating compressors have been adapted and/or developed in response to high-temperature/pressure mechanical vapor compression heat pump cycles using natural refrigerants as ammonia (NH_3), carbon dioxide (CO_2) and, also, HFCs as HFC-410A (e.g., SABROE), and to associated lubricant issues. For example, Danfoss Turbocor developed variable-speed, two-stage compact centrifugal, totally oil-free compressors with HFC-134a and HFO-1234ze (E) as refrigerants, with capacities ranging from 210 to 700 kW focusing air-cooled water chillers for HVAC applications. These compressors have no metal-to-metal contact of rotating components, and include magnetic bearings, variable-speed permanent magnet motors, and intelligent digital electronic solid-state controls. Semi-hermetic piston compressors for supercritical CO_2 heat pumps with maximum allowable pressure on the high pressure side of the compressor of 163 bars are today available on the market (e.g., DORIN products; http://www.dorin.com/en/ News-46/, accessed March 16, 2015). To overcome the large efficiency losses associated with the throttling processes of supercritical CO_2 heat pumps, because of very-high pressure differences required across the compressors, a number of developments have been achieved for various types of positive displacement compressors, mainly of the vane type, which combine compression with some recovery of work from the expansion process (Stošić et al. 2002) (see Section 1.2.2.1). Since 2008, EMBRACO supplies single-stage compressors for CO_2 heat pumps and various light commercial refrigeration systems such as beverage coolers, vending machines, chest freezers, ice cream freezers, and dryers (http://www.embraco.com/Default.aspx?tabid=40, accessed March 16, 2015). The BITZER's OCTAGON subcritical CO_2 compressor, with cooling capacities ranging from 2.7 to 80 kW (−35°C/−5°C), distinguishes by energy efficiency improvements and higher saturation discharge temperatures, as were the admissible pressure loads for high and low pressure (53/30 bar). The BITZER's ECOLINE compressors for R-290/R-1270 are also high efficiency semi-hermetic reciprocating compressors. The four- and six-cylinder models are equipped with capacity control systems (cylinder cut-off), allowing them to run with up to four or six levels of output. Particularly for applications with high load fluctuations, this allows for cost-effective and energy-efficient operation under full- and part-load with all conventional refrigerants. Finally, BITZER's semi-hermetic two-stage compressors are optimized for low-temperature cooling ranges (https://www.bitzer.de/us/ products/technologies/reciprocating-compressors/index-2.jsp, accessed March 16, 2015).

Two-stage compressors are normally combined with intercooling, that is, the process of desuperheating the discharge superheated vapor from the first-stage compressor, for example, by direct-contact with liquid refrigerant maintained at the intercooling (high-stage suction) pressure.

For two-stage compression heat pumps, the optimum intermediate pressure ($P_{opt,int}$) is given by the following equation:

$$P_{opt,int} = \sqrt{P_{discharge,sat} - P_{suction,sat}} \qquad (1.27)$$

where:

$P_{discharge,sat}$ is the absolute saturation pressure corresponding to the discharge or condensing conditions

$P_{suction,sat}$ is the absolute saturation pressure corresponding to the suction saturated conditions

Two-stage compression heat pump systems using ammonia as a refrigerant have been recently developed, because the combination of compression ratio limits of reciprocating or rotary vane compressors and refrigerant discharge superheat limit the ability to provide useful heat pumping in a single-stage compression cycle. Effectively, reciprocating and rotary vane compressors have physical compression ratio (i.e., ratio of absolute discharge to absolute suction pressure) and oil temperature limits depending on the refrigerant discharge temperature. During the compression process, the ammonia pressure increases and, because of its low heat capacity, the temperature increases dramatically. High discharge temperatures tend to increase the rate of compressor lubricating oil breakdown as well as increasing the likelihood of compressor material fatigue. To avoid this, an external source of cooling for the compressor (water- or refrigerant-cooled heads) must be used. To reduce the compressors' energy consumption for a required compression ratio, one approach is to keep the compressor temperature low during running by using suction cold gas or external means, and undergo isothermal compression process by transferring heat from the compression chamber (Chua et al. 2010).

The integral two-stage inverter drive ammonia screw compressors (e.g., KOBELCO; http://www .kobelco.co.jp/english/machinery/products/rotation/refrigerationunit/iza2/index.html, accessed March 16, 2015) are equipped with economizers that regulate and control rotating speed at the first and second stages of the inverter drive. The piston valve traditionally used for capacity control has been replaced to inverter drive capacity control to ensure optimum control of cooling capacity fluctuation.

During the past two decades, the invention of scroll compressors was a major technological break-through in the compressor technology (Wang et al. 2009). Scroll compressors are approximately 10% more efficient than the standard reciprocating compressors because (1) the suction and discharge processes are separate, meaning that no heat is added to the suction gas as it enters the compressor; (2) the compression process is performed slowly and, therefore, fluctuations in driven torque are only 10% of those of reciprocating compressors; (3) the scroll compression mechanism enables the elimination of the suction and discharge valves, which are a source of pressure losses in reciprocating compressors; (4) in addition, scroll compressors have better reliability, because they have fewer moving parts and can operate better under liquid slugging conditions (Winandy and Lebrun 2002). Compact and hermetic scroll CO_2 compressors (e.g., Copeland) allow reducing the refrigerant leakage. They have been designed specifically for use in R-744 (CO_2) refrigeration and heat pump systems. They are 30%–60% lighter than reciprocating piston compressors of similar capacity. A further limitation in the use of CO_2 in normal air conditioning and refrigeration systems is that the range of operating pressures and temperatures required are close to the critical point of CO_2. Hence, the losses associated with throttling are much larger than those associated with conventional refrigerants. It follows that some recovery of power is required from the expansion process in order to achieve an acceptable COP from a CO_2 cycle. In the field of industrial heat pumps and refrigeration, the use of reciprocating and vane compressors is decreasing, while the number of screw compressors is expected to increase. The main advantage of screw compressors over scroll compressors is their fairly large pressure ratio and their excellent part load characteristic. In addition, there are no net radial or axial forces exerted on the main screw or drive shaft components due to the work of compression. Because compression occurs symmetrically and simultaneously on opposite sides of the screw, the compression forces are canceled out. The only vertical loads exerted on the main screw bearings are due to gravity. Because the discharge end of the screw is vented to suction, the suction gas pressure is exerted on both ends of the screw resulting in balanced axial loads (http://www.emersonclimate.com/en-us/brands/vilter/Pages/Vilter.aspx, accessed June 25, 2015).

Screw compressors are simple, reliable, and compact positive displacement rotary volumetric machines, in which all the moving parts operate with pure rotational motion enabling them to operate at higher speeds with less wear than most other types of compressor. They essentially consist of a pair of meshing helical lobed rotors contained in a casing. Together, these form a series of working chambers by means of views from opposite ends and sides of the machine (http://www.carlylecompressor.com). Screw compressors are up to five times lighter than equivalent (same capacity) reciprocating compressors and have a nearly 10 times longer operating life between overhauls. Furthermore, their internal geometry is such that they have a negligible clearance volume and leakage paths within them decrease in size as compression proceeds. Thus, provided that the running clearances, between the rotors and between the rotors and their housing, are small, they can maintain high volumetric and adiabatic efficiencies over a wide range of operating pressures and flows (Stošić 2002). Significant R&D advances and great improvements in performance have been achieved during the past few years in the screw compressors' mathematical modeling and computer simulation of the heat and fluid flow processes, as well as in rolling element bearings and rotor profiles (e.g., with very small tolerances of the order of 5 μm or less at affordable costs allowing reducing the internal leakages) and manufacturing. However, pressure differences across screw compressor rotors impose heavy loads on them and create rotor deformation, which is of the same order of magnitude as the clearances between the rotors and the casing. Consequently, the working pressure differences between entry and exit at which twin screw compressors can operate reliably and economically are limited. These pressure differences create very large radial and axial forces on the rotors whose magnitude and direction is independent of the direction of rotation. Current practice is for a maximum discharge pressure of 85 bars and a maximum difference between suction and discharge of 35 bars (Stošić 2002). Even the development of screw compressors is now advanced, significant improvement through better rotor profiling, housing ports, seals, lubrication systems, and the entire compressor design are still required.

High-pressure single-stage ammonia screw compressors have been developed for industrial heat pumps (http://www.emersonclimate.com/en-us/brands/vilter/Pages/Vilter.aspx, accessed June 25, 2015). This technology allows for sustained operation with ammonia at extremely high saturated condensing temperatures than conventional compressors (see Figure 1.38a and b). Integrated, for example, into existing ammonia refrigeration systems, the high-pressure ammonia heat pumps provide a cost-effective solution to harnessing and converting waste heat of rejection to high grade hot water, up to 90°C. A new high-pressure double screw ammonia compressor for high-temperature heat pumps was developed in Germany in the range from 165 to 2838 kW drive power (IEA 2015b). It allows achieving discharge pressures up to 63 bars and maximum condensing temperature up to 90°C with high COPs. For example, with a heat source at 35°C and a heat sink at 80°C, a single-stage ammonia heat pump using such a compressor can reach a COP of 5.0 at a heating capacity of 14 MW (source: GEA Refrigeration Technologies; http://www.gea.com/global/en/products/gea-refrigeration.jsp; accessed June 25, 2015).

Based on the Danish Rotrex turbo air compressor, a new water vapor (steam) high-speed radial turbo-compressor is under development (Todd et al. 2008). It is dedicated to industrial heat pumps using waste heat at temperature varying between 50°C and 200°C. As working fluid, steam has excellent thermodynamic properties at high temperatures to meet high COP values, is nontoxic and has zero greenhouse potential. The new compressor unit is designed for a pressure ratio up to 3, which is equivalent to a temperature lift about 30°C. To achieve a higher temperature lift, two-stage serial coupling compressors can deliver temperature lifts between 30°C and 60°C. Among focused applications are the high temperature heat pump dryers where the new turbo compressors can be configured in series or parallel to match the operational specification of actual drying systems. Preliminary analyses have shown that the new turbo-compressor-based heat pump drying concepts can achieve COPs between 4 and 7 (IEA 2015b).

1.3.7 Lubricants

To lubricate internal parts of compressors, and to clean and seal the piping/heat exchanger system, heat pumps use lubricating oils mixed with refrigerants. Based on the application types, the selection of lubricants, and the percentage of a given refrigerant that can be dissolved into lubricants, depends on their degrees of miscibility, solubility and volatility, and compatibility with materials used in the heat pump system, such as metals, coatings, and elastomers. Lubricants must be thermally stable enough over the lifetime of the heat pump systems, while contaminants such as water, impurities, and acids must be kept to an absolute minimum, so that the problems such as capillary plugging cannot occur. They also must have a low enough viscosity to provide less drag and good oil return process, as well as a high thermal conductivity. Miscibility and solubility depend on the working pressure of the refrigerant vapor, the temperature of the lubricant and the length of time the two are in contact, play a determinant role in the lubricating oil return to the compressor and, consequently, in the compressor reliability and system performance/efficiency. For example, excessive quantity of oil accumulated in the evaporator reduces the intensity of heat transfer, and thus reduces the efficiency and capacity of the heat pump. Consequently, heat pumps must be designed to promote oil return to the compressor. In the case of large compressors, oil pumps are utilized to continuously supply oil to the compressor crankshaft bearings and shaft seal prior returning to the sump. The percentage of the refrigerant that can be dissolved in oil increases considerably as the temperature of the oil decreases and the pressure of the refrigerant vapor increases. If during a short time, a large quantity of refrigerant vaporizes under the right conditions, then severe oil foaming will result, leading to its migration out of the compressor, thus endangering its operation. The oil loss that occurs as a result of the rotating masses in the crankcase pools the oil from the sump and intensifies mist lubrication to all internal components in the compressor.

CFCs, HCFCs, and HC refrigerants commonly use mineral oils. In order to be used for refrigeration and heat pump applications, mineral oils require special refining to reduce the impurity levels (wax and sulfur content). They are soluble in such lubricants and the viscosity of the refrigerant/lubricant mixture is lower than that of the lubricant alone. Because the attempts to use mineral oils with the HFCs had limited success, synthetic lubricants such as polyalkylene glycols (PAGs) and polyolester oils (POEs) are extensively used today. Such synthetic oils need to be handled more carefully, because they are more hydrophilic than mineral oils. PAGs and POEs are also more miscible with HFC refrigerants. For HFC refrigerants, small amounts of additives are formulated into lubricants in order to boost antiwear protection, metal passivation, acid scavenging, and de-foaming. PAGs and POEs have significantly better miscibility and excellent low-temperature characteristics compared to mineral oils, but they are much more hygroscopic. PAGs are extensively used in most automotive air conditioning applications and, particularly, in direct exchange evaporators. POEs are the lubricants of choice for most stationary refrigeration and heat pump applications. Better performance was seen with PAGs because of the formation of carbonate layers, thus leading to better performance. In contrast, the hydrolysis of the POEs lubricant led to a drop in viscosity of the lubricant and, as a consequence, inferior results. Antiwear additives based on sulfur and phosphorus chemistries used in small quantities can boost the performance of POEs when used with HFC refrigerants. Special types of additives may be required for use with PAGs as conventional types of these additives may not be compatible. Because of their chemical reactivity, they may reduce the lubricant's overall stability and, thus, care must be taken in the antiwear additives selection. Alkylated benzenes (which offer better stability and improved miscibility with HFCs as compared to mineral oils) and polyvinyl ethers (PVEs) are also used in particular applications. The new fluorinated refrigerants, such as HFO-1234yf, are compatible with both PAGs and POEs.

Modern lubricants for ammonia are polyalfaolefine (PAO), PAG (Pearson 2008), and alkylated benzene blends, particularly recommended for low-temperature applications. PAO lubricant provides excellent long life stability and has a high viscosity index, so it remains relatively viscous at high temperature, but still flows freely at low temperature. It is not miscible in ammonia, so some form of oil management is required if it is to be automatically returned to the compressor. To avoid slight shrinkage of O-rings and other elastomers, PAOs are often supplied with a seal additive or in a 50%/50% mix with alkylbenzene lubricant (Pearson 2007, 2008). PAG lubricant is miscible in ammonia and so enables to be used in systems working with fluorocarbon refrigerants. PAGs have a strong affinity for water, so it must be used carefully. It is also relatively expensive (about three times the price of PAO), so there is temptation to keep oil that is left in a drum and not required for top up. This is a bad practice because once a container has been exposed to atmosphere, the oil should be used or sent for disposal (Pearson 2008).

In the case of carbon dioxide (CO_2), because of high compressor discharge temperatures and operating pressures in heat pump applications (up to 160°C and 13 MPa, respectively), special attention must be paid to the stability and wear performance of the lubricant used. For example, end-capped PAG lubricants provide performance improvements over conventional PAGs. POEs and PAOs can also be used in low-temperature applications, even though this base stock is not miscible with CO_2 as a refrigerant. PAGs present the advantage of a better lubricity, because they form carbonates with CO_2. However, POEs have superior miscibility with CO_2 and can provide good performance under heavy duty antiwear/extreme pressure operating conditions when used with additives. More and more attention in the past 5–10 years has been given to lubricant and additive chemistries that can enhance heat transfer and compressor efficiency. Recent studies have shown that the heat transfer of a refrigerant is a strong function of lubricant properties. When a lubricant is added to a refrigerant, either an enhancement or degradation in heat transfer performance is achieved relative to that of the refrigerant, depending upon the lubricant's viscosity, miscibility, and concentration. Addition of copper-based nanoparticles to the lubricant has been shown to boost the heat-transfer properties of the refrigerant.

Future works have to upgrade the hydrolytic stability of POEs. One of the main concerns is the operation of POEs under adverse conditions. Under the right conditions, ester hydrolysis can occur to produce undesirable byproducts such as carboxylic acid that can react with some metals and elastomers. Enhancement of the POEs with improved hydrolytic stability or transition to oils resistant to hydrolysis, such as PVEs, for example, may promote robustness of a refrigeration system, and in some cases, its simplification by removing the filter dryer. Finally, the future use of refrigeration lubricants will initially be dictated by the movement toward new refrigerants such as HFOs that exhibit minimal GWP. The compressor industry anticipates that POEs and PAGs will remain as the base stocks of choice with HFOs.

1.3.8 Microchannel Heat Exchangers

Growing material and energy saving requirements, as well as space limitations for industrial heat pump packaging and ease of unit handling, have focused many R&D activities on the development of miniaturized (or microchannel), light-weight finned-tube evaporators and condensers (i.e., heat exchangers with hydraulic diameters of 1 mm or less) that can provide higher heat transfer coefficients and compactness, as well as lower refrigerant charges and costs compared to conventional devices (Tuckerman and Pease 1981; Detlef et al. 2003; Kandlikar et al. 2006; Hrnjak and Litch 2008). Microchannel heat exchangers reduce heat exchanger size and refrigerant charge of heat pump systems and could aid to their successful commercialization.

Recent advancements in industrial heat exchanger microminiaturization techniques have made possible the fabrication of microchannels in order to reduce the heat pump's weight, improve their compactiveness and size, reduce the refrigerant charge (e.g., up to 20 g/kW, much low compared to ammonia-based conventional water chillers) (Hrnjak and Litch 2008), intensify the overall heat transfer coefficients by using smaller equivalent channel diameters, and optimize/reduce the total manufacturing costs. Compared to the conventional heat exchangers, microchannel heat exchangers provide different heat transfer and flow characteristics as, for example, (1) their reduced scale enhances the fluid compressibility; (2) their increased roughness increases the drag coefficient; (3) the refrigerant is distributed in parallel over a large number of small passages, which reduces the flow velocity in each individual channel; therefore, shorter parallel channel lengths and minimal axial heat conduction, combined with well-distributed flow, result in a low channel side pressure drop, and hence reduce the liquid side pump capacity; and (4) as the surface area-to-volume ratio increases, surface and viscous forces are strengthened, enhancing the axial conduction heat transfer coefficient; for a given heating capacity, the high heat transfer property of microchannels results in shorter channel lengths; as a result, undesired axial heat conduction is minimized, because the channel length and fluid residence time are shortened and also because the entire bulk fluid is in close contact with the microchannel walls. Effectively, the local convective heat transfer coefficient (h) in microchannels depends on the heat transfer surface area (A), which is proportional with the channel equivalent hydraulic diameter (D_{eq}). On the other hand, the refrigerant flow rate inside the microchannel depends on the channel cross-sectional area, which varies linearly with D_{eq}^2. Hence, the heat transfer surface area-to-volume ratio (A/V) varies as $1/D$. Consequently, as D, decreases, A/V increases. In the laminar flow regime, the local heat transfer coefficient (h) varies inversely with the channel diameter, that is, $h \approx (1/D)$. Therefore, as the channel diameter decreases, the heat transfer coefficient increases. As a result, the decreased channel diameter improves both the compactness of the unit as well as the heat transfer rate. According to Shah (1991) and Luo et al. (2007), microchannel heat exchangers may provide heat transfer surface densities of $10,000 \, m^2/m^3$ or more, versus $700 \, m^2/m^3$ achieved by compact heat exchangers. Compared to conventional heat pumps with air-source evaporators and water-cooled condensers, microchannel heat exchangers could minimize the air supply blower capacity and reduce the water supply pump electrical input power for a given flow rate.

The first industrial applications of microchannel heat exchangers were for cooling high-density electronic devices in the 1980s and 1990s. In the early 1920s, microchannel heat exchangers began entering the automotive air conditioning systems, as well as in refrigeration, heat pump, and air conditioning industry. Despite the recent technical achievements in the area of microchannel heat exchangers for heat pump applications, heat transfer processes and fluid flow characteristic, as well as the design methods are not yet well known as for conventional heat exchangers. Even it is possible that some of conventional heat transfer equations/correlations are valid for microchannel evaporators, condensers, and other heat exchangers, more R&D work is required to develop heat transfer and flow theories specific to each particular microchannel heat exchanger. Reducing the air-side flow resistance in order to increase the airflow or reduce the blower electrical power input, or both, are thus some of long-term R&D requirements. Refrigerant flow maldistribution inside the microchannel tubes and condensate retention inside the louvered fins, which may prohibit the air from passing through the evaporator, are two major issues for efficient industrial applications. Particularly, the refrigerant maldistribution is a challenge, especially for heat pump evaporators where the working fluid normally enters in a two-phase state. It may reduce the heat exchange efficiency by 30%, versus 9% for the air-side maldistribution (Choi 1995). Consequently, the refrigerant distribution manifolds must be carefully designed in order to allow an optimum distribution of the liquid–vapor mixture and minimize the pressure drops through both the header and microchannels (Kulkarni et al. 2004; Brix et al. 2010).

1.4 AMMONIA MECHANICAL VAPOR COMPRESSION HEAT PUMPS

Low-grade industrial waste heat rejection at temperatures ranging between 10°C and 35°C represents about 25% of the total primary energy input of many manufacturing industries. Simultaneously, several industrial processes and other usages need hot water or air at temperatures between 645°C and 85°C. To recover such low-grade waste heat rejections in order to heat (or cool) residential as well as nonresidential buildings and industrial processes, ammonia heat pumps are among the most energy-efficient technologies available. Because the production of chemical refrigerants HCFC is prohibited since 2010 and will be total phase-out by 2040, to replace them, energy-efficient and cheap natural fluids with no climate damaging effects, as ammonia (R-717, NH_3), have been focused during the recent years. However, ammonia is still not universally accepted as natural working fluid for single- and/or double-stage industrial heat pumps, especially because of its toxicity and inflammability at high concentrations in the ambient air, even if it is widely used as a refrigerant in large industrial systems as distribution warehouses, food refrigeration, and process cooling.

In addition to ammonia toxicity and inflammability, other barriers are (1) too restrictive national standards and scare of engineering designers and (2) still few incentives for industry to invest in low-GWP alternative technologies. Ammonia liquid is a clear fluid that evaporates quickly when exposed to air at room temperature. A major liquid ammonia spill is thus potentially disastrous, because it creates an explosive fire hazard at high concentrations. However, ammonia's fire hazard rating is usually stated as *slight*, because it is explosive in air at high concentrations of 16%–27% (by volume), even if the fire hazard increases if combined with oil or other combustible materials. Ammonia vapor (gas) is colorless, has a suffocating odor, and is much lighter than air. Any eye protection is mandatory under normal working conditions, that is, below 25 ppm in the ambient air. However, at concentrations >70 ppm in the air, ammonia vapor is irritating to the eyes, nose, and respiratory system, which makes it very easy to detect. Consequently, at concentrations >70 ppm, it is unlikely that any person will remain in a contaminated area, unless the person is trapped or unconscious. At 300 ppm in the air, ammonia causes severe irritation of eyes, nose, or respiratory tract, which becomes intolerable after a few minutes with difficulty breathing and possible burning in lungs. At 2000 ppm or more, ammonia can be fatal after a few breaths. To avoid such concentration in the air in case of leak or emergency, fans must provide cross-ventilation using outside air in the mechanical and storage rooms with at least 15 air changes per hour and a 24-h continuous monitoring of ammonia concentration in air connected to an alarm system. People who work with ammonia have to be instructed in its safe use, storage, handling, and disposal, and be provided with providing protective equipment.

It can be noted that if proper and strict safety regulations are observed, industrial heat pumps using ammonia as refrigerant are as safe, maybe even more than other systems so as ammonia's very distinctive smell provides extremely early warning of leakage. Moreover, today, most manufacturers are able to provide hermetic ammonia refrigeration circuits, as well as reliable detection and evacuation systems in case of massive leakage. It is however suitable to consider a unified approach to safety legislation to ensure that the best combination of efficiency, safety, reliability, and ease of use is achieved (Pearson 2007).

1.4.1 Single-Stage Ammonia Heat Pumps

A typical single-stage ammonia heat pump designed for heat recovery from low-grade industrial waste liquid effluents contains standard components as compressor, desuperheater, condenser, ammonia liquid receiver, and evaporator (Figure 1.30).

1.4.1.1 Residential and Industrial Applications

A 9 kW low-charge (100–120 g) water-to-water ammonia heat pump with microchannel and plate heat exchangers, and desuperheater, produced domestic hot water at close to 60°C with condensing temperatures below 50°C (Palm 2008). By using single-stage ammonia heat pumps with standard 25 bar equipment in nonresidential buildings as office buildings, hospitals, and district networks, the maximum supply water temperature is of about 48°C. However, by using two-stage ammonia heat pumps with standard 40 bar compressors, the maximum supply water temperature can be increased to about 68°C (Stene 2008). Another 7 kW (thermal power supplied) ammonia heat pump with microchannel evaporator and reduced refrigerant charge successfully operated at evaporation temperature of −5°C and condensation temperature of 40°C to provide both tap water preheating space heating of a single-family house (Monfared and Palm 2011). Single-stage mechanical vapor compression ammonia heat pumps using screw compressors can be employed, for example, to recover heat from existing ammonia refrigeration systems by diverting discharge of superheated vapor away from existing condensers to the suction inlet of the heat pump system

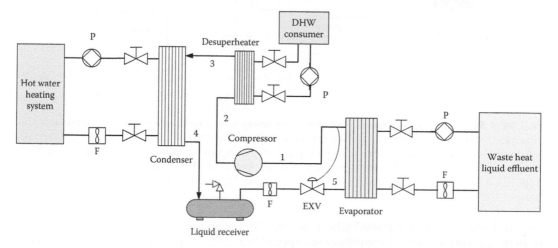

FIGURE 1.30 Schematic representation of a single-stage mechanical vapor compression ammonia heat pump for waste heat recovery. DHW, domestic hot water; EXV, expansion valve; F, flow meter; P, pump.

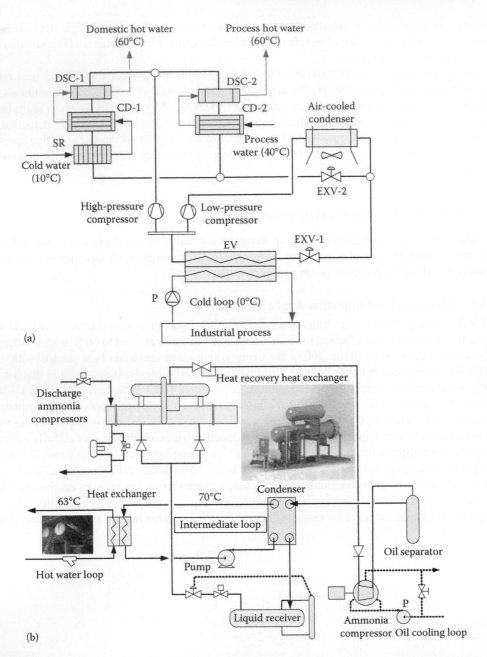

FIGURE 1.31 Single-stage ammonia heat pump recovering heat from existing ammonia compressors. (a) Principle (From http://www.emersonclimate.com/en-us/brands/vilter/Pages/Vilter.aspx; redrawn and reprinted with permission from Emerson Climate Technologies.) (b) As implemented in a Canadian dairy plant (From IEA, Industrial energy-related systems and technologies Annex 13, IEA Heat Pump Program Annex 35, Application of Industrial Heat Pumps, Final Report, Part 1, 2015a; reproduced with permission from IEA HPT Annex 13/35 operating agent.)

(Figure 1.31a). In such an application, heat recovered from the superheated ammonia vapor coming from the existing ammonia compressors in a special heat exchanger is transferred to an intermediate hot water closed loop via a second heat exchanger. Since hot water is not produced and consumed simultaneously, a large hot water storage system is provided. In such a case, the heat pump operates with relatively low differential pressures (up to 24.2 bars) and ammonia may condensate

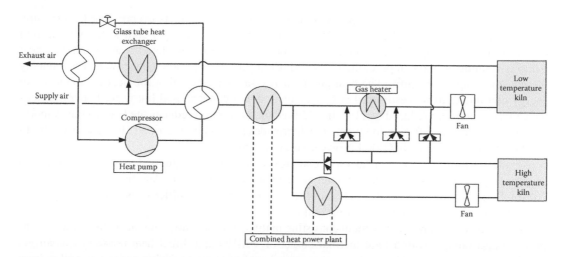

FIGURE 1.32 Ammonia mechanical vapor compression heat pump implemented in a malt production plant in Germany. (From IEA, Industrial energy-related systems and technologies Annex 13, IEA Heat Pump Program Annex 35, Application of Industrial Heat Pumps, Final Report, Part 2, 2015b; redrawn and reprinted with permission from IEA HPT Annex 13/35 operating agent.)

at temperatures up to 71°C in order to produce hot water for domestic or process usage. Such an industrial-scale single-stage ammonia heat pump has been implemented in a Canadian dairy plant (Figure 1.31b) (IEA 2015b).

Another mechanical vapor compression heat pump with ammonia as working fluid has been implemented in a malt production plant in Germany (Figure 1.32). Malt is a major ingredient for beer brewing produced from cereal, which is left to germinate under humid conditions. The germination is stopped by drying the germs in a kiln, a process that typically needs a large amount of hot and dry air at >65°C, while humid exhaust air is released at 28°C. About 2.7 MW of waste heat are recovered by the aid of an ammonia heat pump to preheat the drying inlet air. After the heat pump condenser, the inlet air is further heated by the site's combined heating and power plant. A gas powered auxiliary heater lifts the inlet air temperature up to 65°C, before it enters the low-temperature kiln. The ammonia heat pump operates about 6000 h/year at full load, condensates up to 3000 L/h of water and provides a heating capacity of 3.3 MW with a high COPs (up to 6) because of very low temperature lifts between the heat source and the heat sink (IEA 2015b).

Two ammonia mechanical vapor compression heat pumps equipped with reciprocating compressors (≈26 and 38 kW, respectively) with continuous capacity control have been implemented in 2012 in a French fries producer plant in the Netherlands to provide a part of the energy needed for drying potatoes prior the baking process. The belt-type dryer operates at a maximum temperature of 70°C. The ammonia heat pumps are installed in a mechanical room and the condensers, inside the dryer. The evaporators, installed on the roof, recover heat from the dryer hot and humid exhaust air. By condensing about 1500 kg/h of water, the primary energy savings are estimated at 70% versus the conventional drying system, that is, up to 800,000 Nm³ of natural gas per year. With about 4000 h of operation per year and overall COPs between 5 and 8, the simple payback period was estimated at 4 years (IEA 2015b).

Another 500 kW (thermal capacity) ammonia mechanical vapor compression high-temperature heat pump has been installed (2009) in a veal slaughter house in the Netherlands, in order to provide hot water for cleaning the production rooms and machinery, and removing hair from veal skin, as well as to producing a smaller amount of sterile water at 80°C. The heat pump ammonia reciprocating compressor (maximum discharge pressure of 45 bars) is coupled to the high pressure side of an ammonia 1200 kW refrigeration plant operating at −10°C as evaporating temperature. The heat pump's condenser heats from 15°C up to 62.5°C the municipal cold water with heating COPs of approximately 6.7. The hot water is stored in a 100 m³ insulated water tank. The heat pump is running 16 h/day (approximately 4000 h/year) at almost 100% capacity) and is reducing the gas consumption by 50%. The heat pump energy savings (including the heat pump electrical energy consumption) was estimated at 65% compared to a conventional hot water boiler, whereas the CO$_2$ emissions are reduced by 50%. In addition, the heat pump reduces the load on the evaporative condenser of the refrigeration plant, which provides savings in water and chemical water treatment costs of approximately 6000 Euro/year (IEA 2015b).

An add-on ammonia mechanical vapor compression heat pump equipped with a 71 kW reciprocating compressor has been installed in an artificial ice rink facility in Austria (Figure 1.33) (IEA 2015b). In this application, waste heat from the existing ammonia chiller is used for heating water stored at about 60°C in a storage (buffer) tank.

The combined heat recovery system consisting in three parallel ammonia mechanical vapor compression heat pumps (with a total thermal output of 4 MW) and direct heat recovery exchangers installed in a paper mill in Denmark recovers waste heat from a paper drying process as well as from the natural gas-fired boilers' flue gases (Figure 1.34). The recovered heat is boosted to 70°C and, then, delivered to a district heating network. The ammonia heat pumps recover energy from the hot and moist drying air available at 50°C–55°C and 100% relative humidity. The district heating water is preheated from 37°C to 45°C by using a direct heat exchanger (1.4 thermal MW), and then the three

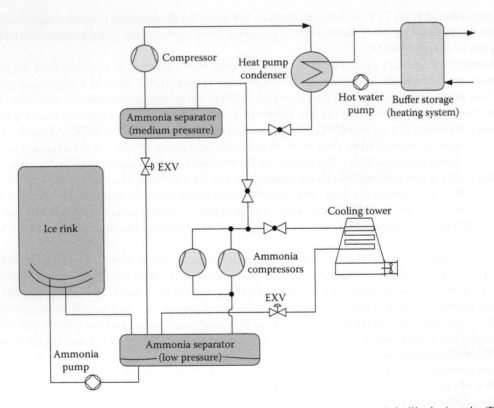

FIGURE 1.33 Process flow sheet of the add-on heat pump for an artificial ice rink facility in Austria. (From IEA, Industrial energy-related systems and technologies Annex 13, IEA Heat Pump Program Annex 35, Application of Industrial Heat Pumps, Final Report, Part 2, 2015b; redrawn and reprinted with permission from IEA HPT Annex 13/35 operating agent.)

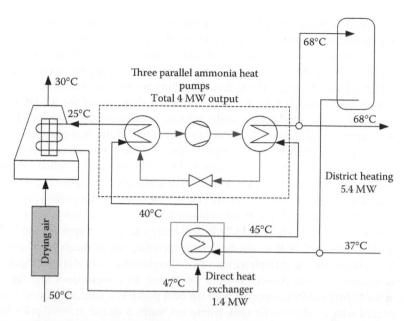

FIGURE 1.34 Principle of the combined ammonia vapor compression heat pumps and direct heat recovery exchangers installed in a paper mill in Denmark. (From IEA, Industrial energy-related systems and technologies Annex 13, IEA Heat Pump Program Annex 35, Application of Industrial Heat Pumps, final report, Part 2, 2015b; redrawn and reprinted with permission from IEA HPT Annex 13/35 operating agent.)

heat pumps increase the water temperature up to 68°C. In periods with very high temperatures of the drying air or low temperatures in the district heating system, the system global COPs have been as high as 11, whereas the total thermal capacity exceeded 5.4 thermal MW (IEA 2015b).

Another mechanical ammonia vapor compression heat pump has been installed in Denmark to recover waste energy from the exhaust air of an industrial washing process (Figure 1.35). The washers

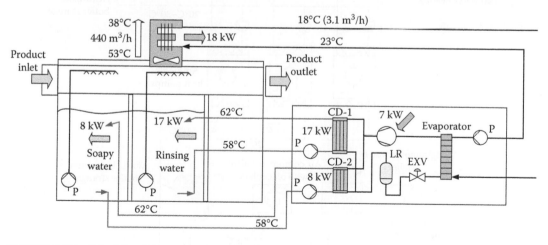

FIGURE 1.35 Principle of the ammonia vapor compression heat pump has been installed in Denmark in an industrial washing plant. CD, condenser; EXV, expansion valve; P, pump. (From IEA, Industrial energy-related systems and technologies Annex 13, IEA Heat Pump Program Annex 35, Application of Industrial Heat Pumps, final report, Part 2, 2015b; redrawn and reprinted with permission from IEA HPT Annex 13/35 operating agent.)

are used to clean metal or plastic devices after machining processes and are typically located where these devices are cleaned to remove oil residues and possible dirt. Such industrial processes generally use electrical energy as heating energy source and thus there is energy saving potentials on energy costs by using heat pumps. In this application, the items to wash and dry are transported through the washer via a belt (or a drum), typically through two washing areas—one with soapy and another with rinsing water, and, finally, through a drying zone. At the top right side of the washer is a ventilation system that removes moist air and keeps a slight negative pressure in the machine. This eliminates unwanted condensation around the machine. Prior to installation of the heat pump as an independent unit connected to washing process via hoses or pipes, the washing plant was equipped with electrical heaters (total 25 kW) to maintain the set temperature of 60°C in both tanks, which consumed considerable amounts of energy (IEA 2015b). Each of the two heat pump's condensers is connected to a tank at the washing plant, while the evaporator is connected to a closed-water loop allowing recovering heat from the hot and humid exhaust air. The results obtained shown that the heat pump reduced the total power demand by 49%, that is, from 31 to 15.7 kW and energy savings of approximately 50%.

Another 800 kW industrial heat recovery system including three mechanical ammonia vapor compression heat pumps has been implemented in a slaughterhouse in Zurich (Figure 1.36) (IEA 2015b). In this system, the heat recovered from oil-cooled air compressors and fan coil units is stored within a waste heat buffer connected with the heat pump evaporators. The warm side of heat pumps is connected with another buffer tank where hot water is stored at 90°C prior being distributed to consumers for slaughtering and cleaning purposes, as feed water for a steam generator, and

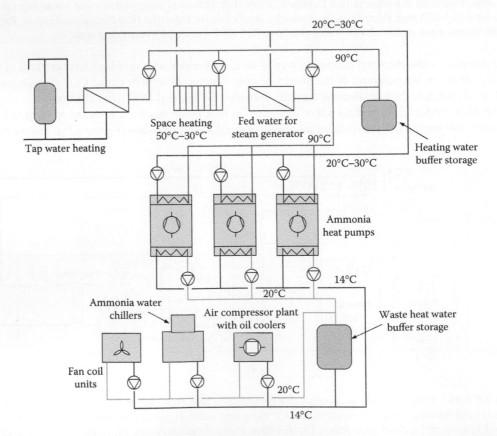

FIGURE 1.36 Schematic of an ammonia vapor compression heat pump system including three units has been implemented in a slaughterhouse in Zurich, Switzerland. (From IEA, Industrial energy-related systems and technologies Annex 13, IEA Heat Pump Program Annex 35, Application of Industrial Heat Pumps, Final Report, Part 2, 2015b; redrawn and reprinted with permission from IEA HPT Annex 13/35 operating agent.)

the heating system. According to the authors, this heat recovery system with ammonia heat pumps allowed to save the equivalent of 2590 MWh from fossil fuels per year, representing an annual reduction in CO_2 emissions of 510 tonnes.

One of the largest mechanical ammonia vapor compression district heat pump in the world has been built by Star Refrigeration Energy (http://www.neatpumps.com/, accessed March 29, 2016) and installed in Drammen, a town located 40 km southwest of Oslo (Norway) (IEA 2015b). The first district heating plant using 8 MW biomass boilers was installed there in 2002. During the second phase of the district heating network, three two-stage heat pumps with series screw compressors, each with a heating capacity of approximately 4.5 MW, and additional 2×30 MW gas fired boiler, as a peak load backup, has been implemented. The heat source for the heat pump is the seawater taken at 40 m depth where the temperature is constant at 8°C–9°C most of the year. After a water intake pipe run of 800 m and return pipes of 600 m, it is returned to the sea at 4°C. The seawater is cooled directly in spray chillers, where ammonia is sprayed across titanium pipes with the seawater inside. The heat pump condenser is split into three sections: the main water stream is heated from 60°C to 69°C through the first section, from 69°C to 78°C in the second section and, finally, from 78°C to 87°C in the third part of the condenser. The return water temperature from the district heating loop is constant at 60°C–65°C year around. Annually, the system equivalent CO_2 emissions were approximately 317 tons, that is, much lower compared to conventional gas-fired boilers, which would give a CO_2 emission of 13,050 tons/year.

A single-stage mechanical ammonia vapor compression heat pump has been built by Star Refrigeration Energy (http://www.neatpumps.com/, accessed March 29, 2016) and implemented in a German chocolate industrial plant for combined heating and cooling (Figure 1.37) (IEA 2015b). The initial existing cooling system used HCFC-22 as a refrigerant and one central coal-fired steam generation plant supplied all end users. The new implemented system aimed at eliminating the HCFC-22 as refrigerant and, also, supplying hot water to end users at 60°C. The chocolate manufacturing requires cooling during several process steps. The simultaneous demands for cooling and heating allowed replacing the existing heating and the cooling system by a combined cooling and heating installation including a single-screw

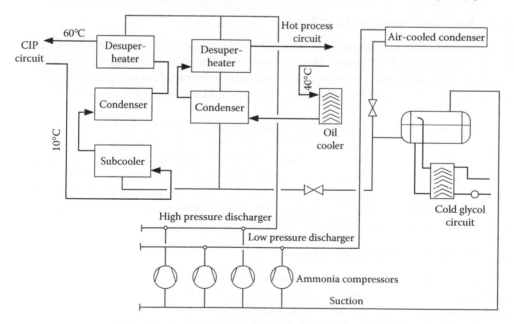

FIGURE 1.37 Configuration of the single-stage ammonia mechanical vapor compression heat pump built by Star Refrigeration Energy (http://www.neatpumps.com/, accessed March 29, 2016) and implemented in a German chocolate industrial plant. (From IEA, Industrial energy-related systems and technologies Annex 13, IEA Heat Pump Program Annex 35, Application of Industrial Heat Pumps, Final Report, Part 2, 2015b; redrawn and reprinted with permission from IEA HPT Annex 13/35 operating agent.)

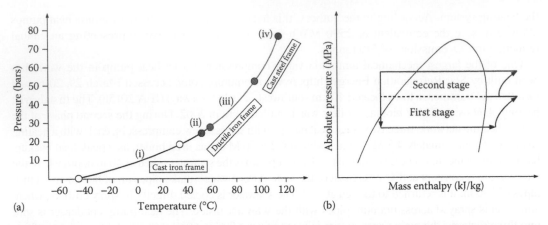

FIGURE 1.38 Ammonia screw compressors for industrial heat pumps. (a) Pressure limits of first and second stage ammonia screw compressors: (i) Normal operating range of ammonia refrigeration compressors; (ii) Design limit for most low-temperature ammonia compressors (26 bars at 60°C); (iii) Compressor duty required for industrial high-temperature ammonia heat pumps (28 bars at 63°C to 52 bars at 96°C); and (iv) Design limit for VILTER cast steel ammonia compressors (76 bars at 110°C). (From http://www. emersonclimate.com/en-us/brands/vilter/Pages/Vilter.aspx; redrawn and reprinted with permission from Emerson Climate Technologies.) (b) Two-stage thermodynamic cycle in p–h diagram.

ammonia compressor heat pump. The heat source (glycol from the cooling process) is cooled from 5°C down to 0°C while ammonia evaporates at −5°C. The temperature lift of the single-stage ammonia heat pump is 61°C and, thus, the process water is heated from 10°C to 60°C. The selected ammonia heat pumps may generate 1.25 MW of high-grade thermal power based on an absorbed thermal power of 346 kW, with a theoretical combined heating and cooling COP estimated at 6.25.

1.4.2 TWO-STAGE AMMONIA HEAT PUMPS

Since today ammonia screw compressors available on the market can work at higher discharge pressures (i.e., up to 52 bar-abs) (Figure 1.38a) and condensing temperatures compared to traditional low-pressure (≈25 bar-abs) ammonia compressors, two-stage ammonia heat pump cycles (Figure 1.38b) are technically feasible (http://www.emersonclimate.com/en-us/brands/vilter/Pages/Vilter.aspx, accessed June 25, 2015).

Such systems, generally provided with desuperheaters, are able to supply hot water at temperatures up to 85°C–96°C by recovering, for example, waste heat from discharge vapor of existing ammonia compressors. Cold water coming from city network at 10°C to 20°C (as shown in Figure 1.39a) or from sea, lake, or river water (i.e., 4°C in the winter and 20°C in the summer), is heated at different temperature levels (60°C, 80°C, and 90°C) to be reused in industrial processes or for district heating (Figure 1.39b) (Source: http://www.emersonclimate.com/en-us/brands/vilter/Pages/Vilter.aspx, accessed June 25, 2015).

1.5 ABSORPTION HEAT PUMPS

In order to simultaneously supply heating and cooling, absorption heat pump systems are directly driven by thermal rather than electrical power sources. As thermal source energies, absorption heat pumps can use natural gas as well as nonconventional sources as solar thermal, deep geothermal reservoirs and industrial waste heat (e.g., from steam boilers of cogeneration systems, ovens, air compressors, and combustion gases), at temperatures between 60°C and 160°C. Thermally driven absorption heat pumps are more suitable in countries where electricity is produced by thermal power plants, because they provide relatively high electricity prices (Ziegler and Riesch 1993). Absorption heat pumps typically provide lower thermal efficiency than mechanical vapor compression heat

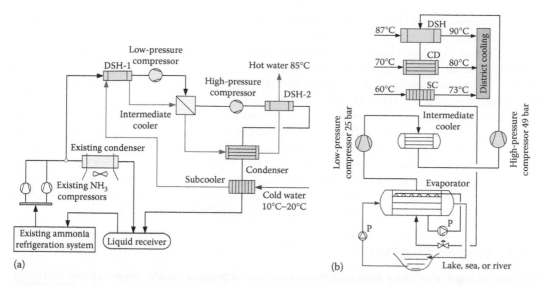

(a)

(b)

FIGURE 1.39 Heat recovery system with high-temperature two-stage ammonia heat pumps. (a) From ammonia refrigeration systems and (b) from lake, sea, or river waters. (From http://www.emersonclimate.com/en-us/brands/vilter/Pages/Vilter.aspx; redrawn and reprinted with permission from Emerson Climate Technologies.)

pumps, but they have some advantages such as few moving parts, no electrical energy requirements (excepting for instrumentation and controls), provide silent operation, long service life, simpler capacity control and low maintenance, as well as high reliability. The main barriers at the industrial implementation of absorption heat pumps, especially for small-scale applications, are their high initial capital costs and the ratio between the cost of electricity and fossil energy sources. Any type absorption heat pump exchanges heat with three external thermal reservoirs available at low, intermediate, and high temperature levels. There are two types of absorption heat pumps.

In the type I (also known as *heat amplifiers,* meaning that the heat pumping process is a heat quality increasing process) (Figure 1.40a), heat (Q_{gen}) is supplied at the generator inlet by waste heat source recovered at a high temperature level (T_{high}) as well as at the evaporator (Q_{evap}) at a much

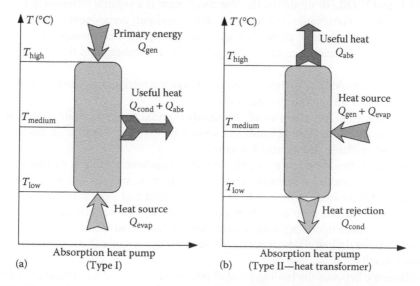

(a)   (b)

FIGURE 1.40 Heat and temperature characteristics of (a) absorption heat pump (Type I) and (b) absorption heat transformer (Type II).

lower temperature level (T_{low}), normally the surrounding ambient air or waste heat. Useful heat is finally supplied at a medium temperature level (T_{medium}) by both condenser (Q_{cond}) and absorber (Q_{abs}). Current two-stage absorption heat pumps using, for example, the mixture $H_2O/LiBr$ as the working fluid are able to deliver heat at 100°C with temperature lifts up to 65°C and COPs of 1.6–1.7. The newest generation of advanced absorption heat pumps will be capable to supply output temperatures up to 260°C (Heat Pump Centre 2007).

In the type II (also referred to as *heat transformers*, meaning the heat pumping process is a temperature increasing process) (Figure 1.40b), heat is recovered at medium temperature levels (T_{medium}), in both the vapor generator (Q_{gen}) and the evaporator (Q_{evap}). One part of the heat recovered (Q_{abs}) is transferred at a high temperature level through the absorber and the other (Q_{cond}) at a low temperature level at the condenser. Such a heat transformer may replace high-temperature primary energy sources by medium-temperature industrial waste heat. Heat transformers using $H_2O/LiBr$ as the working fluid pair are able to deliver heat at 100°C with COPs of 0.45–0.5.

1.5.1 Working Fluid Pairs

Performances of absorption heat pumps are critically dependent on the chemical and thermodynamic properties of the working fluid (Perez-Blanco 1984). This is because, contrary to single-component refrigerants that evaporate and condense isothermally, and the saturation temperature is determined by the pressure, in the case of binary mixtures, saturation temperatures are determined by both pressure and mixture composition. Moreover, the different volatility of components causes one component to evaporate or condense more quickly, leading to concentrations, saturation temperatures, and boiling point changes.

Various working mixtures have been investigated in the past but, today, only ammonia–water (NH_3/H_2O) and water–lithium bromide ($H_2O/LiBr$) are common working fluids commercially available for most of industrial absorption heat pumps and chillers, even if these working fluids have some disadvantages that limit their applications, such as the need for rectification of the refrigerant vapor for the NH_3/H_2O mixture, easy crystallization, solution corrosion problems, and operation at very low pressure for the $H_2O/LiBr$ mixture. Medium and large capacity absorption heat pump and refrigeration systems operating with ammonia–water (for cooling and heating purposes) or water–lithium bromide (for air conditioning) are used in many countries.

In NH_3/H_2O and $H_2O/LiBr$ mixtures, the first component is a volatile (referred to as the *refrigerant*), and the second, a nonvolatile (referred to as the *absorbent*) component. Some 40 refrigerants and 200 absorbents are available today. For usage in absorption heat pumps, the components of working fluid pairs should meet a number of requirements (Holmberg and Berntsson 1990), as

1. The absorbent should have higher boiling point than the refrigerant, strong ability (affinity) to absorb it under the operating conditions.
2. Both refrigerant and absorbent should be highly thermal and chemically stable, nontoxic, noncorrosive (if not, corrosion inhibitors should be used), nonexplosive, environmental-friendly, and available at relatively low costs.
3. The refrigerant should be more volatile than the absorbent, so that it can be easily separated by heating, and have favorable transport properties, such as viscosity, thermal conductivity, and diffusion coefficients to provide high heat and mass transfer coefficients.
4. The refrigerant should have as high as possible enthalpy of vaporization, low saturation pressures at normal operating temperatures and high concentrations within the absorbent in order to maintain low circulation rate between the generator and the absorber per unit of cooling capacity.
5. The difference between the boiling point of pure refrigerant and the mixture boiling point, at the same pressure, should be as large as possible in order to achieve higher temperature lifts compared, for example, to those of mechanical vapor compression heat pumps.

1.5.1.1 Ammonia–Water Mixture

In the ammonia–water mixture (NH_3/H_2O), both of which are highly stable over a wide range of operating temperatures and pressures, environmental friendly, and available at low costs, ammonia is the refrigerant, being more volatile than water, and water is the absorbent. That means that when ammonia is dissolved in water, the boiling point of the mixture is higher than for pure ammonia at the same pressure and temperature, while it is lower than for pure water under the same conditions.

NH_3 has a high latent enthalpy of vaporization and can evaporate at lower temperatures compared to H_2O at the same pressure. However, the disparity in boiling point between NH_3 and H_2O is not very large (i.e., between 4°C and 10°C), which makes it necessary to utilize the distillation equipment. In other words, since both NH_3 and H_2O are volatile, the absorption heat pumps generally require a rectifier to strip away water that normally evaporates with ammonia within the generator. Without such a rectifier, the water would accumulate in the evaporator and offset the overall heat pump performances. On the other hand, ammonia is toxic and has corrosive action to copper and its alloys.

1.5.1.2 Water–Lithium Bromide Mixture

There are two outstanding features of $H_2O/LiBr$ as a working pair fluid in absorption heat pumps: the extremely high enthalpy of vaporization of the refrigerant (H_2O) and the nonvolatility of the absorbent (LiBr), which eliminates the need of refrigerant vapor rectifiers. However, using water as a refrigerant limits the low temperature application to that above 0°C if not, the absorption refrigeration systems must be operated under vacuum conditions. On the other hand, at high concentrations, the $H_2O/LiBr$ solution is prone to crystallization, that is, its most important drawback. It is also corrosive to some metals and, consequently, some additives are usually added as corrosion inhibitors (Iyoki and Uemura 1978) and, also, to improve the heat exchangers' heat and mass transfer performances (Daiguji et al. 1997).

1.5.1.3 New Working Fluid Pairs

Although NH_3/H_2O and $H_2O/LiBr$ mixtures have been widely used for many years and their properties are well known, many research in recent years focused on searching for new working pairs with no corrosion and no crystallization properties, and/or able to eliminate the rectification columns. Different working fluid pairs, such as ammonia–lithium nitrate (NH_3–$LiNO_3$) mixture, where NH_3 is the refrigerant and $LiNO_3$ is the absorbent have been proposed as alternatives to NH_3/H_2O working fluid. Such cycles provide better performance (i.e., COPs as high as 0.5–0.9) than the NH_3/H_2O cycle, not only because of higher COP values but also since lithium nitrate does not evaporate during the vapor generation process; it is not necessary to use a rectifier. Therefore, the absorption system can operate at lower driven temperatures, and it does not have crystallization problems at high heat sink temperatures. However, at evaporating temperatures below −10°C, this cycle cannot operate because of the crystallization possibility (Best et al. 1991; Abdulateef et al. 2008).

Theoretical performances of an intermittent solar absorption refrigeration system with parabolic concentrators and using the ammonia–lithium nitrate (NH_3–$LiNO_3$) as the working mixture varied between 0.78 and 0.33, depending on the day time and the season (Rivera and Rivera 2003). The results also showed that such a system is able to produce up to 11.8 kg of ice at 120°C and 40°C–44°C as generation and condensation (with air or water as cooling agents) temperatures, respectively. Expecting overcoming problems with the oil-lubricated compressor and the water content in the refrigerant vapor, the ammonia–lithium nitrate (NH_3–$LiNO_3$) has been chosen as the working fluid pair for a compression-absorption R&D project in Austria (Hannl and Rieberer 2014). A new absorption chiller of 10 kW of nominal cooling capacity using the ammonia/lithium nitrate mixture ($NH_3/LiNO_3$) as a working pair has been developed by Zamora et al. (2014). Such a chiller does not require a rectifier and his weight and cost are reduced.

Other binary solutions, such as $H_2O/NaOH$ using an inorganic salt absorbent (NaOH) (Best and Holland 1990) or ternary mixtures (Herold et al. 1991; Barragan et al. 1998), such as $LiBr/Zn-Br_2/CH_3OH$, may be successfully used for absorption heat pump and refrigeration systems. Ionic liquids (known as low-temperature molten salts) are room-temperature melting organic salts composed entirely from organic cations and inorganic anions that can remain in the liquid state over a wide range of temperature from near or below room temperature to about 300°C. This property may allow eliminate the crystallization and metal-compatibility problems of H_2O–LiBr absorption system (Marsh et al. 2005; van Valkenburg et al. 2005). Ionic liquids are generally nonflammable, nonvolatile at ambient pressure and temperature (Rogers and Seddon 2003) having excellent thermal stability and very low vapor pressures, as well as melting points below 100°C (Liang et al. 2011). Ionic liquids have also negligible vapor pressure, low combustibility, and good solvating properties. Water, ammonia, or carbon dioxide, all natural fluids, adequately combined with the cations and the anions could be successfully used with ionic liquids as adsorbents, even the latest are moderately to highly viscous, especially at low temperatures, restricting the flow. This issue can be overcome by a careful choice of ionic liquids or by using molecular solvents (additives) to dramatically reduce the viscosity. The low volatility of the ionic liquids enables easy separation of the volatile working fluid (refrigerant) by thermal stratification with minimum harmful impacts on environment. In conclusion, using ionic liquids as the working fluids of absorption cycles can lead to benefit from factors such as less crystallization, less corrosion, low toxicity, and nonflammability in comparison with conventional working fluids including (Khamooshi et al. 2013).

To enhance the heat transfer of current thermal carrier fluids, such as the working pair fluids for absorption heat pump and refrigeration systems, nanofluids appeared to have a huge potential (see Section 1.2.3.5). These are fluids in which nanoparticles below 100 nm in diameter are suspended in order to provide higher (up to 40%) thermal conductivity than the base fluids. Most of R&D works have focused on studying the intensification of absorption rates and heat transfer coefficients by adding nanoparticles (as carbon nanotubes, SiO_2) in the traditional working pairs $H_2O/LiBr$ and NH_3/H_2O. Some experiments on the effect of nanoparticles in the $H_2O/LiBr$ solution in the falling film absorbers, shown that the mass transfer rates substantially increased. The rate of absorption bubble mode without heat removal of the NH_3/H_2O working fluid pair with Cu nanoparticles increased 5.32 times versus the reference values at the highest NH_3 concentration used. Similar experimental studies in bubble mode with NH_3/H_2O and nanoparticles (carbon nanotubes, Al_2O_3, and Ag) and without a chemical surfactant underlined the nanoparticles mass fraction and initial ammonia concentration as key factors.

Although many studies have found enhancement of thermal conductivity in the base fluid by the addition of carbon nanotubes or nanoparticles, the experimental results showed that the thermal conductivity enhancements in working fluids for absorption cycles are significantly smaller. The enhancement of thermal conductivity of the binary working pair $NH_3/LiNO_3$ by adding carbon nanotubes has also been experimentally investigated (Cuenca 2013). A stable distribution of carbon nanotubes in the base fluid was achieved by surface modification of carbon nanotubes through oxidation. The key parameters are the concentrations of NH_3 and carbon nanotubes in weight, ranging from 30% to 50% for NH_3), and 0.005%–0.2% for carbon nanotubes.

1.5.2 Single-Stage Ammonia Absorption Heat Pumps

Single- or two-stage ammonia–water absorption heat pumps, generally used to upgrade industrial waste heat, present some advantages as (1) operation at high temperatures corresponding to relatively low pressures; (2) better match of the heated fluid temperature within the absorber because of the working fluid temperature glides; (3) higher thermal performances compared to conventional mechanical vapor compression cycles for specific operating conditions; and (4) compared to conventional single-stage mechanical vapor compression cycles, where temperature lifts are of maximum 30°C–40°C, ammonia–water absorption heat pumps can provide temperature lifts as

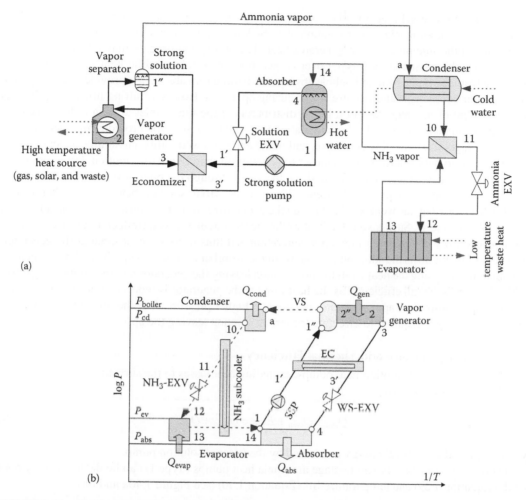

(a)

(b)

FIGURE 1.41 Single-stage ammonia–water heat pump (type I) (a) principle and (b) thermodynamic process in log P–$1/T$ diagram. abs, absorption; cd, condensing; ev, evaporating; EXV, throttling device; P, pressure; Q, heat; T, absolute temperature; VS, vapor separator.

high as 110°C–165°C. A typical single-stage ammonia absorption heat pump contains four heat exchangers: vapor generator, condenser, absorber, and economizer (Figure 1.41a). The vapor generator and absorber assembly replaces the compressor in the actual vapor-compression systems to achieve a thermochemical-type compression heat pump. In such a system, a relatively high temperature heat source (such as gas, thermal solar, or industrial waste) is supplied to the vapor generator, where ammonia vapor is boiled out at high pressure and separated from the ammonia–water strong solution. As the absorbent (water) is highly volatile, it will be partially evaporated together with ammonia (refrigerant) (state a) (Figure 1.41b). Thus, the vapor separator (VS) will contribute to purify the ammonia–water vapor mixture before it enters the condenser. Without this vapor separation, the water will be condensed and accumulate inside the evaporator, dropping the system performance.

The high-pressure ammonia vapor at state a is condensed in the condenser by transferring heat to the heated water stream. The high-pressure condensate leaving the condenser at state 10 is then throttled to a lower pressure prior entering the evaporator at state 12. Inside the evaporator, low-temperature waste heat is recovered to vaporize the low-pressure ammonia and generate ammonia vapor at state 13. After a superheating process (13–14), the ammonia vapor enters the absorber where it is absorbed by the weak solution coming from the vapor generation at state 4 via an exothermic process. In the absorber, one of the important/critical components of absorption heat pumps, the weak ammonia–water solution absorbs low-pressure vapor coming from the evaporator by an exothermic process that generates heat. The absorption process may take place by both falling film (the most common type) over cooled horizontal tubes, while heat is simultaneously removed from the liquid film and bubble absorption. Falling film absorption provides relatively high heat transfer coefficients and is more stable during operation. However, falling film absorption has wettability problems, requires good liquid distributors at the liquid inlet and, in order to achieve better performances, requires high recirculation rate. Bubble absorption provides high heat transfer coefficients, good mixing between vapor, and liquid and good wettability. Generally, vapor distribution required in bubble absorption is easier to accomplish than liquid distribution, but it requires a pressure difference on the vapor side to drive the vapor through the pool of liquid, which rules out bubble absorption in low-pressure systems. The resulted strong solution from the absorber at state 1 is further heated by the weak solution inside the economizer prior re-entering the vapor generation via the VS at state 1″. Such a solution heat exchanger (economizer) is introduced in order to reduce the irreversibility of the ammonia vapor generation and thus decrease heat input at the generator. The economizer allows the solution coming from the absorber at state to be preheated prior entering the generator by using the heat from the hot solution leaving the generator and thus improving (up to 60%) the cycle's overall efficiency as the heat input at the generator is reduced. Because the separation process within the vapor generator occurs at a higher pressure than the absorption process, a circulation pump is required to circulate the strong solution leaving the absorber.

1.5.2.1 Energy Balance and Thermal Efficiency

The heat balance of absorption heat pumps, excluding heat losses to the surroundings, is expressed as follows (see Figure 1.41b):

$$Q_{gen} + Q_{evap} + E_{SSP} = Q_{cond} + Q_{abs} \tag{1.28}$$

where E_{SSP} is the electrical energy consumed by the (strong) solution pump.

Theoretical efficiency of single-stage ammonia heat pumps (Type I) can be determined by both COP (Equation 1.29) and temperature lift (Equation 1.30) (see Figure 1.40a and b):

$$COP = \frac{\text{Useful heat delivered at } T_{medium}}{\text{Heat supplied at } T_{high} + \text{Electrical energy consumed}} = \frac{Q_{cond} + Q_{abs}}{Q_{gen} + E_{SSP}} \tag{1.29}$$

$$\Delta T_{lift}^{AHP} = T_{medium} - T_{low} \tag{1.30}$$

The possible temperature lift is nearly constant over the whole operating temperature range. A rough rule of thumb for the H_2O/LiBr absorption heat pumps is

$$\left(T_{high} - T_{medium}\right) < 2\left(T_{medium} - T_{low}\right) \tag{1.31}$$

Maximum temperature lifts for H_2O/LiBr absorption heat pumps is 45–50 K, and the maximum possible temperature of delivered heat (T_{high}) is 100°C because of the risk of crystallization. Typical values of H_2O/LiBr absorption heat pumps are 1.3–1.4, including reasonable thermal efficiencies of the vapor generator.

1.5.3 Two-Stage Ammonia Absorption Heat Pumps

Figure 1.42a schematically represents a two-stage ammonia–water heat pump. In such a system, strong ammonia–water solutions are boiled in vapor generators 1 and 2 by the aid of a high-temperature (such as combustion exhaust gas, thermal solar, or industrial waste) heat source.

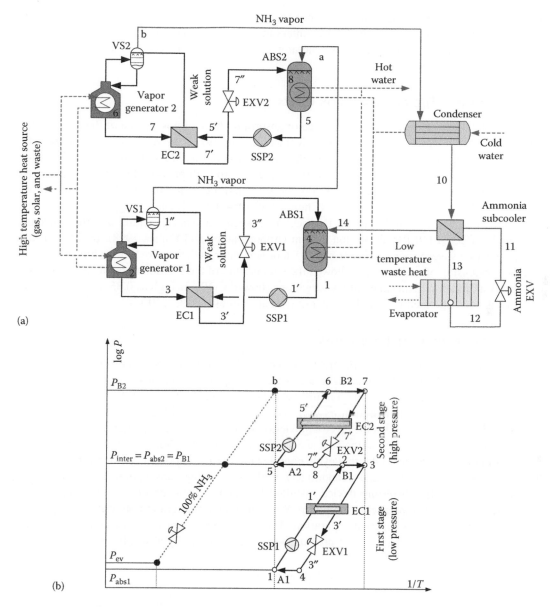

FIGURE 1.42 Two-stage ammonia–water heat pump. (a) Schematic representation and (b) thermodynamic process of the ammonia–water solution in the log P–$1/T$ diagram. ABS, absorber; B, boiler; EC, economizer; EXV, expansion (throttle) valve; SSP, strong solution pump; T, temperature; VS, vapor separator.

Ammonia quasi-pure vapor produced by the vapor generator 2 is condensed in the condenser and the ammonia liquid enters the evaporator via a SC and a throttling device EXV. The low pressure, pure ammonia is absorbed inside the absorber (ABS1) by the weak solution coming at the same pressure from the vapor generator 1. The resulted strong solution is pumped through the economizer EC1, where it is heated by the weak solution leaving the vapor generator 1, to the vapor generator 1 where a boiling process takes place. After separation in the VS1, the ammonia vapor is absorbed in the second-stage absorber ABS2 by the weak solution flowing from the second-stage vapor generator 2 at higher pressure. The strong solution is then pumped back to the vapor generator 2 via the economizer EC2 where it is also heated by the hot weak solution. The high pressure vapor resulted from the vapor generator 2 is condensed in a water-cooled condenser. As noted, the weak ammonia solution from the second-stage vapor generator 2, after having passed through the econo-mizer EC2 and the throttling valve EXV2, enters the absorber ABS2 at the same pressure as the vapor coming from the first-stage vapor generator 1 (see also the solution thermodynamic process in Figure 1.42b).

1.5.4 COMBINED EJECTOR–ABSORPTION CYCLES

Ejectors can be used to improve performance of absorption heat pumps. In one of the first concept proposed with two-stage vapor generators (Kuhlenschmidt 1973), the refrigerant vapor from the second-stage generator was used as a motive fluid for the ejector in order to produce a cooling effect within the evaporator. A similar concept used an ejector with high-pressure liquid solution coming from the vapor generator as motive fluid in order to maintain the pressure inside the absorber at a higher level than the evaporator pressure (Chung et al. 1984; Chen 1988).

Another concept proposed to locate the ejector between the vapor generator and the condenser of a single-effect H_2O/LiBr absorption system (Aphornratana and Eames 1998). In this concept, the ejector used water vapor coming from the vapor generator as the motive fluid at higher pressure that the condensing pressure, which allowed the temperature of the solution to be increased without risk of crystallization. Experimental investigations showed COP's as high as 0.86 to 1.04, when the system has been operated with high-temperature heat sources (190°C–210°C).

1.5.5 ABSORPTION HEAT TRANSFORMERS

The absorption heat transformers (Figure 1.43) (see also Figure 1.40b) use heat from a thermal res-ervoir at medium (intermediate) temperature as the driving heat (normally, from industrial waste heat or thermal solar captors at temperatures varying between 60°C and 100°C), supplies useful heat output at a higher temperature level by recovering up to 50% of the waste heat input without any other energy input except some electrical energy required for circulating the working fluid pair and, finally, rejects out the remaining heat at a low temperature level, normally to the surroundings.

As can be seen in Figure 1.43, this cycle contains the main components (vapor generator, con-denser, evaporator, absorber, and economizer) as the single-stage absorption heat pumps, with the difference that the throttling device installed between the condenser and the evaporator is substituted by a refrigerant circulating pump. Waste heat at a medium temperature is supplied to the vapor generator for refrigerant separation. The liquid refrigerant from the condenser is then pumped to the evaporator at a higher pressure where it vaporizes by absorbing waste heat at the

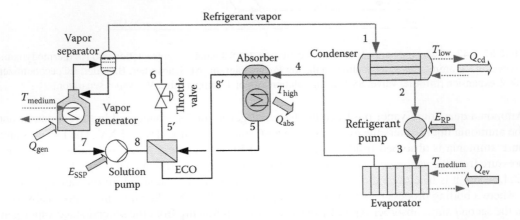

FIGURE 1.43 Schematic representation of an absorption heat transformer (type II absorption heat pump, see Figure 1.40). E, energy; ECO, economizer.

same medium temperature as that used to drive the vapor generator. The vapor refrigerant is then absorbed into the weak solution in the absorber via an exothermic process, which rejects useful heat at a high temperature level. Absorption heat transformers are thus devices with the unique capability of raising the temperature from low or moderately warm waste heat sources to much higher useful temperature levels. The refrigerant vapor at state 4 is produced in the evaporator, heated by the low–medium grade heat source. The refrigerant vapor 4 is absorbed in the refrigerant–absorbent weak solution entering the absorber at state 8′ and leaves it as a strong solution at state 5. The heat of absorption is transferred to the fluid to be heated. The strong solution enters the vapor generator at state 6 after having absorbed heat from the weak solution within the economizer (process 5–5′) and lowered its pressure via a throttling valve (isenthalpic process 5′–6). In the vapor generator, some refrigerant vapor is removed from the strong boiling solution (process 6–7), and the remaining weak solution is returned to the absorber at state 8′ after increasing its pressure (process 7–8) and transferring some heat to the strong solution within the economizer (process 8–8′).

The vaporized refrigerant at state 1 is condensed in the condenser up to state 2 then pumped to a higher pressure state 3 prior entering the evaporator. The evaporated refrigerant at state 4 is finally absorbed in the absorber (exothermic process $4 + 8' = 5$) at a higher temperature (Figure 1.43).

1.5.5.1 Energy Balance and Thermal Efficiency

The heat balance of absorption heat transformers, excluding heat losses to the surroundings, is expressed as follows (see Figure 1.43):

$$Q_{gen} + Q_{evap} + E_{SSP} + E_{RP} = Q_{cond} + Q_{abs} \tag{1.32}$$

where E_{SSP} and E_{RP} are the electrical energy consumed by the strong solution pump and refrigerant pump, respectively.

Theoretical efficiency of heat transformers can be determined by both COP (Equation 1.33) and temperature lift (Equation 1.34) (see Figure 1.43):

$$\text{COP}_{heating}^{AHT} = \frac{\text{Useful heat delivered at } T_{high}}{\text{Heat supplied at } T_{medium} + \text{Electrical energy consumed}} = \frac{Q_{cond} + Q_{abs}}{Q_{gen} + Q_{evap} + E_{RP} + E_{SSP}}$$

$$\Delta T_{lift}^{HT} = T_{high} - T_{medium} \tag{1.33}$$

The possible temperature lift is nearly constant over the whole operating temperature range. A rough rule of thumb for the $H_2O/LiBr$ absorption heat transformers is

$$\left(T_{high} - T_{medium}\right) < 0.8\left(T_{medium} - T_{low}\right) \tag{1.34}$$

Typical $\text{COP}_{heating}^{AHT}$ values of single-stage $H_2O/LiBr$ absorption heat transformers are 1.45–1.49, including reasonable thermal efficiencies of the vapor generator. This means that almost 50% of the waste heat can be brought back to the process at a higher temperature and useful temperature level, without needing to supply any primary energy. Maximum temperature lifts for single-stage $H_2O/LiBr$ absorption heat transformers is 45–50 K, and the maximum possible temperature of delivered heat (T_{high}) is 150°C because of the risk of crystallization.

1.5.6 Industrial Applications of Absorption Heat Pumps

A two-stage ammonia–water absorption thermally driven absorption system, similar to the system represented in Figure 1.42, but for refrigeration purposes, could be implemented in industrial plants in order to produce cold brine at −10°C with waste heat at 95°C and cooling water at maximum

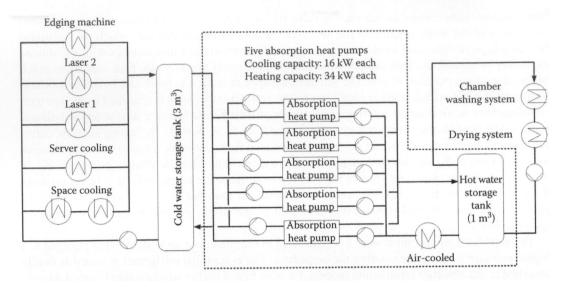

FIGURE 1.44 Heating and cooling system with ammonia–water absorption heat pumps in a metal sheet manufacturing plant in Germany. (From IEA, Industrial energy related systems and technologies Annex 13, IEA Heat Pump Program Annex 35, Application of Industrial Heat Pumps, Final Report, Part 2, 2015b; redrawn and reprinted with permission from IEA HPT Annex 13/35 operating agent.)

30°C during the summer. The waste heat carrier (hot water in this application), coming from the industrial plant, enters the vapor generators 1 and 2, and the rest of the thermodynamic cycle is similar to that previously explained (Minea 2010c). With a refrigeration efficiency of 33.6%, such a system may contribute at improving the competitiveness of any industrial plant by reducing primary energy consumption.

A centralized process cooling system was installed in 2007 in a metal sheet manufacturing plant in Germany (IEA 2005b) in order to cool two new production machines. The central mechanical room (Figure 1.44) includes five parallel ammonia–water absorption heat pumps equipped with circulating pumps on both heat source and heat sink sides, each having heating and cooling capacities of 34 and 16 kW, respectively. Cold water at 20°C is recovered from several industrial processes (edging machine, laser, server, and space cooling) and stored inside a 3 m³ storage tank as a heat source for the absorption heat pumps. Hot water produced at 60°C by these heat pumps is stored inside a 1 m³ storage tank prior being distributed to the plant's hot air drying and washing systems. In the case of a heat surplus, the heat can be evacuated to the environment via an air cooler. The simple payback period for the investment was estimated at 4 years and compared to the old system about 40% of the CO_2 emissions have been eliminated.

1.5.7 COMPRESSION–ABSORPTION (RESORPTION) HEAT PUMPS

Compression–absorption (resorption) heat pump, also known as *hybrid ammonia–water heat pump*, combines mechanical vapor compression and absorption heat pump thermodynamic cycles.

It is well known that mechanical vapor compression heat pumps provide limited temperature lifts and absorption heat pumps have limited temperature ranges. By combining the two systems in a hybrid compression–absorption heat pump, these individual limitations can be reduced. By replacing the condenser and the evaporator of conventional mechanical vapor compression heat pumps with a resorber (vapor absorber) and a desorber (vapor generator), respectively, compression–absorption (resorption) cycle is especially suited for heat pumping processes requiring large temperature lifts. It allows recovering industrial waste heat at relatively low temperatures and supply heat at much higher temperatures compared to standard mechanical vapor compression heat pumps, because the saturation pressures of ammonia–water mixture are significantly lower than those of pure ammonia. Consequently, high heat sink outlet temperatures are possible at moderate pressure levels compared to mechanical vapor compression heat pumps working with pure refrigerants. For example, temperatures above 100°C at the heat sink can be reached with a high-pressure level below 20 bars for ammonia–water instead of a high-pressure level higher than 62 bars for pure ammonia heat pumps.

There are two variants of compression–resorption (absorption) heat pumps: (1) with solution loop, where a liquid pump and single-phase (saturated vapor) compressor are used to overcome the pressure differentials and (2) with two-phase (wet) compression, where the compressor simultaneously works as a vapor compressor and as a liquid pump (Infante Ferreira and Zaytsev 2002).

1.5.7.1 Single-Phase (Dry) Compression–Absorption Heat Pumps

In the single-phase (or dry) compression–absorption (resorption) heat pump, also known as the *Osenbrück cycle* a strong (i.e., high ammonia concentration) ammonia–water solution boils in a special heat exchanger (desorber) by absorbing heat from a (waste) heat source (Figure 1.45a). Such a system leads to large ammonia vapor superheating temperatures.

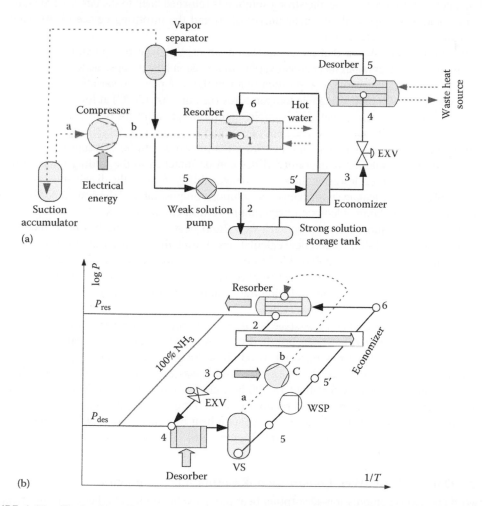

FIGURE 1.45 Single-phase (dry) compression-resorption ammonia–water heat pump. (a) Schematic and (b) thermodynamic process of ammonia–water solution in log P–$1/T$ diagram. C, ammonia compressor; EXV, expansion (throttle) valve; VS, vapor separator; WSP, weak solution pump.

The first theoretical comparison of this thermodynamic cycle, involving nonisothermal absorption–desorption processes, with conventional isothermal evaporation/condensation processes was provided by Altenkirch (1950) (Groll 1997).

Because the fluid exiting the desorber at state 5 as a result of the boiling process is mostly a two-phase mixture, the saturated pure ammonia vapor must be separated from the ammonia saturated liquid within a VS, prior entering the ammonia electrically driven compressor at state a via the SA. The pressure ratio across the compressor is much lower than in conventional mechanical vapor compression systems, which may result in higher COPs. The evaporation process of the strong ammonia–water mixture entering the desorber at state 4, results in a lower solution concentration and in a high ammonia vapor concentration with a significant temperature glide (Figure 1.45b). In other words, as the solution composition decreases during desorption process, the ammonia saturation temperature increases resulting in a mixture temperature glide.

In the ammonia compressor, the pressure and temperature of the ammonia vapor are increased, and then it enters the resorber at state b. In the resorber, the vapor is absorbed through an exothermic process by the weak solution 6 coming from the VS via the weak solution pump (process 5–5′) and the economizer (process 5′–6). The heat generated by the resorption process (6 + 1 = 2) is transferred to a hot water loop, whereas the strong solution is returned back to the desorber at state 6 via the strong solution storage tank, the economizer (2–3) and the throttling device (3–4). Then, the ammonia and solution cycles repeat.

Several studies have shown that the compression–absorption heat pump exhibits some advantages compared to conventional mechanical vapor compression heat pumps, as, for example, (1) the possibility of varying the mixture composition to adapt the heat pump to variations in temperature levels and capacities; (2) high attainable temperatures (i.e., at least 150°C) at lower pressures compared to pure ammonia refrigerant; for example, saturation pressure for pure ammonia at 100°C is 62.6 bars, whereas the saturation pressure of a liquid mixture with 90% (weight) of ammonia concentration at 100°C is 54.4 bars, and the saturation pressure with 50% (weight) of ammonia is 23.6 bars; and (3) the solution temperature glides can be matched to the gliding temperatures of heat sinks and heat sources, lowering the system irreversibility. Among the disadvantages, it can be noted that any ammonia leakage will change the solution composition and the system performances, and that ammonia is a flammable/toxic refrigerant in certain circumstances.

A two-stage dry compression-absorption (resorption) heat pump cycle has been theoretically studied by Sveine et al. (1998). Theoretical calculations showed that, with a heat source at 53°C, heat can be provided at 117°C with a COP of 3.8. Moreover, the desuperheater 1, acting as an intercooler, between the compressors, has a significant impact on the system performance because, while the ammonia vapor is cooled down, thus reducing the compressors' discharge temperatures, the liquid solution is heated to its saturation temperature.

A 7.5 MW (heating capacity) single-stage H_2O/LiBr absorption heat pump with internal solution heat exchanger (economizer) has been installed in an Austrian industrial company that produces cellulose and bioenergy from raw material wood (IEA 2005b). The biomass cogeneration power plant, fired by 77% of wood and 23% of in-house solid waste, supplies steam (30 MWth) to the company, as well as electricity for about 15,000 households and heat for the local district heating network. The H_2O/LiBr absorption heat pump uses as driving heat source steam from the biomass cogeneration plant at about 165°C and recovers latent heat by condensation of hot and humid flue gases exhausted by the biomass power plant (Figure 1.46). The system operates with a seasonal performance factor (SPF) of about 1.6 during a significant number of hours per year (IEA 2005b).

1.5.7.2 Double-Phase (Wet) Compression–Resorption (Absorption) Heat Pumps

The two-phase (wet) compression–absorption heat pump cycle provides higher operating temperatures and higher energy efficiencies (Itard 1998; Infante Ferreira and Zaytsev 2002).

The most critical component of the double-phase (wet) compression–absorption cycle is the wet compressor. It compresses a liquid-vapor ammonia–water mixture and thus has to be, as much

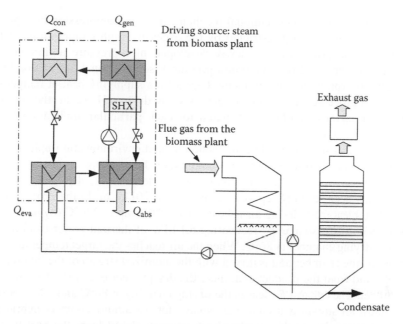

FIGURE 1.46 Process flow sheet of the single-stage H₂O/LiBr absorption heat pump for the flue gas condensation of an Austrian biomass plant. (From IEA, Industrial energy related systems and technologies Annex 13, IEA Heat Pump Program Annex 35, Application of Industrial Heat Pumps, Final Report, Part 2, 2015b redrawn and reprinted with permission from IEA HPT Annex 13/35 operating agent.)

as possible, oil-free, and able to provide acceptable isentropic compression efficiencies. As in the single-phase compression–absorption cycle, within both desorber and resorber, the two-phase heat transfer for ammonia–water strong and weak solutions, respectively, differs from evaporation or condensation of a pure fluid, because the phase changes are not isothermal, the vapor and liquid phases have different concentrations, and the concentration of both phases change during the respective processes.

An experimental prototype (50 kW) with oil-free twin-screw compressor and falling film vertical shell-and-tube type resorber, desorber, and economizer has been designed to upgrade a water flow from 110°C to 130°C, using heat source (wastewater effluent) at 80°C (Infante Ferreira and Zaytsev 2002). Compared to a dry compression–resorption heat pump, the experimental wet compression–resorption heat pump achieved power input reductions, eliminated the vapor superheating and improved the system COPs by up to 20%. According to van de Bor et al. (2014), optimal performances (i.e., low compressor power input and short payback periods) would be obtained if the ammonia concentration at the resorber inlet is 100%.

1.5.7.3 Industrial Applications of Compression–Absorption Heat Pumps

There are many sectors where compression–absorption (resorption) heat pumps could be implemented, as, for example, on large metallurgical industrial platforms aiming at recovering low-temperature heat from air compressors to heat domestic hot water for large district networks (Pop et al. 1983; Minea and Chiriac 2006). To recover heat from high-temperature combustion gases, ammonia–water hybrid absorption–compression heat pumps could be competitive if the maximum concentration of ammonia is kept between 0.7 and 0.9 (Ommen et al. 2011). Spray drying processes, as large energy consumers, traditionally using fossil fuels as primary energy sources to provide high amounts of greenhouse gas emissions are also eligible for improving their energy efficiency by using (dry or wet) compression–absorption heat pumps. This is mainly because the implementation of conventional mechanical vapor compression heat pumps in such an industrial sector is technically restricted by the excessive high temperatures of the exhaust air (80°C–100°C). The availability of newly developed high-pressure ammonia compressors and components (i.e., up to 52 bar maximum pressure) (see Section 1.3.2) in combination with the compression–absorption heat pump technology allows supplying heat at temperatures up to 150°C in industrial processes as processes such as spray drying. However, the actual economic and environmental savings should be estimated for each particular application (Brunin et al. 1997, Jensen et al. 2014a, b).

A detailed theoretical study on technical, economic, and environmental implications of implementing ammonia–water hybrid absorption–compression heat pumps in energy-intensive spray drying industrial processes used for dry solids, as powdered milk, detergents, and dyes from liquid feedstock has been recently reported (Jensen et al. 2014a, b). In the proposed application, 100,000 m³/h of ambient fresh drying air is heated from 20°C up to 200°C (by using a gas-burned air heater as a backup heat source) by recovering 6.1 MW of thermal power from the exhaust air leaving the spray drying chamber at 80°C. When the air reaches the target temperature of 200°C, it enters the spray drying chamber and is mixed with the atomized stream of the liquid product. This evaporates the liquid from the product and, then, the dry product is extracted from the bottom of the drying chamber. The exhaust air leaves the drying chamber at 80°C and a MC of 0.045 kg/kg. Half of the exhaust air stream is used as heat source for the ammonia–water hybrid absorption–compression heat pump. Inside the resorber heat recovery closed loop, the heat transfer fluid is water. It was theoretically found that an 865-kW ammonia–water hybrid compression–absorption heat pump with an ammonia mass fraction of 0.81 and a circulation ratio of 0.45 could reduce the facility's CO_2 emissions by 210 tons/year (Jensen et al. 2014a, b).

Among the most promising applications for compression–resorption heat pumps with high temperature lifts are distillation columns, which can generate half of the operating costs of petrochemical plants. Typically, the temperature of distillation columns is low at the top and increases when moving to the lower sections, reaching a maximum in the re-boiler. In distillation, two or more components are separated based on their difference in boiling point. A mixture is boiled up in the re-boiler of a column, stripping off most of the light component and a part of the heavy component. The remaining flow leaves the column as product stream. The vapor created is relatively rich in low boiling component and moves toward the condenser in the top of the column. Here, the overhead vapor is condensed and leaves the system as product, or flow back into the column as reflux. Between the re-boiler and condenser there are trays to increase the separation efficiency. Again starting at the re-boiler the vapor moves up each tray, loses some of the heavier component, therefore becoming richer in the light component. The liquid flowing downward becomes richer in the heavy component. The temperature gradually decreases from the bottom of the column to the top of the column (van de Bor et al. 2014). Such installations achieve low thermodynamic efficiency (e.g., about 12% for crude distillation), requiring high qualities of energy in the re-boiler, while rejecting a similar amount of heat in the condenser, at lower temperature. Compression–resorption heat pumps can thus be used to upgrade the low quality energy in the condenser in order to upgrade the condenser heat to re-boiler temperature level and thus reduce the consumption of valuable utilities. By matching the absorption cycle with the heat loads of distillation columns it is possible to minimize the consumption of primary energy (van de Bor et al. 2014).

1.6 MECHANICAL VAPOR RECOMPRESSION HEAT PUMPS

Basically, mechanical vapor recompression devices are special cases of closed-loop mechanical vapor compression heat pumps operating as open (or semi-open) systems. They are generally used to recover waste heat from very low-pressure (vacuum) water vapor (steam) rejected from energy-intensive industrial processes, such as concentration of aqueous, saline and sugar solutions, and milk, and evaporation/distillation in the chemical industry.

1.6.1 Principle

Mechanical vapor recompression is a technique by which a compressor or a high-pressure blower is used to compress (i.e., increase the pressure and temperature) a process vapor (generally, steam) and, then, condense it within a special evaporator–condenser heat exchanger in order to heat the vapor's *mother* liquid or solution being concentrated by an evaporation process. This process may reduce by 10%–15% the energy normally consumed by conventional concentration, evaporation, and distillation processes.

The most common single-stage mechanical vapor recompression system is represented in Figure 1.47a where a process steam is directly compressed (Minea 2014c). In such a mechanical vapor recompression heat pump, the saturated vapor at state 1 (e.g., at a vacuum pressure p_1 and temperature T_1) is recompressed by an electrically driven compressor or a fast revolving high pressure blower ($\approx$3000 rpm) (both able of operating under vacuum) up to state 2 (pressure $p_2 > p_1$ and temperature $T_2 > T_1$) by absorbing electrical energy and transferring it to the recovered and recompressed vapor as an additional energy input. The vapor leaving the evaporator at saturated state 1 is (or could be) superheated by a small amount (1.5°C, not shown in Figure 1.47b). A vacuum pump, together with a small amount of cooling water, maintains the desired vacuum pressure in the system. Compression process raises the pressure and saturation temperature of the steam, so that it may be returned to the evaporator/condenser heat exchanger to be reused as a heating source to evaporate more water instead of being rejected outdoor through a conventional condenser. This reduces the steam needed to meet the evaporative load of the industrial system concerned.

The compression process sufficiently increases the vapor pressure to allow the vapor condensation at a temperature high enough to boil the incoming liquid inside the evaporator–condenser. During the quasi-isentropic compression process 1–2, the steam temperature increases, for example, from 70°C–80°C up to 110°C–150°C and, in some cases, up to 200°C, depending on compression ratios (Heat Pump Centre 2007). However, because the energy consumed by the compressor is utilized most efficiently at low compression ratios (generally, lower than 2), the increase of vapor pressure and temperature is limited. The vapor recompression process can be performed directly from a two-phase wet mixture, from a dry saturated steam (state 1) to the final superheated state 2 or by two-stage compression with liquid injection between stages (Becker and Zakak 1986). To compress wet water vapor, screw compressors must be used, whereas to compress dry steam, centrifugal compressors must be employed to avoid erosion damages, even though they provide lower efficiency compared to that of screw compressors. For compression ratios equal to (or lower than) 2, COPs as

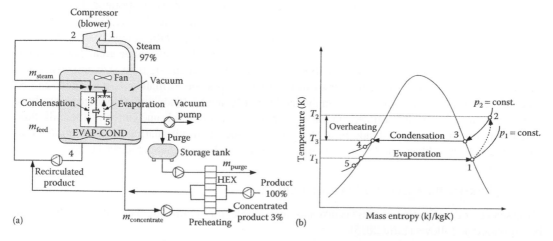

(a) (b)

FIGURE 1.47 Principle of a mechanical vapor recompression heat pump. (a) Schematic and (b) thermodynamic process in *T–s* diagram. COND, condenser; EVAP, evaporator.

high as 12 (or higher) can be achieved (Becker and Zakak 1986). The superheated steam leaving the compressor at state 2 is first cooled up to the saturated state 3 (usually, by using a desuperheater) and then condensed and subcooled up to state 4 within the condensing side of the evaporator–condenser, simultaneously transferring the recovered sensible and latent heat to the liquid being preheated evaporated (process 5–1 in Figure 1.47a). On the condensing side, by subcooling the condensate, the vapor enthalpy (latent heat) of vaporization is recovered. In other words, the high-temperature vapor at state 2 serves as the heating agent for the inlet (colder) solution (also named *mother* solution) being concentrated by water evaporation. As it can be seen in Figure 1.47a, the hot side (vapor) of the system is separated from the cold side (liquid or solution) by the heat exchange surface of the evaporator–condenser that, in such open vapor recompression, heat pumps basically replaces the function of EXVs of closed-loop mechanical vapor compression heat pumps. In practice, small amounts of additional energy may be required to achieve the system heat balance, thereby allowing constant pressure ratios and stable operating conditions. After the vapor compression and subsequent condensation, the concentrate condensate (shown at 3% in Figure 1.47a) leaves the cycle and, with the purge stream, is used to preheat the product close to the evaporating temperature. Because of this heat recovery process, the make-up steam consumption is nearly zero, being required only a small quantity for the unit start up. At this point, the cycle restarts.

1.6.1.1 Compressors

Mechanical vapor recompression systems typically use electrically driven volumetric rotating machines, as positive displacement (reciprocating), screw, centrifugal, and axial compressors, or industrial high-pressure fans (blowers) working at constant volumetric flow rates at a given speed, regardless of the suction pressure. Selection of compressors or blowers depends on their nominal capacity, reliability, efficiency, and cost. Single-stage centrifugal compressors are usually equipped with liquid separators in the suction line, because they are highly susceptible to erosion from entrained liquid droplets, leading to lower efficiencies, dynamic instability from rotor imbalance and possible blade mechanical failures. Axial multistage compressors are used for very large volumetric flow rates.

1.6.1.2 Evaporator–Condenser Heat Exchangers

The concentrated product enters at the top of the evaporator–condenser (see Figure 1.47a) and then is distributed as a thin film flowing down under a slight negative pressure across a very large heat transfer surface area, assisted by an induced draft. Concentrated product (feed) with a small internal liquid volume is separated from the vapor in the bottom liquid chamber of the evaporator/condenser, recombined with additional feed and re-circulated through the top of the evaporator. The temperature difference between the heating media (compressed vapor entering the evaporator–condenser at state 2) and the boiling liquid is very low and, therefore, the evaporator bundle is subject to very low fouling or scaling risks, but it requires a very large internal heat exchange area. The evaporators are operated at high temperature and pressure, which allow reducing their size and cost, but it may promote scaling on heat exchanger surfaces. To address scale formation, five methods can be used, namely, (1) remove carbonates from the feed water by acidification and stripping the resulting carbon dioxide; (2) remove sulfates via ion exchange; (3) promote salt nucleation in the bulk fluid rather than on surfaces; (4) abrade heat exchanger surfaces with circulating *cleaning balls* commonly made from rubber; and (5) apply nonstick coatings to heat exchanger surfaces.

1.6.2 THERMAL BALANCES AND ENERGY EFFICIENCY

If the seawater desalination process is taken as an example, the steam-side thermal power (kW) can be expressed as follows (Lara 2015):

$$\dot{Q} = \dot{m}_s h_{fg} \qquad (1.35)$$

$$\dot{Q}_{\text{steam}} = \dot{m}_{\text{steam}} h_{fg} \tag{1.36}$$

where:

$\dot{m}_s$ and $\dot{m}_{\text{steam}}$ are the steam flow rate (kg/s)

h_{fg} is the condensing enthalpy (latent heat) (kJ/kg)

The product (seawater)-side thermal energy will be

$$\dot{Q} = \dot{m}_v h_v - \dot{m}_f h_f + \dot{m}_b h_b = (\dot{m}_f - \dot{m}_b) h_v - \dot{m}_f h_f + \dot{m}_b h_b \tag{1.37}$$

$$\dot{Q}_{\text{product}} = (\dot{m}_{\text{feed}} - \dot{m}_{\text{con}}) h_{\text{purge}} - \dot{m}_{\text{feed}} h_{\text{feed}} + \dot{m}_{\text{con}} h_{\text{con}} \tag{1.38}$$

where:

$\dot{m}_{\text{feed}}$ is the saltwater feed flow rate (kg/s)

$\dot{m}_{\text{con}}$ is the flow rate of exiting concentrate (brine) (kg/s)

$\dot{m}_{\text{purge}} = \dot{m}_{\text{feed}} - \dot{m}_{\text{con}}$, which is the flow rate if vapor to the next effect (purge) (kJ/kg)

h_{purge} is the specific enthalpy of vapor going to the next effect (purge) (kJ/kg)

h_{feed} is the specific enthalpy of saltwater feed (kJ/kg)

h_{con} is the specific enthalpy of exiting concentrate (brine) (kJ/kg)

Using the boiling temperature as a reference, the enthalpy h_{feed} can be calculated from the specific heat of saltwater c_{pfeed} (kJ/kgK):

$$h_{\text{feed}} = c_{\text{pfeed}} (T_b - T_f) \tag{1.39}$$

where:

T_b is the temperature of the concentrate (brine) leaving the evaporator/condenser heat exchanger (°C)

T_f is the saltwater (feed) temperature entering the evaporator/condenser heat exchanger (°C)

Using the boiling point temperature as a reference (i.e., $h_{\text{con}} = 0$), the specific enthalpy h_{purge} (h_v) of the leaving vapor equals the latent heat of evaporation plus sensible superheat. However, the sensible superheat is small, so it is approximately true that $h_{\text{purge}} (h_v)$ is the latent heat of evaporation, which is h_{fg}. With this simplifying assumption, the steady-state evaporator energy balance derived using Equations 1.40 and 1.41 becomes:

$$\dot{m}_s h_{fg} = (\dot{m}_f - \dot{m}_b) h_{fg} - \dot{m}_f c_{pf} (T_b - T_f) + 0 \tag{1.40}$$

$$\dot{m}_{\text{steam}} h_{fg} = (\dot{m}_{\text{feed}} - \dot{m}_{\text{con}}) h_{fg} - \dot{m}_{\text{feed}} c_{\text{pfeed}} (T_b - T_f) + 0 \tag{1.41}$$

The energy efficiency of a mechanical vapor recompression heat pump depends on the compressor (or blower) efficiency and energy consumption, and on the overall heat transfer coefficient of the steam evaporator–condenser installed inside the original colder *mother* solution or liquid in order to simultaneously boil it and condense the vapor (steam).

The energy input (E_c, kW) of a single-stage compressor can be theoretically expressed as follows:

$$E_{\text{compr}} = \dot{m}_{\text{steam}} (h_2 - h_1) \tag{1.42}$$

where:

$\dot{m}_{\text{steam}}$ is the vapor (steam) mass flow rate (kg/s)

h_1 and h_2 are the mass enthalpy of vapor at the compressor inlet and outlet, respectively (kJ/kg)

The actual electrical power required by a single-stage centrifugal compressor will be

$$\dot{W}_{compr} = \dot{m}_{steam}\left(h_2 - h_1\right)/\eta_s \tag{1.43}$$

where:

$(h_2 - h_1)$ is the specific isentropic compression work (kJ/kg)

$\eta_s = (h_{2s}^* - h_1)/(h_2 - h_1) \approx 0.8$ is the compressor internal (isentropic) efficiency (see Figure 1.2b)

h_2^* is the actual specific enthalpy of the steam leaving the compressor after the polytropic compression process (kJ/kg)

The compression ratio depends on (1) compression ratio (p_2/p_1) and, therefore, it must be maintained to the lowest value required. (2) The boiling point elevation of the liquid to be evaporated; higher the boiling point rise, higher will be the compression ratio required. (3) The minimum differential temperature gradient required for effective heat transfer; indirect condensers require a minimum temperature gradient across the fluids exchanging heat, and the condensers should be designed for least temperature lift operation. (4) The total system pressure drop in the piping and valves; adequate size of piping and valve selection should be done for minimum pressure drop during transfer of fluid through them.

The energy efficiency of a mechanical vapor recompression heat pump is expressed by the COP defined as follows:

$$COP_{heating}^{MVR} = \frac{Useful\,thermal\,power\,delivered}{Compressor\,electrical\,power\,input} = \frac{\eta_{cd-ev} * \dot{m}_{steam} *\left(h_2 - h_4\right)}{\dot{W}_{compr}} \tag{1.44}$$

where:

η_{cd-ev} is the thermal efficiency of the condenser–evaporator heat exchanger

h_2 and h_4 are the steam specific enthalpy (kJ/kg) at the inlet and the outlet of the condensing process, respectively (Figure 1.47b)

In Equation 1.45, the electrical power input of the vacuum pump is neglected. Compared to mechanical vapor compression heat pumps, but at compression ratios restricted at between 1.8 and maximum 2, which corresponds to an absolute saturated steam temperature lifts between 12°C and 18°C and maximum 30°C, depending on the compressor suction pressure, mechanical vapor recompression heat pumps provide much higher COPs, typically between 10 and 30, sometimes up to 100 or even higher. If higher quality materials, such as titanium, are used for the compressor, compression ratios up to 2.5 or higher can be achieved today.

1.6.3 ADVANTAGES AND LIMITATIONS

Some of technological benefits of mechanical vapor recompression systems can be summarized as follows: (1) recover waste vapor (generally, steam) and use it as an environmental-friendly working fluid within a very simple thermodynamic process; (2) require fewer components than conventional closed-loop mechanical vapor compression heat pumps; and (3) reasonable operation and maintenance costs, almost no steam consumption and excellent partial load behavior.

Mechanical vapor recompression systems present also some disadvantages as (1) higher capital cost compared to conventional steam heated systems; (2) high electrical power input and energy consumption requiring high voltage power supply, which, from an economic point of view, may limit the number of industrial applications; and (3) relatively low evaporating (boiling) temperatures require very large vapor volumes and, thus, therefore, large heat transfer surfaces which increase the initial capital costs of industrial applications.

1.6.4 Industrial Applications

As one of the most efficient energy saving systems, mechanical vapor recompression heat pumps are currently applied in some energy-intensive processes, such as separation (evaporation) and distillation in chemical industry where between 25% and 40% of the total energy consumed is used, while a lot of low-pressure steam is rejected to the atmosphere as waste heat, or condensed in cooling towers.

During the last 15 years, mechanical vapor recompression have been also installed in dairy, food and beverage (e.g., evaporation of milk, whey, juice, and refining of sugar solution before final crystallization), brine concentration, ethylene glycol (antifreeze) refortification, car wash recycling, chemical solution, and environmental technology (concentration of wastewater and generation of dry effluents), where large amounts of water must be evaporated (DOE, eere. energy.gov). Mechanical vapor recompression systems could also be integrated (combined) even with high-temperature large-scale industrial dryers. Since 1969, circa 260 of single- and multieffect mechanical vapor recompression units have been installed worldwide stations (Lokiec and Ophir 2007). In Spain alone, for example, over 70 units have been installed in sites as tourist resorts, municipalities, ports, refineries, and power plants. A mechanical vapor recompression system has been implemented in 2000 in the United States in a sugar plant (Pope et al. 2015). In this system, the molasses desugarization process takes soft molasses from the factory sugar end and converts it into at least two streams, that is, product (extract) and nonsugar byproduct (raffinate). Because of limits on available steam supply for the molasses desugarization facility, a single- or multiple-effect steam heated evaporator system alone was not an acceptable option. Boiler capacity to meet the process demands could not be met without the addition of new equipment and costly infrastructure upgrades. Thus, the mechanical vapor recompression evaporation, system was selected to meet the required criteria.

The mechanical vapor recompression heat pumps include recirculating evaporators with vertical tubes operating under a falling film heat transfer mode. The generated vapor enters the VSs from which the liquid is returned to the evaporator chest and the evaporated vapor collected then compressed and finally supplied to the steam chests to condense and, thus, heat the raffinate. The condensate is collected and used as preheating agent. By concentrating the raffinate stream in the evaporator, 90% of the required evaporation load is provided by the mechanical vapor recompression system. The concentrated raffinate (~28%) is finally collected in a storage tank and then pumped to a new multiple-effect evaporation plant (Pope et al. 2015). A mechanical vapor recompression system has been used for distillation (desalination) of saline water (Tleimat 2015).

A new mechanical vapor recompression desalination system operating at high temperatures has been developed in the United States (Lara et al. 2015). It uses sheet-and-shell latent (evaporator–condenser) heat exchangers that achieve efficient dropwise condensation allowing small temperature and pressure lifts between the saturated boiling liquid and the condensing steam, hence reducing the energy requirements. This system consists of a train of nonscaling counter flow evaporators–condensers with hydrophobic copper plates recovering heat from both the condensate stream and the concentrated discharge brine with ultra-efficient heat transfer coefficients. The forced convective boiling in the water chamber is produced by an internal jet ejector powered by a pump. A high-efficiency compressor provides the compression work required to return saturated steam to the initial stage of the evaporator train. To preheat the feed prior entering to the evaporators–condensers, a sensible heat exchanger is employed, which recovers thermal energy from the incoming feed water and transfers it to the discharged distilled water and concentrated brine (Lara et al. 2015).

A two-stage mechanical vapor recompression system has been implemented in a sugar refinery in Korea in order to improve the conventional process, in which a double-effect concentration rising film-type tube was used as a concentration process under vacuum of sugar solution. In the initial

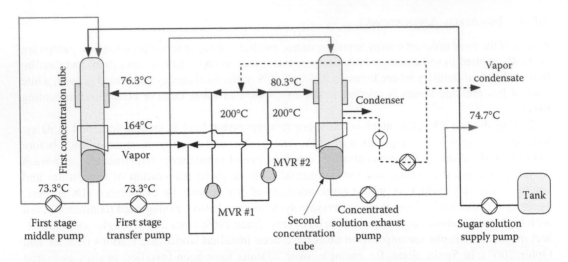

FIGURE 1.48 Diagram of improved process with mechanical vapor recompression in a Korean sugar refinery. (From IEA, Industrial energy-related systems and technologies Annex 13, IEA Heat Pump Program Annex 35, Application of Industrial Heat Pumps, Final Report, Part 2, 2015b; redrawn and reprinted with permission from IEA HPT Annex 13/35 operating agent.)

process, the waste steam from the concentration tube was employed to prepare process hot water, whereas the condensate was used as make-up supply water to a steam boiler, or rejected as wastewater (IEA 2015b). In the improved process, a first-stage mechanical vapor recompression system recompresses the waste steam generated by the first concentration tube and supplies it again into the first concentration tube (Figure 1.48). A part of the waste steam from first concentration tube is recompressed by the second mechanical vapor recompression system and then used as heat source for the second concentration tube.

Propylene is a key material in the production of a number of chemical products including polypropylene (PP) and solvents. It is obtained by distillation separation of propylene and propane in a so-called PP-splitter column. In conventional distillation processes, the re-boiler is heated by low-pressure steam or hot condensate and the overhead vapors are cooled with cooling water. The operating pressure is one of the key variables for the separation of propylene and propane by distillation. The volatility ratio is significantly greater at pressures in the range of 3–10 bars, compared to the traditional values at 15–20 bars. Moreover, the operation at low pressure prevents the use of cooling water. This problem was solved in the Netherlands (IEA 2015b) by using a 5.8 MW mechanical vapor recompression system (Figure 1.49). In this improved system, the overhead top vapor is used to heat the column at the bottom. An electrically driven two-stage fixed speed compressor increases the pressure of the top vapor, which is then condensed in a condenser/re-boiler. The main part of the condensed overhead vapor is returned to the column as reflux, the remainder providing feed stock to downstream chemical plant. Because the column can operate without cooling, the column pressure has been reduced resulting in a better split between propylene and propane, increase of relative volatility and production of propylene with a purity of minimum 99.5%. This mechanical vapor recompression system saved 1.2 PJ/year of primary energy and reduced the CO_2 emissions by 67 ktons/year, with a simple payback period estimated at 2 years.

A combined mechanical and thermal vapor recompression system has been implemented in Japan in an ethanol distillation tower process aiming at reducing the recompression electrical power of the low-pressure steam recycled for an ethanol rectifying tower (Figure 1.50) (IEA 2015b). Thermal vapor recompression is a thermodynamic process based on the principle of

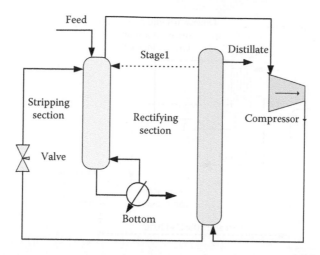

FIGURE 1.49 Mechanical vapor recompression system integrated in a propylene and propane distillation process in the Netherlands. (From IEA, Industrial energy-related systems and technologies Annex 13, IEA Heat Pump Program Annex 35, Application of Industrial Heat Pumps, Final Report, Part 2, 2015b; redrawn and reprinted with permission from IEA HPT Annex 13/35 operating agent.)

converting the pressure (i.e., kinetic) energy contained in a motive fluid stream into velocity energy to entrain suction fluid (Elbel and Hrnjak 2008). For that, a high-pressure live (motive) vapor (steam, refrigerant, etc.) enters the ejector consisting of a nozzle, a suction chamber, a mixing section, and a diffuser. Inside the converging–diverging nozzle, the high-speed motive vapor expands at supersonic speed and, thus, its static pressure sharply decreases. This static pressure drop allows to a secondary fluid to be entrained toward the suction chamber where mixing process takes place. At the exit of the mixing section, the mixed fluid still has a high kinetic energy (i.e., high flow velocity), which is recovered within the diffuser by converting it into potential energy, thereby increasing the static pressure. In the initial conventional process, 95% (vol.) ethanol solution (80 kL/day) was cooled down and then condensed at 75°C to generate a large amount of hot water effluent at less than 75°C, which was released to the atmosphere via a cooling tower. With a mechanical vapor recompression system alone (Figure 1.50a), the compression of low-pressure steam with a compression ratio of 3.5 would require a 700 kW compressor, that is, a too large capacity and, thus, low energy efficiency.

Another solution was to combine the mechanical vapor recompression system with a thermal vapor recompression, steam-driven system in order to reduce the compression ratio to 1.7 (Figure 1.50b). The thermal vapor recompression system uses as motive drive steam from a boiler at 1.5 MPa and 197°C to extract vapor at 0.039 MPa and 75°C and, finally, supply 4.45 tons/h of steam at 0.066 MPa and 88°C. The subsequent mechanical vapor recompression system recompresses the vapor at a ratio of 2.1 and, therefore, the total compression ratio of the system is 3.5, whereas the electrical power requirement for the mechanical vapor recompression compressor has been reduced by 50%. The mechanical vapor recompression system, installed after the thermal vapor recompression system, recovers 4.45 tons/h of steam by the aid of a 250 kW (shaft power) recompression compressor. The steam enters the compressor at 0.066 MPa and 88°C and leaves it at 0.137 MPa and 140°C, thus operating with a compression ratio of 2.1.

The combined thermal and mechanical recompression system reduced by 56.3% and 43% the primary energy consumption for steam production and CO_2 emissions (vs. crude oil CO_2 emissions of 2.62 kg/kL), respectively, for both rectifying and methyl towers. The simple payback period was estimated at 3 years.

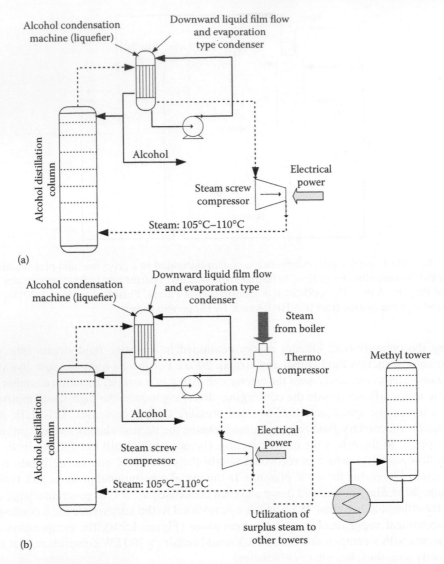

FIGURE 1.50 Mechanical vapor recompression system implemented in an alcohol distillation process in Japan. (a) Distillation column with mechanical vapor recompression and (b) distillation column with combined mechanical and thermal vapor recompression. (From IEA, Industrial energy-related systems and technologies Annex 13, IEA Heat Pump Program Annex 35, Application of Industrial Heat Pumps, Final Report, Part 2, 2015b; redrawn and reprinted with permission from IEA HPT Annex 13/35 operating agent.)

Another combined mechanical and thermal vapor recompression system has been implemented in Korea (IEA 2015b) (Figure 1.51) in order to reduce the thermal losses by recycling exhausted gas otherwise wasted. The combined system collects exhausted low-pressure steam from the solvent separation of a synthetic rubber production process.

Another combined use of a conventional mechanical vapor compression heat pump with a mechanical vapor recompression system has been theoretically evaluated (Richard and Labrecque 2014). In this concept, the mechanical vapor compression heat pump recovers sensible and latent heat from a hot and humid drying air stream (e.g., from a pulp and paper drying industrial process) at 60°C and produces steam at very low-pressure (under) vacuum conditions. This steam is

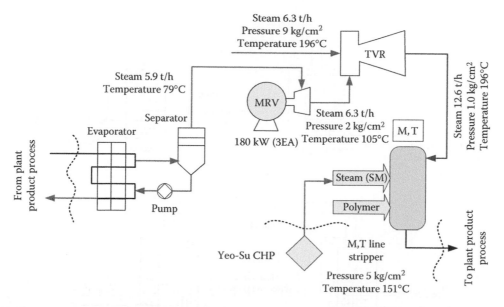

FIGURE 1.51 Diagram of energy saving process with combined mechanical and thermal vapor recompression system in Korea. (From IEA, Industrial energy-related systems and technologies Annex 13, IEA Heat Pump Program Annex 35, Application of Industrial Heat Pumps, Final Report, Part 2, 2015b; redrawn and reprinted with permission from IEA HPT Annex 13/35 operating agent.)

afterward recompressed above atmospheric pressure and up to 110°C by an open-loop mechanical vapor recompression system as a steam utility. According to these authors, such a combined heat recovery system could achieve COPs higher than three.

1.7 CHEMICAL HEAT PUMPS AND TRANSFORMERS

Heat pumps recovering heat at low- or medium-temperatures are able to upgrade it to higher temperature levels. Such a process may be achieved by consuming electricity, with mechanical vapor compression (Section 1.2), or thermal energy, with absorption heat pumps and heat transformers (Section 1.3).

Similar devices are the chemical heat pumps and transformers, defined as systems that utilize organic or inorganic substances, with relatively high thermochemical energy storage densities, as well as reversible chemical reactions to upgrade the quality of heat recovered to higher temperatures by absorbing heat (via endothermic) and then releasing heat (via exothermic reactions) for useful consumption (Spinner 1996; Kato et al. 1996; Kawasaki et al. 1999). Chemical heat pumps and transformers directly utilize various energy sources, as low-, medium-, and high-temperature industrial waste heat (e.g., from cogeneration power plants and air convective dryers) or solar energy, thus eliminating irreversibility associated with traditional heat conversion to electrical power and reduce the global primary energy use. Similar to conventional absorption heat pumps, chemical heat pumps may operate in two modes depending on the available input and required output temperatures, that is, in the heat pump (where heat is stored at high temperature, recovered at low temperature and supplied [intermediate] medium-level temperature) (see Figure 1.52) and in the heat transformer mode (where heat is recovered and stored at medium temperature, supplied at high temperature and released at low temperature) (see Figure 1.53). Among the advantages of chemical heat pumps and transformers can be noted: (1) operation in batch or continuous mode with any moving components, (2) storage of energy by using chemical reactants with high storage density and capacity, and (3) release of any harmful gases with negative environmental impacts (Daghigh et al. 2010).

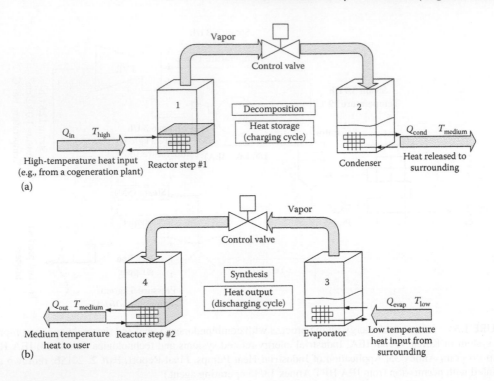

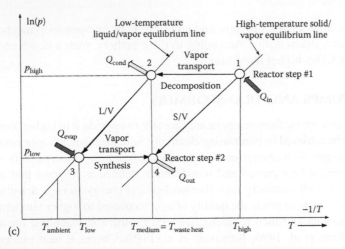

FIGURE 1.52 Schematic principle of a chemical heat pump. (a) Decomposition (heat storage/charging) step; (b) synthesis (heat output/discharging) step; and (c) process representation in Clausius–Clapeyron *p-T* diagram. L/V, liquid–vapor; S/V, solid–vapor.

1.7.1 Chemical Heat Pumps

Chemical heat pumps use organic or inorganic working pair substances as solid– and liquid–vapor (Wongsuwan et al. 2001). The temperature and pressure ranges in which such devices may operate depend on the reaction equilibrium parameters of the chosen working couple (Mujumdar 2009).

A batch-type chemical heat pump (Figure 1.52a and b) contains a reactor and a condenser–evaporator heat exchanger where condensation and evaporation processes alternately take place. The reactor contains a solid material (e.g., a salt) that interacts with a vapor present in its liquid form in condenser–evaporator. Such a system works at three temperature, that is, high (T_{high})

medium–intermediate (T_{medium}) and low (T_{low}) and two pressure levels, that is, low (p_{low}) and high (p_{high}) (Figure 1.52c) (Goetz et al. 1993; Aristov 2008). Batch chemical heat pumps can be applied, for example, to load leveling in cogeneration plants by chemically storing the surplus heat (e.g., from diesel, gas engines, and fuel cells) during the low demand periods and supplying heat during the high load demand periods (Kato et al. 1999). As working pairs, various chemical substances can be used as, for example, alkaline- or chloride salts-ammonia vapor (Neveu and Castaiang 1993), metal hydride-hydrogen vapor, oxides/carbon dioxide vapor (Saito et al. 2007) and magnesium oxide–water vapor (Kato et al. 2005). The most used working pairs, having zero ODP and GWP, are metal chloride-ammonia vapor and metal hydride–hydrogen vapor (Wang et al. 2009; Yu et al. 2008). In such systems, the waste heat can be upgraded, for example, from 150°C to 250°C, that is, with temperature lifts up to 100°C or higher and achieve energy efficiencies typically of around 30%.

Liquid–vapor chemical heat pumps, more appropriate for continuous heat recovery processes where the reactants and products could be fed or removed continuously, include two (endothermic and exothermic) reactors. In these systems, while the endothermic reactor uses heat for decomposition, for example, a metal hydride, the exothermic reactor simultaneously supplies heat at higher temperature.

This section refers particularly to solid–vapor batch-type chemical heat pumps.

1.7.1.1 Principle

The working principle of chemical heat pumps is based on two batch-step reversible, discontinuous (alternate) chemical transformations, that is, decomposition (heat charging/storage cycle) and synthesis (heat output/discharge cycle) of chemical bonds existing between a solid (e.g., metal salts) and a vapor (e.g., ammonia, hydrogen, carbon dioxide, or water) used as a working pair (Figure 1.52).

During the first step, the reactor, linked to the condenser, which impose its pressure, is heated at a high temperature, for example, by industrial waste heat or surplus heat from a cogeneration plant. The input thermal heat (Q_{in}) is converted at high temperature T_{high} and high pressure p_{high} to chemical energy by decomposing the solid–vapor working pair by an endothermic process (1–2 in Figure 1.52c) and, thus, storing it into the working couple. When the high pressure p_{high} is attained, the control valve opens and the vapor generated by the decomposition reaction flows to the condenser as a result of the pressure differential established and as the vapor is condensed. By this process, a large amount of condensation latent heat (Q_{cond}) is released at medium temperature (T_{medium}), and it can be used for heating domestic or process water or air. As a rule, for space air heating purposes, the condensing temperature must be higher than 40°C, while for space floor heating, it must be between 30°C and 35°C (Aristov 2008). When complete decomposition process is achieved, the chemical heat pump is ready for a new production phase (Neveu and Lebrun 1991). During the second step of operation, the condenser linked to the reactor becomes evaporator. The condensed liquid present there is evaporated by absorbing heat (Q_{evap}) from a surrounding (e.g., waste heat source) at low temperature (T_{low}) (Figure 1.52c). The evaporator is thus the source of volatile compound (vapor) from condense generated during the first (decomposition) step. As the vapor equilibrium state in the evaporator is higher than the actual pressure in the reactor, the vapor is driven to the reactor, in which it reforms the initial solid–liquid compound by means of an exothermic synthesis chemical reaction (process 3–4 in Figure 1.52c). This process releases heat by absorbing the vapor within solid compound existing in the reactor (Aristov 2008). The heat released (Q_{out}) by the exothermic synthesis process is transferred from the reactor to an industrial process or for space or domestic hot water heating at a medium temperature (T_{medium}).

In solid–vapor chemical heat pumps with dual reactors, the decomposition and synthesis processes operate in counter phase, that is, when the medium-temperature reactor is in the decomposition mode, the high-temperature reactor is in the synthesis mode, and vice versa. In other words, although one set is charged and recovers heat at low temperature, the other one releases heat at higher temperature compared to that of heat recovered.

1.7.1.2   Energy Efficiency

The maximum (Carnot) heating COP ($COP_{Carnot}^{heating}$) of conventional mechanical vapor compression heat pumps is defined by Equation 1.5.

In the case of chemical heat pumps, the maximum (ideal) heating COP depends only on the cycle temperature boundaries and is defined as the ratio of the useful heat supplied to the heat recovered (Aristov 2008):

$$COP_{heating,max}^{ChemHP} = \frac{\text{Useful heat supplied}}{\text{Recovered heat}} = \frac{T_{high}}{T_{medium}} \frac{T_{medium} - T_{low}}{T_{high} - T_{low}} \tag{1.45}$$

However, the actual COP of chemical heat pumps is not dependent on the temperature levels, but on the changes of specific enthalpy of chemical reactions, that is, based on the thermophysical characteristics of the solid–vapor working pair. By neglecting the system thermal mass and sensible heat changes of the reactants, as well as auxiliary energy consumptions (fluid pumping), and by assuming that the heat released in the reactor during the synthesis step and in the condenser are used for heating purposes, the actual $COP_{heating,actual}^{ChemHP}$ can be calculated as follows (Aristov 2008) (see Figure 1.52):

$$COP_{heating,actual}^{ChemHP} = \frac{Q_{out} + Q_{cond}}{Q_{in}} = \frac{\left(\Delta h_{reactor}^{exotherm} + \Delta h_{cond}\right)}{\Delta h_{evap}} \tag{1.46}$$

where:
$\Delta h_{reactor}^{exotherm}$ is the specific enthalpy (kJ/kg) variation corresponding to the useful heat released at T_{medium} during the high-temperature synthesis exothermic reaction

Δh_{evap} is the specific enthalpy (kJ/kg) variation corresponding to the heat entering the evaporator at T_{low}

The $COP_{cooling}$ of a chemical heat pump can similarly be defined as the ratio of the specific heat of evaporation (Δh_{evap}) of the working fluid and the heat of the decomposition chemical reaction (Aristov 2008):

$$COP_{cooling} = \frac{Q_{evap}}{Q_{in}} = \frac{\Delta h_{evap}}{\Delta h_{reactor}^{endotherm}} \tag{1.47}$$

where $\Delta h_{reactor}^{endotherm}$ is the specific enthalpy (kJ/kg) variation corresponding to the heat input at T_{high} during the endothermic chemical reaction.

1.7.2   Chemical Heat Transformers

A solid–vapor chemical heat transformer also consists of a reactor and an evaporator–condenser (Figure 1.53). A transport process of a mobile phase (vapor) takes place between these components. The working principle of a batch-type chemical heat transformer is based on the alternate, reversible decomposition and synthesis of a solid–vapor working pair. The decomposition process is endothermic absorbing heat, while the synthesis is exothermic releasing heat.

During the first step, the solid-vapor reactor absorbs thermal energy (Q_{in}) under equilibrium conditions from an external heat source, such as industrial waste heat or solar energy at a medium-temperature (T_{medium}). This heat drives inside the reactor the decomposition endothermic process 1–2 at a low pressure (p_{low}), determined by the condenser operating temperature (T_{low}) (Figure 1.53c). When this pressure is attained, the control valve opens and the vapor generated by the decomposition reaction flows to the condenser as a result of the pressure differential established. The vapor is condensed at lower temperature (T_{low}) and the condensing heat (Q_{cond}) is released to surroundings.

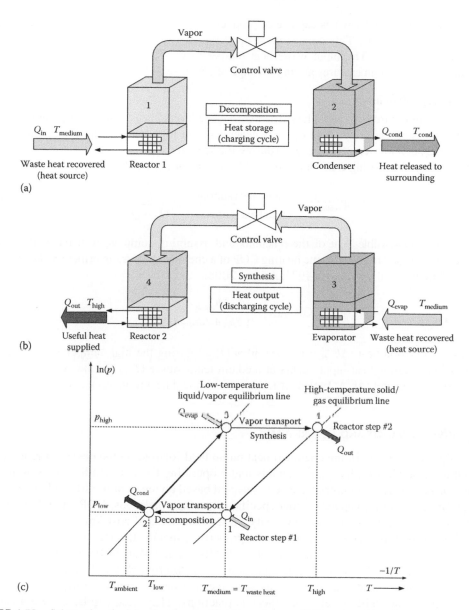

FIGURE 1.53 Schematic principle of a chemical heat transformer. (a) Decomposition (heat storage/charging) cycle; (b) synthesis (heat output/discharge) cycle; and (c) process representation in Clausius-Clapeyron *p-T* diagram.

The next step is the synthesis (heat discharging) phase (process 3–4 in Figure 1.53c). During this step, the condenser becomes evaporator where an evaporating process takes place by using waste heat source at medium temperature (T_{medium}). As the evaporator temperature increases to T_{medium}, the vapor pressure increases according to the equilibrium $p–T$ relation. When the vapor reaches its (saturation) high pressure p_{high}, the control valve opens to allow it entering the reactor. Within the synthesis process, vapor molecules are concentrated onto the working material's solid surface, and lose some energy. Thus, as the vapor diffuses through the reactor's bed, an exothermic reaction 3–4 takes place at a high temperature (T_{high}) corresponding to the vapor high pressure (p_{high}). The heat generated (Q_{out}) by such an exothermic reaction is released outside the system for industrial process usage at high-temperature (T_{high}).

A chemical heat transformer is capable of producing temperature lifts up to 100°C and output temperatures up to 250°C. This creates more opportunities for industrial applications and more energy saving potential. The input waste heat temperatures (T_{medium}) can be between 80°C and 150°C and the output temperatures are in the range of 150°C–250°C (T_{high}).

1.7.2.1 Energy Efficiency

As can be seen in Figure 1.53, waste heat ($Q_{in} + Q_{evap}$) is recovered at a medium temperature (T_{medium}) during the first step and, then, heat (Q_{out}) is generated at a high temperature (T_{high}) and released (rejected, Q_{cond}) at a low temperature (T_{low}) during the second step.

Consequently, the heating COP of a chemical heat transformer can be defined as follows:

$$\text{COP}_{heating}^{ChemHT} = \frac{\text{Useful heat supplied}}{\text{Input heat recovered}} = \frac{Q_{out}}{Q_{in} + Q_{evap}} \tag{1.48}$$

By neglecting the sensible heat of the reactants and assuming complete heat recovery, reaction enthalpies can be used to estimate the heating COP of a chemical heat transformer can be expressed as follows (Raldow and Wentworth 1979; Aristov 2008):

$$\text{COP}_{thermal}^{ChemHT} = \frac{\Delta h_{useful}}{\left(\Delta h_{in} + \Delta h_{evap}\right)} < 1 \tag{1.49}$$

where useful reaction heat (Δh_{useful}) is released at (T_{high}) during the high-temperature exothermic reaction while reaction heat input occurs at medium temperature (T_{high}) in the evaporator (Δh_{evap}) and the first step reactor (Δh_{in}). Typical efficiency of chemical heat transformers turns around 30%.

1.7.3 R&D Advances and Future Needs

Historically, R&D works in the chemical heat pump field focused on the design of reactors and new and more stable working pair reactants under operating thermal conditions. Among other criteria, the new working couples have been selected based on temperature lifts achievable, thermal stability and storage capacity under specific operating conditions, ODP and GWP, specific mass/volume, thermal conductivity, viscosity, surface tension and cost, transportability, corrosion and inflammability, external heat transfer properties, specific thermal power output per unit of mass/area, and so on (Lebrun and Neveu 1991; Maizza and Maizza 1996; Wongsuwan et al. 2001; Wang et al. 2008; Yu et al. 2008; Jangam 2011). Meirovitch and Segal (1990) reported that low-temperature heat sources, such as solar thermal energy, could be upgraded to satisfy the requirements at higher level by exothermic reactions. They used a solar collector directly integrated to a chemical reactor for methane reforming. An ammonia-based thermochemical solar energy storage and transport system has been developed and experimentally studied by Lovegrove and Luzzi (1996). Even several studies noted that combined solar-assisted chemical heat pump dryers may offer opportunities for lucrative drying systems, few of them have demonstrated interesting energetic and economic efficiency allowing large-scale industrial applications. Consequently, major improvements at the level of the system overall COP and the control strategy of the whole drying cycle and of the final quality of heat sensitive products must be provided in the future. Reducing global energy consumption and environmental impact of working fluids cannot be enough to promote solar-assisted chemical heat pump dryers for industrial implementations. It is also necessary to perform more theoretical and experimental works to improve the overall performance and cost-effectiveness of such systems, as well as studies on market potential.

Future R&D works must however focus on the design of new reactors, and must also develop better heat transfer equipment and more efficient control strategies, especially in drying applications, using efficient renewable energies, such as solar and geothermal, along with industrial waste heat either to upgrade it to higher temperatures or for thermal energy storage, combine chemical heat pumps with mechanical vapor compression heat pumps, and scale-up laboratory prototypes to industrial-scale applications.

1.7.4 INDUSTRIAL APPLICATIONS

Commercial applications of chemical heat pumps are relatively rare mainly due to their low heating and cooling COPs, safety issues and relatively high initial costs.

The majority of reported studies are on small-scale applications and include refrigeration, air conditioning (e.g., automobile), space heating and cooling, waste heat recovery (e.g., from power plants, wind turbines, distillation, evaporation and condensation, and drying), high capacity and long-term thermal energy storage, upgrade of solar energy, cooling of electronic devices, and some applications in drying processes for heat recovery and dehumidification (Howerton 1978; Mbaye et al. 1998; Ogura and Mujumdar 2000; Wongsuwan et al. 2001; Wang et al. 2008; Fadhel et al. 2011; Jangam 2011).

1.7.4.1 Chemical Heat Pump-Assisted Drying

According to several researchers, chemical heat pump potentially represents a promising technology for effective energy utilization in drying.

Rolf and Corp (1990) developed a chemical heat pump dryer for drying of bark and lumber, and concluded that this drying method is easy to adapt to other industrial drying processes, in particular to drying in the pulp and paper industry. Ogura and Mujumdar (2000) proposed a chemical heat pump drying concept using reversible reactions with calcium oxide/calcium hydroxide ($CaO/H_2O/Ca(OH)_2$) as working materials because of their safety, high reaction enthalpy, equilibrium temperature–pressure level, and relatively low cost. These authors focused on the heat and mass transfer processes and suggested that integration of solar thermal panels would help extending the concept utilization, especially in tropical regions. Chemical heat pump-assisted batch air convective dryers, can efficiently dehumidify and then heat the drying air. Ogura et al. (2004, 2005) found that the amounts of input and output heats of a 100 $kg_{dry-air}$/h compact batch chemical heat pump with constant drying rate in a repetitive operation can be controlled by the special pressure adjustment procedure such as instant vacuuming of the reactor during both heat-storing and heat-releasing steps. In this concept, two chemical heat pumps operate concurrently, that is, when the first chemical heat pump is in the heat storing step, the second is in the heat releasing phase. They then reverse their operating modes with each other every hour or so. According to Ogura et al. (2004, 2005), the most critical issue has been the controllability of chemical reactions, heat transfer processes, air dehumidification process, and dry air production.

A solid–vapor ($CaCl_2–NH_3$) solar-assisted chemical heat pump for drying agricultural products as lemon grass has been proposed, constructed and tested under Malaysia's weather conditions (Ibrahim et al. 2008, 2009, 2010). The system consisted of a solar collector, a storage tank, a chemical heat pump (reactor and condenser–evaporator), and a continuous (tray) dryer chamber. In this system, solar heat was supplied to the reactor at high temperature to decompose the solid–vapor working pair, store heat, and generate ammonia vapor to be then condensed. This solar-assisted chemical heat pump system provided 85% of the total energy required to maintain the drying air temperature at 55°C. The rest (15%) of energy has been provided by an auxiliary heater. Experimental efficiency of solar tube collector was of 74%, and the reported experimental heating COP.

In the future, solar-assisted chemical heat pump drying must improve the solar collector conception (thermal or combined *PV/T*, air, or water-based), thermal performance, system integration and control in order to be acceptable for commercial applications.

1.8 SOLID-STATE HEAT PUMPS

Solid-state devices, as thermoelectric and thermoacoustic heat pumps may efficiently provide useful heating and cooling (Hochbaum et al. 2008).

1.8.1 THERMOELECTRIC HEAT PUMPS

A thermoelectric heat pump is a heating and cooling device based on Peltier effect that consists in transferring heat between two junctions (*n*- and *p*-type) of two dissimilar electrically conductive materials (Figure 1.54a), as doped bismuth telluride and silicon germanium semiconductors (Duck et al. 1997). Such materials form modules thermally placed in parallel and electrically in series to each other, connected side by side and then sandwiched between two thermally conducting ceramic plates (Figure 1.54b) (https://en.wikipedia.org/wiki/Thermoelectric_cooling, accessed June 15, 2015). These modules are configured so that all junctions on one side heat and those on the other side cool. In practice, they are multiplied by more than 100 times, and organized into arrays and the electrical current passes from one segment to another through connecting shunts.

When a direct current (DC) potential is applied to the free ends of such a module, a temperature difference, depending on the current direction and semiconductors' electron densities, is generated. When the applied potential is reversed, the opposite effect is produced, similar to a conventional heat pumps working between two (cold and hot) thermal reservoirs. But, unlike conventional heat pumps, which use refrigerants, thermoelectric devices use the electrical current as the working fluid. Contrarily, if a temperature gradient is applied between *n*- and *p*-type junctions, a thermoelectric DC electric current between the hot and cold ends of junctions is generated, a phenomenon known as *Seebeck effect*. It is the basic principle of thermocouples and thermoelectric DC power generators able to convert waste heat into electricity.

This section refers only to thermoelectric heat pumps based on the Peltier effect that can achieve today temperature lifts as high as 70°C with maximum operating temperatures of 80°C. However, the most advanced multistage thermoelectric heat pumps operate at temperatures up to 175°C and provide heat pumping densities of up to 14 W/cm², that is, twice as high as standard thermoelectric modules, and cooling capacities from 100 to 300 W (http://www.lairdtech.com/products).

Compared to traditional mechanical vapor compression heat pumps, the advantages of thermoelectric heat pumps are: (1) simple and compact (very small size), light weight, flexible shape, silent and vibration-free, robust (durable) in demanding environments, rapid response time; (2) precise temperature control (tolerances of ±0.1°C); (3) capacity controllable via changing the applied electrical tension; (4) lack of moving parts or potentially hazardous working fluid as synthetic refrigerants; (5) high reliability, maintenance-free operation over 20 years technical life time with little performance degradation; and (6) near-infinite technical lifetime exceeding 100,000 h.

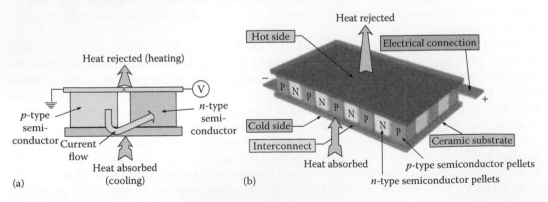

FIGURE 1.54 (a) Peltier cooling/heating effect and (b) schematic of a typical thermoelectric heat pump. (Redrawn from https://en.wikipedia.org/wiki/Thermoelectric_cooling, accessed June 15, 2015.)

Disadvantages can be listed as high cost, low thermal capacities and efficiency, particularly in large-scale industrial applications (Bell 2008).

1.8.1.1 Energy Efficiency

The operation of a thermoelectric module with N pairs of thermoelectric legs implies Peltier and Joule effects and thermal conduction.

The thermal fluxes absorbed ($Q_{abs,cold}$) and rejected ($Q_{rej,hot}$) by a thermoelectric element can be respectively expressed as follows (Rowe 1995):

$$\dot{Q}_{abs,cold} = 2N*\left(\alpha*I*T_{cold} - 0.5*R*I^2 - K*\Delta T\right) \tag{1.50}$$

$$\dot{Q}_{rej,hot} = 2N*\left(\alpha*I*T_{hot} - 0.5*R*I^2 - K*\Delta T\right) \tag{1.51}$$

where:

α, R, and K are the Seebeck coefficient, electrical resistance, and thermal conductance of thermoelectric legs at the mean temperature, respectively

T_{cold} and T_{hot} are the temperatures of cold and hot junctions, respectively (Kim et al. 2013)

The electrical power supplied is the difference between these two thermal fluxes:

$$\dot{W}_{elec} = \dot{Q}_{rej,hot} - \dot{Q}_{abs,cold} = 2*n\left(\alpha*I*\Delta T + R*I^2\right) \tag{1.52}$$

The amount of heat absorbed by a thermoelectric heat pump can be also expressed as follows:

$$Q = P*I*\tau \tag{1.53}$$

where:

P is the Peltier coefficient that depends on the temperature and the type of joined semiconductors

I is the current intensity

τ is the time

If the auxiliary electrical input power of circulating devices (fans and pumps) of secondary thermal carriers (air and water) are neglected, the heating COP of a thermoelectric heat pump can be expressed as follows:

$$\mathrm{COP}_{heating}^{TEhp} = \frac{\dot{Q}_{rej,hot}}{\dot{W}_{elec}} \tag{1.54}$$

Today, the maximum theoretical efficiency that may be achieved by thermoelectric coolers turns around 10%–15% of the ideal Carnot cycle, compared to 40%–60% achieved by conventional mechanical vapor compression heat pumps (Rama et al. 2001). On the other hand, the efficiency of thermoelectric heat pumps (or coolers) and power generators with p-type and n-type materials, is determined by the dimensionless thermoelectric material figure of merit defined as follows:

$$zT = \frac{\sigma S^2 T}{\lambda} \tag{1.55}$$

where:

Z is a parameter equal to the square of the Seebeck coefficient per unit of temperature

T is the absolute temperature

σ is the electrical conductivity

S is the Seebeck coefficient of the thermoelectric material, a measure of the magnitude of the induced thermoelectric voltage in response to a temperature difference across that material (μVolt/Kelvin)

λ is the thermal conductivity. During the last few decades, zT of commercial materials has been limited to around 1

However, in the case of waste heat recovery from vehicle exhaust operating with about 350°C temperature differential, the efficiency needs to be 10% and the corresponding average zT should be 1.25. To double the efficiency of an ideal thermoelectric system the figure of merit has to increase to about 2.2. It is estimated that materials with $zT > 3$ (i.e., about 20%–30% Carnot efficiency) are required to replace traditional heat pumps and coolers in most applications (as air conditioners) and even around nine (Harman et al. 2002), but no known thermoelectric devices have today figures of merit $zT > 3$ (Tritt and Subramanian 2011).

1.8.1.2 R&D Advances

Active R&D works in material sciences have been deployed during the past 20 years in order to improve the modules' energy conversion efficiency, optimize the design, reducing parasitic losses, weight and costs, demonstrate and implement the most promising concepts (Angrist 1965; Vining 2007), and develop advanced low-cost, nontraditional materials, such as alloys, crystals, and multiphase nanocomposites, with high electric and low thermal conductivities, respectively, and with sufficiently high figures of merit.

Average figures of merit between 1.5 and 2 are considered today enable for waste-heat harvesting, as in residential and commercial solid-state heating and cooling applications. At figures of merit of 2, cooling and temperature control of microprocessors, as well as for industrial waste-heat recovery and to replace small internal combustion engines, become attractive. Gains in material figure of merit of 30%–40% have been reported by using nanostructured silicon materials and, in the case of thermoelectric modules with nanoscale inclusions, figures of merit up to 3.2 (at temperatures of about 300°C) have been achieved. Such materials enhanced the Seebeck coefficient, and dramatically reduced the thermal conductivity. However, despite such promising results, the scaling of the nanomaterials has proven to be difficult and, thus, the efficiency gains have yet to be demonstrated (Bell 2008). Recently discovered, tin selenide has figures of merit of 2.6, the highest value reported till date attributed to an extremely low thermal conductivity (0.23 W/mK) at high temperatures of about 650°C (Zhang and Talapin 2014; Zhao et al. 2014).

Other thermoelectric materials under research are magnesium and oxide compounds, silicides, skutterudites, half Heusler alloys (Culp et al. 2006), organic materials, silicon-germanium alloys (currently, the best thermoelectric materials at around 1000°C), sodium cobaltate, and nanocrystalline materials (Voneshen et al. 2013). A theoretical concept integrating a thermoelectric heat exchanger within a conventional mechanical vapor compression air-to-water heat pump cycle has been developed in order to enhance the system heating capacity and improve its overall efficiency (Zhu and Yu 2015). In this concept, a multi-module thermoelectric element and two microchannel heat exchangers were integrated into a thermoelectric heat exchanger module inserted at the condenser outlet in order to transfer heat from the main refrigerant circuit to the bypass refrigerant circuit. In addition, an ejector is used to drive the refrigerant through the bypass circuit. Because of the cooling and heating function of the thermoelectric heat exchanger, the two-phase refrigerant stream from the EXV is condensed in the cold-side microchannel heat exchanger, while the refrigerant liquid from the condenser is evaporated in the hot side microchannel heat exchanger. Thus, the thermoelectric heat exchanger enables heat to be transferred from the main circuit to the bypass circuit, resulting in the increase of the amount of rejected heat in the condenser. The simulation results obtained showed that improvements can be achieved in both heating COP and capacity between 16.4% and 21.7% under thermoelectric-assisted vapor compression cycle.

1.8.1.3   Applications

Different thermoelectric coolers are available for specific industrial applications of thermoelectric devices, as, for example, (1) air-to-air, with air heat exchangers equipped with fans to absorb and dissipate heat; (2) cold plate-to-air (or to-liquid) where heat is absorbed through a cold plate, transferred through the thermoelectric modules and dissipated to air or water via air or water heat exchangers; and (3) liquid-to-air (or to-liquid) with liquid heat exchangers to absorb heat and transfer it through the thermoelectric modules to air or water heat exchangers.

For industrial applications, cooling capacity from 10 to 100 W, footprint less than 13 × 13 mm (for less than 10 W cooling capacity), multistage (cascade), and high power density (with heat pumping densities up to 14 W/cm^2) thermoelectric modules are available (http://www.lairdtech.com/products).

Currently, thermoelectric devices are used in industrial sectors with tight space constraints, low weight requirements and requiring cooling capacities from 10^{-3} W to several thousand of W, as space (satellites and spacecraft), as modules that can operate in the absence of sunlight or as mini-electrical plants to convert heat into electricity using the Seebeck effect and cool astronomical telescopes and spectrometers, submarines and railroad trucks, and heat-seeking missiles with night-vision operating at temperatures as low as −80°C. Such devices can be competitive with conventional mechanical vapor compression heat pumps when used in small-scale applications, such as cooling/heating car seats, night-vision systems, electrical enclosures, laser diodes, and low-wattage power generators.

Many other applications can be mentioned: (1) small food, portable beverage and picnic (camping) ice-free coolers, wine-storage cabinets, hotel room minirefrigerators, office water coolers, medical, telecom, military, cooling microprocessors, electronic instruments and enclosures, optic fibers, remote data communication systems for oil and gas pipelines, polar weather station power generators, and thermal cycles for DNA synthesizers; (2) temperature control of hybrid car seats, that is, rapid seat cooling in the summer and fast heating in the winter; and (3) air dehumidifiers for digital color printing systems precise temperature control and low relative humidity.

1.8.2   Thermoacoustic Heat Pumps

Thermoacoustic heat pumps are near-solid-state devices that have no moving parts requiring sealing or lubrication and that use high-amplitude sound waves, equivalent to an external power input, to transfer heat from one thermal reservoir to another. The lack of moving parts gives thermoacoustic refrigerators the advantages of simplicity, reliability, and low cost. Because the sound waves are confined in sealed cavities, the devices are fairly quiet.

Thermoacoustic heat pumps contain any ozone-depleting or toxic working fluid and are intending to eliminate the compressors of conventional heat pumps to provide maximum temperature lifts up to 50°C.

1.8.2.1   Principle

As known, a sound wave is associated with changes in pressure, temperature, and density of the medium through which the sound wave propagates. In addition, the medium itself is moved around an equilibrium position, but these fluctuations are too small to be noticed in the sounds. Thermoacoustic effect, which converts sound to heat, has been known for over a 100 years. The most significant barrier to rapid commercial deployment of thermoacoustic devices was the large number of precision components required. Thermoacoustic heat pumps (Figure 1.55) use standing sound waves to take the working fluid (a gas) through a thermodynamic cycle. They rely on the heating and cooling that accompany the compression and expansion of a gas in a sound wave in a pressure range of 8 bars.

In this concept, an electrically driven loudspeaker generates high-amplitude standing sound (acoustic) (≈300 Hz) waves resulting in noticeable fluctuations through an inert gas (air, noble gas)

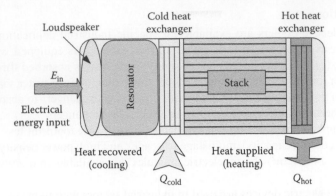

FIGURE 1.55 Principle of a thermoacoustic device. (From https://en.wikipedia.org/wiki/Thermoacoustic _heat_engine, accessed June 30, 2015.)

present in the resonator (Spoelstra and Tijani 2005). These waves interact with an array of small parallel solid plate channels (stacks) placed inside the resonator (Figure 1.55). When extremely loud standing sound waves exist inside resonator, a temperature difference is generated across it (Malone Refrigeration 2015). This phenomenon can be explained by the fact that as the gas oscillates back and forth it changes in pressure, mainly because of its adiabatic compression and expansion produced by the sound pressure, as well as by the heat transfer between the gas and the stack. Because the pressure changes simultaneously, when pressure reaches a maximum or a minimum, the temperature changes (Source: Malone Refrigeration). In other words, the acoustic waves generate temperature differences and transfer heat from one place to another, as a heat pump. By placing heat exchangers at each side of the stack, heat is transferred from the cold to the hot sides by absorbing and suppling heat. Each element of gas goes through a thermodynamic cycle in which the element is compressed and heated, rejects heat at the right end of its range of oscillation, is depressurized and cooled, and absorbs heat at the left end. For standing wave devices, the process can be described using the ideal reverse Brayton cycle that consists of four processes, that is, *adiabatic gas compression, isobaric heat transfer, adiabatic gas expansion, and isobaric heat transfer*. Consequently, each element of gas moves heat from left to right, from cold to hot, during each cycle of the sound wave. The combination of the cycles of all the elements of gas transports heat from the cold heat exchanger to the hot heat exchanger much as a bucket brigade transports water (https://www.fas.org /sgp/othergov/doe/lanl/pubs/00326024.pdf, accessed July 31, 2015).

Although the working principle of thermoacoustic technology is quite complex, the practical implementation is relatively simple, offering great advantages with respect to the economic feasibility of this technology. The following are some of the other advantages noted: (1) no moving components or noise, (2) long life span, (3) environmental-friendly working medium (air and noble gas), without phase changes; (4) high temperature lifts; and (5) low maintenance costs. Their principal disadvantages are very low power density and energy efficiency yet to be proven.

1.8.2.2 Energy Efficiency

The heating COP of a thermoacoustic heat pump can be expressed as follows:

$$\text{COP}_{\text{heating}}^{\text{ThAC}} = \frac{Q_{\text{hot}}}{E_{\text{in}}} \tag{1.56}$$

where:
    Q_{hot} is the thermal energy supplied to user (kWh)
    E_{in} is the electrical energy input to loudspeaker (kWh)

1.8.2.3 Applications and R&D Activities

Common applications of thermoacoustic heat pumps (coolers) are in cryogenic laboratories for deep cooling processes without using cryofluids (Ceperley 1979).

Most of R&D works focus on developing (1) coolers able to cool up to −25°C with 600 W thermal capacity and efficiency of 33%, (2) intense sound generators and resonators, (3) compact (microchannel) heat exchangers, (4) solar thermoacoustic coolers with parabolic captors and helium gas as the working fluid (http://ipnwww.in2p3.fr/Thermo-acoustic, accessed July 31, 2015), and (5) air-conditioning devices and cooking stoves for use in developing countries.

1.9 FUTURE R&D AND APPLICATION CHALLENGES

Most countries will continue in the future to be net primary energy importers exposed to supply security risks. In the building sector, for example, global demand for heating and air-conditioning will increase, while increasing prices for fossil energy resources will provide higher production costs in many industries. In such a context, heat pumps may partially contribute at reducing these risks by more efficient use of primary energy sources in both building and industrial sectors.

1.9.1 RESIDENTIAL, INSTITUTIONAL/COMMERCIAL HEAT PUMPS

The ambient air and ground will remain the most common heat sources for mechanical vapor compression heat pumps in the residential and institutional/commercial sectors.

Future R&D works on air-source heat pumps have to improve the defrosting techniques in order to reduce the demand for auxiliary heating and, thus, increase the seasonal coefficients of performance. In cold climates, double-source (hybrid) (e.g., coupled with gas boilers), multifunction (e.g., simultaneous or alternate space and domestic hot water heating, and radiant floor and air heating and cooling) and vapor injection heat pumps will be further developed and/or continuously improved. Ground-source heat pumps, that use renewable energy stored in the soil or bedrock, or surface waters and groundwater as heat and sink sources for building heating and cooling, will remain among the most efficient and clean systems available for building heating and cooling. In the future, ground-source heat pumps will continue to efficiently reduce energy consumption by more than 45% compared to air-source heat pumps and up to 75% compared to electric resistance heating. The design optimization and the reduction of the capital cost of vertical ground heat exchangers will be among the most challenging tasks.

1.9.2 INDUSTRIAL HEAT PUMPS

As for any kind of heat pumps, future R&D targets for industrial heat pumps include improvements of existing components and development of new types of compressors, heat exchangers, and controls.

To improve overall heating performances, industrial heat pumps have to be equipped with inverter controlled or two-stage compressors with or without refrigerant injection at intermediate pressures in order to better match heating and/or cooling capacities and demand loads aiming at reducing the overall energy consumption. Sustained efforts will be deployed in the future to develop new, long-term natural or synthetic refrigerants with very low or zero GWPs and ODPs. For using ammonia on a large scale, safer installation and operation measures as well as new national standards will have to be implemented. Components that use CO_2 are already on the market, but the target of future R&D activities is to improve their performances and to expand the application field by developing new types of compressors, heat exchangers, and controls. HCs are flammable; therefore, it will be necessary to take measures to reduce the system refrigerant charges and prevent major accidental leakages. Further research will be also required to develop new user-friendly lubricants to replace the existing synthetic ones.

In the field of industrial heat pumps, other R&D challenges are (1) develop cost-effective and more efficient microchannel evaporators, condensers, and gas coolers with little temperature approaches and lower refrigerant charges; (2) recover the expansion work especially in the case of high-pressure refrigerants as CO_2; (3) develop mechanical vapor compression and absorption heat pump cycles with single or cascade ejectors in order to improve their energy performances by replacing some of the input power; (4) improve the control strategies by involving forecast technologies, intelligent on-board fault detection and diagnostics, smart user interfaces, bidirectional connectivity, and demand response readiness; and (5) efficiently integrate the heat pumps in whole industrial processes by using advanced pinch analysis technics. Mechanical vapor compression and recompression, absorption heat pumps, and transformers will find in the future more interesting applications in large-scale industrial processes such as drying pulp and paper and wood, distillation, and evaporation.

In the field of chemical heat pumps, difficult tasks as continuous or simultaneous operation and control strategies of chemical reactors must be addressed. Future R&D works will focus on the design of new reactors, stability of the selected working pairs under operating conditions, development of better heat transfer equipment and more efficient control strategies, especially in drying applications, use efficient renewable energies, such as solar and geothermal, along with industrial waste heat either to upgrade it to higher temperatures or for thermal energy storage, combine chemical heat pumps with mechanical vapor compression heat pumps, and scale-up laboratory prototypes to industrial-scale applications.

1.9.2.1   Heat Pump-Assisted Drying

Heat pump-assisted dryers based on the mechanical vapor compression cycle are the most used today. Even the heat pump-assisted drying in agro-food and wood drying processes is sometimes considered as a mature technology, it still require further theoretical and experimental R&D works, and demonstration projects, in order to improve the system integration and control, and thus provide higher energy and dehumidification performances.

Obviously, without improving the integration of dryers and heat pumps and their coupled operation throughout the whole drying process, no real technology advancement can be achieved. The lack of reliable integrated systems may compromise the future applicability of standard and/or innovative heat pump drying concepts. In order to make relevant decisions about using or not using heat pumps as dehumidification devices, the drying mechanism of each product has to be relatively well known. Data on the physical properties of the dried material is also required to develop new drying heat pump concepts. Inadequate integration and/or control strategy may lead to operation troubles and low dehumidification efficiencies. According to materials' drying mechanisms, optimum air convective dryers must be selected, as well as specific drying schedules based on the time-drying curve and drying characteristics of each dried product. Research studies also have to consider that air velocity on the product surface is a critical parameter that must be high enough to produce rapid air changes, avoid the formation of death zones, and provide uniform drying. The control of the dehumidification processes is the next major challenge, strongly interconnected with the system integration. Continuously or intermittently matching the material thermal dewatering rate with the heat pump dehumidification capacity is thus a mandatory task. Increasing attention must also be given to the prevention of any hazards or industrial risks that can occur during drying. This goal could be achieved by providing safety parameters for the heat pump compressor, as well as prevention methods through adequate drying controls. Troubleshooting specific to drying heat pumps, such as short on/off compressor cycles, excessively low or high refrigerant pressures and temperatures, and evaporator frosting, has to be carefully considered and, if possible, avoided by appropriate system integration and control. There is also a need for developing specific type of heat pump dryers per product or group of similar products, that is, agro-food, wood, heat-sensitive, biological, solid, and liquid, and for each specific range of drying temperatures.

In the future, more attention will have to be paid to the environmental and safety aspects of heat pump-assisted drying systems. For many dried materials, such as wood, the water extracted from the dryers will need to be treated before being rejected into the environment, because some condensates may contain acid and/or organic loads as formaldehyde and, to a lesser degree, acetaldehyde. Options to consider for treatment are the UV peroxidation and reverse osmosis. Drying heat pump technology is an interdisciplinary technology involving drying, heat pump, and control experts. As a consequence, future R&D studies provide numerous R&D challenges in terms of inter- and multi-disciplinary interference of various scientific and engineering fields. Academia must develop modeling and simulation techniques for optimizing dryer operation, whereas industry has to consider more complex approaches for the selection and implementation of drying technologies.

Finally, periodical analysis of market potential and economic issues, development of new standards, simulations, laboratory tests and *field* demonstrations of various prototypes will help heat pump dryers attain future technical, economic, and environmental targets.

REFERENCES

Abdulateef, J.M., K. Sopian, M.A. Alghoul, M.Y. Suliman, A. Zaharim, I. Ahmad. 2008. Solar absorption refrigeration system using new working fluid pairs. In: *3rd IASME/WSEAS International Conference on Energy and Environment*, February 23–25, University of Cambridge, Cambridge, UK.

Adapa, P.K., G.J. Schoenau. 2005. Re-circulating heat pump assisted continuous bed drying and energy analysis. *International Journal of Energy Research* 29:961–972.

Aikins, K.A., S.H. Lee, J.M. Choi. 2013. Technology review of two stage vapor compression heat pump system. *International Journal of Air-Conditioning and Refrigeration* 21:1330002. http://www.worldscientific.com/doi/pdf. Accessed August 2015.

Altenkirch, E. 1950. Vapor compression refrigeration machine with solution circuit. (Original language German: Kompressionskältemachine mit lösungskreislauf). *Kältetechnik* 2(10–12):251–259, 310–315, 279–284.

Angrist, S.W. 1965. *Direct Energy Conversion*. Allyn and Bacon, Boston, MA, Chapter 4, pp. 144–150.

ANSI/ASHRAE Standard 34. 2013. *Designation and Safety Classification of Refrigerants*. American Society of Heating, Refrigerating and Air-Conditioning Engineers, Atlanta, GA.

Anstett, P. 2006. Measurement of the performance of an air/water heat pump using CO_2 or R744 for the production of hot water for use in a hospital. *IEA Heat Pump Centre Newsletter* 25(2): 35–38.

Aphornratana, S., I.W. Eames. 1998. Experimental investigation of a combined ejector absorption refrigerator. *International Journal of Energy Resources* 22:195–207.

Aristov, Y.I. 2008. Chemical and adsorption heat pumps: Cycle efficiency and boundary temperatures. *Theoretical Foundations of Chemical Engineering* 42(6):873–881.

ASHRAE. 2011. *ASHRAE Handbook: HVAC Applications*, SI Edition. ASHRAE, Atlanta, GA.

ASHRAE. 2013. *Fundamentals Handbook*. American Society of Heating, Refrigerating and Air Conditioning Engineers, Atlanta, GA.

Auer, W.W. 1980. Solar energy systems for agricultural and industrial process drying. In: Mujumdar, A.S. (ed.) *Drying'80*. Hemisphere, New York, pp. 280–292.

Baek, C., J. Heo, J. Lung, E. Lee, Y. Kim. 2014. Effects of vapor injection techniques on the heating performance of a CO_2 heat pump at low ambient temperatures. *International Journal of Refrigeration* 43:26–35.

Baek, J.S., E.A. Groll, P.B. Lawless. 2005. Piston-cylinder work producing expansion device in a transcritical carbon dioxide cycle. Part I: Experimental investigation. *International Journal of Refrigeration* 28:141–151.

Baker, C.G.J. 1995. *Industrial Drying of Foods*, 1st ed. Springer-Verlag, New York.

Bannister, P., B. Bansal, C.G. Carrington, Z.F. Sun. 1998. Impact of kiln losses on a dehumidifier drier. *International Journal of Energy Research* 22:515–522.

Bannister, P., G. Carrington, G. Chen, Z.F. Sun. 1999. Guidelines for operating dehumidifier timber kilns. Energy Group's Heat Pump Dehumidifier Research Programme Report, EGL-RR-02.

Bansal, P.K., J.E. Braun, E. Groll. 2001. Improving the energy efficiency of conventional tumbler clothes drying systems. *International Journal of Energy Research*. 25:1315–1332.

Barber, J., D. Brutin, L. Tadrist. 2011. A review on boiling heat transfer enhancement with nanofluids. *Nanoscale Research Letters* 6:280. http://www.nanoscalereslett.com/content/6/1/280. Accessed May 22, 2015.

Barragan, R.M., V.M. Arellano, C.L. Heard, R. Best. 1998. Experimental performance of ternary solution in an absorption heat transfer. *International Journal of Energy Research* 22:73–83.

Basunia, M.A., T. Abe. 2001. Thin-layer solar drying characteristics of rough rice under natural convection. *Journal of Food Engineering* 47:295–301.

Becker, F.E., A.I. Zakak. 1986. Heat recovery in distillation by mechanical vapor recompression. In: *Proceedings of the 8th Annual Industrial Energy Technology Conference*, June 17–19, Houston, TX.

Bell, L.E. September 12, 2008. Cooling, heating, generating power, and recovering waste heat with thermoelectric systems. *Science* 321:1457–1461. http://www.sciencemag.org. Accessed June 18, 2015.

Bertsch, S.S., E.A. Groll. 2008. Two-stage air-source heat pump for residential heating and cooling applications in northern U.S. climates. *International Journal of Refrigeration* 31:1282–1292.

Best, R., F.A. Holland. 1990. A study of the operating characteristics of an experimental absorption cooler using ternary systems. *International Journal of Energy Research* 14:553–561.

Best, R., L. Porras, F.A. Holland. 1991. Thermodynamic design data for absorption heat pump system operating on ammonia-nitrate: Part I cooling. *Heat Recovery System and CHP* 11(1):49–61.

Best, R., W. Soto, I. Pilatowsky, L.J. Gutlerrez. 1994. Evaluation of a rice drying system using a solar assisted heat pump. *Renewable Energy* 5:465–468.

Bi, S., L. Shi, L. Zhang. 2008. Application of nanoparticles in domestic refrigerators. *Applied Thermal Engineering* 28:1834–1843.

Bourke, G., P. Bansal. 2010. Energy consumption modeling of air source electric heat pump water heaters. *Applied Thermal Engineering* 30:1769–1774.

Braun, J.E. P.K. Bansal, E.A. Groll. 2002. Energy efficiency analysis of air cycle heat pump dryers. *International Journal of Refrigeration* 25(7):954–965.

Brauwelt. 2010. Energieeinsparung in Mälzereien. *Brauwelt* 1–2:9–10.

Brian, T.A., K. Sumathy. 2011. Transcritical carbon dioxide heat pump systems: A review. *Renewable and Sustainable Energy Reviews* 15:4013–4029.

Brix, W., R.M. Kærn, B. Elmegaard. 2010. Modelling distribution of evaporating CO_2 in parallel minichannels. *International Journal of Refrigeration* 33:1086–1094.

Brown, J.S. 2009. HFOs: New, low global warming potential refrigerants. *ASHRAE Journal* 51:22–29.

Brown, J.S., C. Zilio, A. Cavallini. 2009. The fluorinated olefin R-1234ze(Z) as a high-temperature heat pumping refrigerant. *International Journal of Refrigeration* 32:1412–1422.

Brunin, O., M. Fiedt, B. Hivet. 1997. Comparison of the working domains of some compression heat pumps and a compression-absorption heat pump. *International Journal of Refrigeration* 20(5):308–318.

Byrne, P., J. Miriel, Y. Lenat. 2009. Design and simulation of a heat pump for simultaneous heating and cooling using HFC and CO_2 as a working fluid. *International Journal of Refrigeration* 32:1711–1723.

Calm, J.M. 2008. The next generation of refrigerants—Historical review, considerations, and outlook. *International Journal of Refrigeration* 31:1123–1133.

Carrington, C.G., P. Bannister, Q. Liu. 1995. Performance analysis of a dehumidifier using HFC-134a. *International Journal of Refrigeration* 18:477–488.

Cavallini, A., L. Cecchinato., M. Corradi., E. Formasieri., C. Zilio. 2005. Two-stage transcritical carbon dioxide cycle optimization: A theoretical and experimental analysis. *International Journal of Refrigeration* 28:1274–1283.

Cecchinato, L., M. Chiarello, M. Corradi, E. Fornasieri, S. Minetto, P. Stringari et al. 2009. Thermodynamic analysis of different two-stage transcritical carbon dioxide cycles. *International Journal of Refrigeration* 32(5):1058–1067.

Cecchinato, L., M. Corradi, E. Fornasieri, I. Zamboni. 2005. Carbon dioxide as refrigerant for tap water heat pumps: A comparison with traditional solution. *International Journal of Refrigeration* 28(8):1250–1258.

Cech, M.J., F. Pfaff. 2000. *Operator Wood Drier Handbook for East of Canada*, edited by FORINTEK Corp., Canada's Eastern Forester Products Laboratory, Québec, Canada.

Ceperley, P. 1979. A pistonless Stirling engine—The travelling wave heat engine. *Journal of the Acoustical Society of America* 66: 1508–1513. DOI: 10.1121/1.383505.

Cervantes, J.G., E. Torres-Reyes. 2002. Experiments on a solar-assisted heat pump and an exergy analysis of the system. *Applied Thermal Engineering* 22:1289–1297.

Chaturvedi, S.K., J.Y. Shen. 1984. Thermal performance of a direct expansion solarassisted heat pump. *Solar Energy* 33(2):155–162.

Chen, L.T. 1988. A new ejector-absorber cycle to improve the COP of an absorption refrigeration system. *Applied Energy* 30:37–51.

Cho, H., C. Baek, C. Park, Y. Kim. 2009. Performance evaluation of a two-stage CO_2 cycle with gas injection in the cooling mode operation. *International Journal of Refrigeration* 32:40–46.

Cho, H., C. Ryu, Y. Kim. 2007. Cooling performance of a variable speed CO_2 cycle with an electronic expansion valve and internal heat exchanger. *International Journal of Refrigeration* 30:664–671.

Cho, H., C. Ryu, Y. Kim, H.Y. Kim. 2005. Effects of refrigerant charge amount on the performance of transcritical CO_2 heat pumps. *International Journal of Refrigeration* 28:1266–1273.

Choi, S.U.S. 1995. Enhancing thermal conductivity of fluids with nanoparticles. In: *Proceedings of the 1995 ASME International Mechanical Engineering Congress and Exposition*, November 12–17, San Francisco, CA.

Chou, S.K., K.J. Chua. 2006. Heat pump drying systems. In: Mujumdar, A.S. (ed.) *Handbook of Industrial Drying*. Taylor & Francis Group, Boca Raton, FL, pp. 1122–1123.

Chua, K.J., S.K. Chou, J.C. Ho, M.N.A. Hawlader. 2002. Heat pump drying: Recent developments and future trends. *Drying Technology* 20(8):1579–1610.

Chua, K.J., S.K. Chou, W.M. Wang. 2010. Advances in heat pump systems: A review. *Applied Energy* 87(12):3611–3624.

Chung, H., M.H. Huor, M. Prevost, R. Bugarel. 1984. Domestic heating application of an absorption heat pump, directly fired heat pumps. In: *Proceedings of the International Conference*, University of Bristol, Bristol, UK, paper 2.2.

Claussen, I.C., T.S. Ustad, I. Strommen, P.M. Walde. 2007. Atmospheric freeze drying—A review. *Drying Technology* 25:957–967.

Cox, N., V. Mazur, D. Colbourne. 2009. The development of azeotropic ammonia refrigerant blends for industrial process applications. In: *International Conference on Ammonia Refrigeration Technology*, Ohrid, Macedonia.

Cuenca Y.M. 2013. Experimental study of thermal conductivity of new mixtures for absorption cycles and the effect of the nanoparticles additions, Doctoral Thesis, Universidad ROVIRA I VIRGILI, Department of mechanical Engineering, Spain.

Culp, S.R., S.J. Poon., N. Hickman, M. Tritt, J. Blumm. 2006. Effect of substitutions on the thermoelectric figure of merit of half-Heusler phases at 800°C. *Applied Physics Letter* 88:042106. http://dx.doi.org/10.1063/1.2168019 (accessed July, 2015).

Daghigh, R., M.H. Ruslan, M.Y. Sulaiman, K. Sopian. 2010. Review of solar assisted heat pump drying systems for agricultural and marine products. *Renewable and Sustainable Energy Reviews* 14:2564–2579.

Daiguji, H., E. Hihara, T. Saito. 1997. Mechanism of absorption enhancement by surfactant. *International Journal of Heat and Mass Transfer* 40(8):1743–1752.

Detlef, W., W.R. Kurt, B. James. 2003. Microchannel heat exchangers—emerging technologies. *ASHRAE Journal* 45(12); 107–109.

Diamante, L.M., P.A. Munro. 1993. Mathematical modeling of the thin layer solar drying of sweet potato slices. *Solar Energy* 51:271–276.

Dincer, I. 2003. *Refrigeration Systems and Applications*, 2nd ed., John Wiley & Sons, London.

Dokandari, D.A., A.S. Hagh, S.M.S. Mahmoudi. 2014. Thermodynamic investigation and optimization of novel ejector-expansion CO_2/NH_3 cascade refrigeration cycles (novel CO_2/NH_3 cycle). *International Journal of Refrigeration* 46:26–36.

Duck, Y.C., T. Hogan, J. Schindler, L. Iordarridis, P. Brazis, C.R. Kannewurf, B. Chen, C. Uher, M.G. Kanatzidis. 1997. Searching for new thermoelectrics in chemically and structurally complex bismuth chalcogenides. In: *Proceedings of ICT'97, 16th International Conference on Thermoelectrics*, 459p.

Elbel, S., P. Hrnjak. 2004. Flash gas bypass for improving the performance of transcritical R744 systems that use microchannel evaporators. *International Journal of Refrigeration* 27(7):724–735.

Elbel, S., P. Hrnjak. 2008. Experimental validation of a prototype ejector designed to reduce throttling losses encountered in transcritical R744 system operation. *International Journal of Refrigeration* 31(3):411–422.

Fadhel, M.I., K. Sopian, W.R.W. Daud, M.A. Alghoul. 2011. Review on advanced of solar assisted chemical heat pump dryer for agriculture produce. *Renewable and Sustainable Energy Reviews* 15:1152–1168.

Fan, S., Q. Liu, S. He. 2008. Scroll compressor development for air-source heat pump water heater applications. In: *Proceedings of International Refrigeration and Air Conditioning Conference at Purdue*, Purdue University, West Lafayette, IN.

Federation of American Scientists. 2014. Project on Government Secrecy, Malone Refrigeration thermoacoustic engines and refrigerators. https://www.fas.org/sgp/othergov/doe/lanl/pubs/00326024.pdf. Accessed June 2015.

Fischer, S.K., P.J. Hughes, P.O. Fairchild, C.L. Kusik, J.T. Dieckmann, E.M. McMahon, N. Hobay. 1991. Energy and global warming impacts of CFC alternative technologies. http://www.afeas.org/tewi.htm. Accessed March 20, 2015.

Fronk, B.M., S. Garimella. 2011. Water-coupled carbon dioxide microchannel gas cooler for heat pump water heaters: Part II-model development and validation. *International Journal of Refrigeration* 34:17–28.

Fukuda, S., C. Kondou, N. Takata, S. Koyama. 2014. Low GWP refrigerants R1234ze(E) and R1234ze(Z) for high temperature heat pumps. *International Journal of Refrigeration* 36:161–173.

Goetz, V., F. Elie, B. Spinner. 1993. The structure and performance of single effect solid–gas chemical heat pumps. *Heat Recovery Systems and CHP* 13(1):79–96.

Goodman, C., B. Fronk, S. Garimella. 2010. Transcritical carbon dioxide microchannel heat pump water heaters: Part II - System Simulation and Optimization. *International Refrigeration and Air Conditioning Conference*, Paper 1159, July 12–15. http://docs.lib.purdue.edu/iracc/1159 (accessed July 15, 2015).

Groll, E.A. 1997. Current status of absorption/compression cycle technology. *ASHRAE Transactions* 103(1): 1136p.

Guilbeault, P. 1987. Case study: Residential heat pump. In: *Association Québécoise Pour la Maitrise de L'énergie (AQME)*, April, Montréal, Canada.

Haiqing, G., M. Yitaai, L. Minxia. 2006. Some design features of a CO_2 swing piston expander. *Applied Thermal Engineering* 26:237–243.

Hannl, D., R. Rieberer. 2014. Absorption/compression cycle for high temperature heat pumps—Simulation model, prototype design and initial experimental results. In: *11th IEA Heat Pump Conference*, Montreal, Québec, Canada.

Harman, T.C., P.J. Taylor, M.P. Walsh, B.E. Laforge. 2002. Quantum dot superlattice thermoelectric materials and devices. *Science* 297(5590):2229–2232.

Harrell, G.S., A.A. Kornhauser. 1995. Performance tests of a two-phase ejector. In *Proceedings of the 30th Intersociety Energy Conversion Engineering Conference*, Orlando, FL, pp. 49–53.

Hawlader, M.N.A., T.Y. Bong, Y. Yang. 1998. A simulation and performance analysis of a heat pump batch dryer. In: *Proceedings of the 11th International Drying Symposium*, August 19–22, Halkidiki, Greece, vol. A, pp. 208–215.

Hawlader, M.N.A., S.K. Chou, K.A. Jahangeer, S.M.A. Rahman, K.W.E. Lau. 2003. Solar-assisted heat-pump dryer and water heater. *Applied Energy* 7(1):185–193(9).

Hawlader, M.N.A., K.A. Jahangeer. 2006. Solar heat pump drying and water heating in the tropics. *Solar Energy* 80:492–499.

Hawlader, M.N.A., S.M.A. Rahman, K.A. Jahangeer. 2008. Performance of evaporator-collector and air collector in solar assisted heat pump dryer. *Energy Conversion and Management* 49:1612–1619.

He, S., W. Guo, M. Wu. 2006. Northern China heat pump application with the digital heating scroll compressor. In: *Proceedings of the 18th Compressor Engineering Conference at Purdue* and *11th International Refrigeration and Air Conditioning Conference at Purdue*.

Heat Pump Centre. 2007. http://www.heatpumpcentre.org/en/aboutheatpumps/Sidor/default.aspx. Accessed June 2015.

Henderson, K., Y.G. Park, L. Liu, A.M. Jacobi. 2010. Flow-boiling heat transfer of R-134a-based nanofluids in a horizontal tube. *International Journal of Heat and Mass Transfer* 53(5–6):944–951.

Heo, J., M.W. Jeong, C. Baek, Y. Kim. 2011. Comparison of the heating performance of air-source heat pumps using various types of refrigerant injection. *International Journal of Refrigeration* 34:444–453.

Hepbasli, A., Y. Kalinci. 2009. A review of heat pump water heating systems. *Renewable & Sustainable Energy Reviews* 13:1211–1229.

Herold, K.E., R. Radermacher, L. Howe, D.C. Erickson. 1991. Development of an absorption heat pump water heater using an aqueous ternary hydroxide working fluid. *International Journal of Refrigeration* 14:156–167.

Higashi, Y. 2010. Thermo-physical properties of HFO-1234yf and HFO-1234ze(E). In: *International Symposium on Next-Generation Air Conditioning and Refrigeration Technology*, February 17–19, Tokyo, Japan.

Hihara, E. 2006. Performance test of a carbon dioxide heat pump for combined domestic hot water and floor heating. *IEA Heat Pump Centre Newsletter* 4:21–23.

Hochbaum, A.I., R. Chen, D.R. Diaz, W. Liang, E.C. Garnett, M. Najarian, A.S. Majumdar, P. Yang. 2008. Enhanced thermoelectric performance of rough silicon and nanowires. *Nature* 451(7175):163–167.

Holmberg, P., T. Berntsson. 1990. Alternative working fluids in heat transformers. *ASHRAE Transactions* 96:1582–1589.

Honma, M., T. Tamura, Y. Yakumaru, F. Nishiwaki. 2008. Experimental study on compact heat pump system for clothes drying using CO_2 as a refrigerant. In: *7th IIR Gustav Lorentzen Conference on Natural Working Fluids*, Trondheim, Norway.

Howerton, M.T. 1978. Thermochemical energy storage system and heat pumps, Martin Marietta Aerospace, Report SAE/P78/75:935–940.

Hrnjak, P., A.D. Litch. 2008. Microchannel heat exchangers for charge minimization in air-cooled ammonia condensers and chillers. *International Journal of Refrigeration* 3:658–668.

Hsu, C.T. 1984. *Investigation of an Ejector Heat Pump by Analytical Methods*. Oak Ridge National Laboratory, Oak Ridge, TN.

Hua, T., M. Yitai., L. Minxia, W. Wei. 2010. Study on expansion power recovery in CO_2 transcritical cycle. *Energy Conversion and Management* 51:2516–2522.

Huang, B.J., W.Z. Ton, C.C. Wu, H.W. Ko, H.S. Chang, H.Y. Hsu, J.H. Liu, J.H. Wu, R.H. Yen. 2014. Performance test of solar-assisted ejector cooling system. *International Journal of Refrigeration* 39:172–185.

Hwang, Y., A. Celik, R. Radermacher. 2004. Performance of CO_2 cycles with a two-stage compressor. In: *Proceedings of International Refrigeration and Air Conditioning Conference at Purdue*, Purdue University, West Lafayette, IN.

Ibrahim, M., W.R.W. Daud, K. Ibrahim, A. Zaharim, K. Sopian. 2010. Performance of solid-gas chemical heat pump sub-system of a solar dryer. *Recent Advances in Applied Mathematics.* http://www.wseas.us/e-library/conferences/2010/Harvard/MATH/MATH-081.pdf. Accessed May 15, 2015.

Ibrahim, M., K. Sopian, W.R.W. Daud, M.A. Alghoul, M. Yahya, M.Y. Sulaiman. 2009. Solar chemical heat pump drying system for tropical region. *WSEAS Transactions on Environment and Development Journal* 5(5):404–413.

Ibrahim, M., K. Sopian, W.R.W. Daud, M. Yahya, M.A. Alghoul, A. Zaharim. 2008. Performance predication of solar-assisted chemical heat pump drying system in tropical region. In: *2nd WSEAS/IASME International Conference on Renewable Energy Sources,* October 26–28, Corfu Island, Greece, pp. 122–126.

IEA. 2014. Key world energy statistics. http://www.iea.org/publications/freepublications. Accessed May 15, 2015.

IEA. 2015a. Industrial energy-related systems and technologies Annex 13, IEA Heat Pump Program Annex 35. Application of Industrial Heat Pumps, Final Report, Part 1.

IEA. 2015b. Industrial energy-related systems and technologies Annex 13, IEA Heat Pump Program Annex 35. Application of Industrial Heat Pumps, Final Report, Part 2.

Imre, L. 2006. Solar drying. In: Mujumdar, A.S. (ed.) *Handbook of Industrial Drying.* Taylor & Francis Group, Boca Raton, FL, pp. 317–319.

Imre, L., L.I. Kiss, K. Molnar. 1982. Design of convective dryers. In: Ashworth, J.C. (ed.) *Proceedings of the Third International Drying Symposium*, Drying Research Limited, Wolverhampton, England, pp. 370–395.

Infante Ferreira, C.A., D. Zaytsev. 2002. Experimental compression—Resorption heat pump for industrial applications. In: *International Refrigeration and Air Conditioning Conference*, School of Mechanical Engineering, Purdue University Purdue e-Pubs, Purdue University, West Lafayette, IN.

Islam, M.R., A.S. Mujumdar. 2008. Heat pump-assisted drying. In: Mujumdar, A. S. (ed.) *Guide to Industrial Drying: Principles, Equipment, and New Development; International Drying Symposium*, Hyderabad, India.

Itard, L. 1998. Wet compression-resorption heat pump cycles: Thermodynamic analysis and design. PhD thesis, Delft University of Technology, Delft, the Netherlands.

Iyoki, S., T. Uemura. 1978. Studies on corrosion inhibitor in water-lithium bromide absorption refrigerating machine. *Reito* 53(614):1101–1105.

Jangam, S. 2011. Guest editorial: Chemical heat pumps to enhance *drying* efficiency—Some thoughts for new R&D opportunities. *Drying Technology: An International Journal* 29(6):610–611.

Jensen, J.K., W.B. Markussen, L. Reinholdt, B. Elmegaard. 2014b. Exergoeconomic optimization of an ammonia-water hybrid heat pump for heat supply in a spray drying facility. In: *Proceedings of ECOS 2014—The 27th International Conference on Efficiency, Cost, Optimization, Simulation and Environmental impact of Energy Systems*, June 15–19, Turku, Finland.

Jensen, J.K., L. Reinholdt, W.B. Markussen, B., Elmegaard. 2014a. Investigation of ammonia/water hybrid absorption/compression heat pumps for heat supply temperatures above 100°C. In: *International Sorption Heat Pump Conference*, March 31–April 2, University of Maryland, Washington, DC.

Jia, X., S. Clements, P. Jolly. 1993. Study of heat pump assisted microwave drying. *Drying Technology* 11(7):1583–1616.

Jolly, P., X. Jia, S. Clements. 1990. Heat pump assisted continuous drying part 1: Simulation model. *International Journal of Energy Resources* 14:757–770.

Jones, P. 1992. Electromagnetic wave energy in drying processes. In: Mujumdar, A.S. (ed.) *Drying 92*, Elsevier Science Publisher BV, Amsterdam, the Netherlands, pp. 114–136.

Kandlikar, S.G., S. Garimella, D. Li, S. Colin, M.R. King. 2006. *Heat Transfer and Fluid Flow in Minichannels and Microchannels.* Elsevier, Oxford, UK.

Kara, O., K. Ulgen, A. Hepbasli. 2008. Exergetic assessment of direct-expansion solarassisted heat pump systems: Review and modeling. *Renewable and Sustainable Energy Reviews* 12:1383–1401.

Kato, Y., Y. Harada, Y. Yoshizawa. 1999. Kinetic feasibility of a chemical heat pump for heat utilization of high-temperature processes. *Applied Thermal Engineering* 19:239–254.

Kato, Y., Y. Sasaki, Y. Yoshizawa. 2005. Magnesium oxide/water chemical heat pump to enhance energy utilization of a cogeneration system. *Energy* 30:2144–2155.

Kato, Y., N. Yamashita, K. Kobayashi, Y. Yoshizawa. 1996. Kinetic study of the hydration of magnesium oxide for chemical heat pump. *Applied Thermal Engineering* 16(11):853–862.

Kawasaki, H., T. Watanabe, A. Kanzawa. 1999. Proposal of a chemical heat pump with paraldehyde depoly-merization for cooling system. *Applied Thermal Engineering* 19:133–143.

Khamooshi, M., K. Parham, U. Atikol. 2013. Overview of ionic liquids used as working fluids in absorp-tion cycles. *Advances in Mechanical Engineering*. http://www.uk.sagepub.com/aboutus/openaccess.htm (accessed September 2015).

Kiang, C.S., C.K. Jon. 2006. Heat pump drying systems. In: Mujumdar, A.S. (ed.) *Handbook of Industrial Drying*. Taylor & Francis Group, Boca Raton, FL, pp. 1104–1105.

Kim, H.J., J.M. Ahn, S.O. Cho, K.R. Cho. 2008. Numerical simulation on scroll expander-compressor unit for CO_2 transcritical cycles. *Applied Thermal Engineering* 28:1654–1661.

Kim, M.H., J. Pettersen, C.W. Bullard. 2004. Fundamental process and system design issues in CO_2 vapor compression systems. *Progress in Energy and Combustion Science* 30:119–174.

Kim, S.G., Y.J. Kim, G. Lee, M.S. Kim. 2005. The performance of a transcritical CO_2 cycle with an internal heat exchanger for hot water heating. *International Journal of Refrigeration* 28:1064–1072.

Kim, S.J., T. McKrell, J. Buongiorno, L.W. Hu. 2010. Subcooled flow boiling heat transfer of dilute alu-mina, zinc oxide, and diamond nanofluids at atmospheric pressure. *Nuclear Engineering and Design* 240(5):1186–1194.

Kim, Y.W., J. Ramousse, G. Fraisse, P. Dalicieux, P. Baranek. 2013. Optimal performance of air/air thermo-electric heat pump (THP) coupled to energy-efficient buildings coupling in different climate conditions. In: *Proceedings of the 13th Conference of International Building Performance Simulation Association*, August 26–28, Chambéry, France.

Klöcker, K., E.L. Schmidt, F. Steimle. 2001. Carbon dioxide as a working fluid in drying heat pump. *International Journal of Refrigeration* 24(2):100–107.

Klöcker, K., E.L. Schmidt, F. Steimle. 2002. A drying heat pump using carbon dioxide as working fluid. *Drying Technology* 20(8):1659–1671.

Kornhauser, A.A. 1990. The use of an ejector as a refrigerant expander. In: *Proceedings of the 1990 USNC/IIR-Purdue Refrigeration Conference*, Purdue University, West Lafayette, IN, pp. 10–19.

Kowalski, S.J., A. Pawlovski. 2008. Drying of wet materials in intermittent conditions. In: *Proceedings of the 16th International Drying Symposium*, Hyderabad, India, pp. 951–957.

Kuang, Y.H., R.Z. Wang, L.Q. Yu. 2003. Experimental study on solar assisted heat pump system for heat supply. *Energy Conversion and Management* 44:1089–1098.

Kudra, T., A.S. Mujumdar. 2001. *Advanced Drying Technologies*. Marcel Dekker, New York, 457p.

Kuhlenschmidt, D. 1973. Absorption refrigeration system with multiple generator stages, U.S. Patent No. 3717007.

Kulkarni, T., C.W. Bullard, K. Cho. 2004. Header design tradeoffs in microchannel evaporators. *Applied Thermal Engineering* 24:759–776.

Kumar, R.R., K. Sridhar, M. Narasimha. 2013. Heat transfer enhancement in domestic refrigerator using R600a/mineral oil/nano-Al_2O_3 as working fluid. *International Journal of Computational Engineering Research* 3(4):42–51. http://www.ijceronline.com. Accessed June 3, 2015.

Laipradit, P., J. Tiansuwan, T. Kiatsiriroat, I. Aye. 2007. Theoretical performance analysis of heat pump water heaters using carbon dioxide as refrigerant. *International Journal of Energy Research* 32(4):356–366.

Lara, J.R., O. Osunsan, M.T. Holtzapple. 2015. Advanced mechanical vapor-compression desalination system. http://cdn.intechopen.com/pdfs-wm/13757.pdf (accessed on April 16, 2015).

Lawton, J. 1978. Drying: The role of heat pumps and electromagnetic fields. *Physics in Technology* 9:214–220.

Lazzarin, R.M. 2012. Dual source heat pump systems: Operation and performance. *Energy Buildings* 52:77–85.

Lebrun, M., P. Neveu. 1991. Conception, simulation, dimensioning and testing of an experimental chemical heat pump. *ASHRAE Transaction* 98:420–429.

Lee, J., I. Mudawar. 2007. Assessment of the effectiveness of nanofluids for single-phase and two-phase heat transfer in micro-channels. *International Journal of Heat and Mass Transfer* 50(3–4):452–463.

Leimbach, J.G., J.H. Heffner. 1992. Injection valve for a refrigeration system. U.S. Patent No. 5148684.

Liang, S., W. Chen, K. Cheng, Y. Guo, X. Gui. 2011. The latent application of ionic liquids in absorption refrig-eration. Applications of Ionic Liquids in Science and Technology. http://www.intechopen.com. Accessed April 14, 2015.

Lifson, A., M.F. Taras, T.J. Dobmeier. 2006. Flash tank for heat pump in heating and cooling modes of opera-tion. U.S. Patent, No. 7137270B2.

Lokiec, F., A. Ophir. 2007. The mechanical vapor compression: 38 years of experience. In: *IDA World Congress-Maspalomas*, October 21–26, Gran Canaria, Spain.

Longo, G.A., C. Zilio, G. Righetti, J.S. Brown. 2014. HFO1234ze(Z) saturated vapor condensation inside a brazed plate heat exchanger. In: *International Refrigeration and Air Conditioning Conference*, School of Mechanical Engineering, Purdue e-Pubs, Purdue University, West Lafayette, IN.

Lorentzen, G. 1989. Method of operating a vapour compression cycle under supercritical conditions. European Patent 0424474B2.

Lorentzen, G. 1993. Revival of the carbon dioxide as a refrigerant. *International Journal of Refrigeration* 17(5):292–301.

Lovegrove, K., A. Luzzi. 1996. Endothermic reactors for an ammonia based thermochemical solar energy storage and transport system. *Solar Energy* 56(4):361–371.

Luo, L., Y. Fan, D. Tondeur. 2007. Heat exchanger: From micro- to multi-scale design optimization. *International Journal of Energy Research* 30:1266–1274.

Ma, G., Q. Chai, Y. Jiang. 2003. Experimental investigation of air-source heat pump for cold regions. *International Journal of Refrigeration* 26:12–18.

Ma, G.Y., Q.H. Chai. 2004. Characteristics of an improved heat-pump cycle for cold regions. *Applied Energy* 77:235–247.

Ma, G-Y., H.-X. Zhao. 2008. Experimental study of a heat pump system with flash-tank coupled with scroll compressor. *Energy and Buildings* 40(5):697–701.

Maizza, V., A. Maizza. 1996. Working fluids in non-steady flows for waste energy recovery systems. *Applied Thermal Engineering* 16(7):579–590.

Marsh, K.N., J.A. Boxall, R. Lichtenthaler. 2005. Room temperature ionic liquids and their mixtures—A review. *Fluid Phase Equilibria* 219(1):93–98.

Marshall, M.G., A.C. Metaxas. 1998. Modeling the radio frequency electric field strength developed during the RF assisted heat pump drying of particulates. *International Microwave Power Institute* 33(3):167–177.

Marshall, M.G., A.C. Metaxas. 1999. Radio frequency assisted heat pump drying of crushed brick. *Applied Thermal Engineering* 19(4):375–388. DOI: 10.1016/S1359-4311(98)00058-1.

Mbaye, M., Z. Aidoun, V. Valkov, A. Legault. 1998. Analysis of chemical heat pumps (CHPS): Basic concepts and numerical model description. *Applied Thermal Engineering* 18(3/4):131–146.

Meirovitch, E., L.M. Segal. 1990. Theoretical modelling of a directly heated solar-driven chemical reactor. *Solar Energy* 45(3):139–148.

Menegay, P., A.A. Kornhauser. 1996. Improvements to the ejector expansion refrigeration cycle. In: *Proceedings of the 31st Intersociety Energy Conversion Engineering Conference*, Washington, DC, pp. 702–706.

Metaxas, A.C., R. Meredith. 1983. *Industrial Microwave Heating*. Peter Peregrinus, London.

Meyers, S., V. Franco, A. Lekov, L. Thompson, A. Sturges. 2010. Do heat pump clothes dryers make sense for the U.S. market? Presented at ACEEE Summer Study on Energy Efficiency in Buildings, August.

Minea, V. 2004. Heat pumps for wood drying—New developments and preliminary results. In: *Proceedings of the 14th International Drying Symposium*, August 22–25, Sao Paulo, Brazil, vol. B, pp. 892–899.

Minea, V. 2010a. Improvements of high-temperature drying heat pumps. *International Journal of Refrigeration* 33(1):180–195.

Minea, V. 2010b. Industrial drying heat pumps. In: Mikkel, E. L. (eds.), *Refrigeration: Theory, Technology and Applications*. Nova Science Publishers, New York, pp. 1–70.

Minea, V. 2010c. Thermally-driven ammonia-water absorption system. In: *9th IIR Gustav Lorentzen Conference*, Sydney, Australia.

Minea, V. 2011. Dual-energy source heat pump. In: *10th IEA Heat Pump Conference*, May 16–19, Tokyo, Japan.

Minea, V. 2012. Efficient energy recovery with wood drying heat pumps. *Drying Technology* 30:1630–1643.

Minea, V. 2013. Valorisation des rejets thermiques dans les PMI—étape 3, rapport LTE-RT-2013-0081, Août (in French).

Minea, V. 2014a. Overview of heat pump-assisted drying systems—Part I: Integration, control complexity and applicability of new innovative concepts. *Drying Technology*. DOI: 10.1080/07373937.2014.952377.

Minea, V. 2014b. Overview of heat pump-assisted drying systems—Part II: Data provided vs. results reported. *Drying Technology*. DOI: 10.1080/07373937.2014.952378.

Minea, V. 2014c. Efficient process integration and cooling & heating energy performance of supercritical CO_2 heat pumps. In: *11th IEA Heat Pump Conference*, May 12–16, Montréal, Québec, Canada.

Minea, V. 2015. High-temperature heat pump-assisted softwood dryer: Sizing and control requirements & energy performances. In: *24th International Congress of Refrigeration*, August 16–22, Yokohama, Japan.

Minea, V., F. Chiriac. 2006. Hybrid absorption heat pump with ammonia/water mixture—Some design guidelines and district heating application. *International Journal of Refrigeration* 29(7):1080–1092.

Minor, B., M. Spatz. 2008. HFO-1234yf Low GWP refrigerant update. In: *International Refrigeration and Air Conditioning Conference*, Purdue University, West Lafayette, IN. http://docs.lib.purdue.edu/iracc/937.

Miyara, A., Y. Onaka, S. Koyama. 2012. Ways of next generation refrigerants and heat pump/refrigeration systems. *International Journal of Air-Conditioning and Refrigeration* 20:1130002.

Molina, M.J., F.S. Rowland. June 28, 1974. Stratospheric sink for chlorofluoromethans: Chlorine atom-catalysed destruction of ozone. *Nature* 249(5460):810–812.

Monfared, B.A., B. Palm. 2011. Design and test of a domestic heat pump with ammonia as refrigerant. In: *Proceedings of the 4th IIR Conference Ammonia Refrigeration Technology*, Ohrid, Republic of Macedonia.

Morrison, G.L. 1994. Simulation of packaged solar heat-pump water heaters. *Solar Energy* 53(3):249–257.

Mujumdar, A.S. 1987. *Handbook of Industrial Drying*, 2nd ed. Marcel Dekker, New York.

Mujumdar, A.S. 1995. *Handbook of Industrial Drying*, vol. 2, Marcel Dekker, New York, pp. 1241–1272.

Mujumdar, A.S. 1996. Innovation in drying. *Drying Technology*, 14(6):1459–1475.

Mujumdar, A.S. 2002. Drying research-current state and future trends. *Developments in Chemical Engineering and Mineral Processing* 10(3–4):225–246.

Mujumdar, A.S. 2006. *Handbook of Industrial Drying*, 3rd ed. CRC Press, Boca Raton, FL.

Mujumdar, A.S. 2008. Guide to Industrial Drying. Principles, Equipments & New Developments. In: *International Drying Symposium*, Hyderabad, India.

Mujumdar, A.S. 2009. *Advanced Drying Technologies*, 2nd ed. CRC Press, Boca Raton, FL, 520p.

Mujumdar, A.S., Z. Wu. 2008. Thermal drying technologies-cost-effective innovation aided by mathematical modeling approach. *Drying Technology* 26:145–153.

Navarro-Esbri, J., J.M. Mendoza-Miranda, A. Mota-Babiloni, A. Barragan-Cervera, J.M. Belman-Flores. 2013a. Experimental analysis of R1234yf as a drop-in replacement for R134a in a vapor compression system. *International Journal of Refrigeration* 36:870–880.

Navarro-Esbri, J., F. Moles, A. Barragan-Cervera. 2013b. Experimental analysis of the internal heat exchanger influence on a vapor compression system performance working with R1234yf as a drop-in replacement for R134a. *Applied Thermal Engineering* 59:153–161.

Neksa, P., P. Hrnjak, J. Pettersen, R. Rieberer, E.L. Schmidt, J. Suss. 1999. CO_2 as a working fluid. In: *20th International Congress of Refrigeration*, Sydney, Australia.

Neksa, P., H. Rekstad, G.R. Zakeri, P.A. Schiefloe. 1998. CO_2 heat pump water heater: Characteristics, system design and experimental results. *International Journal of Refrigeration* 21(3):172–179.

Neksa, P., H.T. Walnum, A. Hafner. 2010. CO_2—A refrigerant from the past with prospects of being one of the main refrigerants in the future. In: *9th IIR Gustav Lorentzen Conference*, Sydney, Australia.

Neveu, P., J. Castaiang. 1993. Solid-gas chemical heat pumps: Field of application and performance of the internal heat recovery process. *Heat Recovery System and CHP* 13(3):233–251.

Neveu, P., M. Lebrun. 1991. High efficiency process for gas-solid thermochemical heat pumps. In: *4th Congress of Chemical Engineering*, Karlruhe, Germany.

Nguyen, V.M., S.B. Riffat, P.S. Doherty. 2001. Development of a solar-powered passive ejector cooling system. *Applied Thermal Engineering* 21:157–168.

Nickl, J., G. Will, W.E. Kraus, H. Quack. 2003. Third generation CO_2 expander. In: *Proceedings of the 21th International Congress of Refrigeration*, Washington, DC.

Nickl, J., G. Will, H. Quack, W.E. Kraus. 2005. Integration of a three-stage expander into a CO_2 refrigeration system. *International Journal of Refrigeration* 28(8):1219–1224.

Nipkow, J., E. Bush. 2009. Promotion of energy-efficient heat pump dryers, Swiss Agency for Efficient Energy Use (S.A.F.E.), Topten International Group. In: *Proceedings of EEDAL*, Berlin, Germany.

Nipkow, J., E. Bush. 2010. Promotion of energy-efficient heat pump dryers. In: *Energy Efficiency in Domestic Appliances and Lighting Proceedings of the 5th International Conference EEDAL'09*, June 16–18, Berlin, Germany, vol. 3, pp. 1215–1225, EUR 24139 EN/1–2.

Ogura, H., A.S Mujumdar. 2000. Proposal for a novel chemical heat pump dryer. *Drying Technology* 18(4):1033–1053.

Ogura, H., T. Yamamoto, Y. Otsubo, H. Ishida, H. Kage, A.S. Mujumdar. 2004. Controllability of a chemical heat pump dryer. In: *Drying 2004—Proceedings of the 14th International Drying Symposium*, August 22–25, São Paulo, Brazil, vol. B, pp. 998–1004.

Ogura, H., T. Yamamoto, Y. Otsubo, H. Ishida, H. Kage, A.S. Mujumdar. 2005. A control strategy for a chemical heat pump dryer. *Drying Technology* 23:1189–1203.

Ommen, T.S., C.M. Markussen, L. Reinholdt, B. Elmegaard. 2011. Thermoeconomic comparision of industrial heat pump. In: *ICR 2011*, August 21–26, Prague, Czech Republic.

Ozgur, A.E., A. Kabul, O. Kizilkan. 2014. Exergy analysis of refrigeration systems using an alternative refrigerant (HFO-1234yf) to R134a. *International Journal of Low-Carbon Technologies* 9:56–62. http:// dx.doi. org/10.1093/ijlct/cts054. Accessed March 12, 2015.

Paakkonen, K., J. Havento, B. Galambosi, M. Pyykkonen. 1999. Infrared drying of herbs. *Agricultural and Food Science in Finland* 8:19–27.

Palm, B. 2008. Ammonia as refrigerant in small-capacity systems. *IEA Heat Pump Centre Newsletter* 26(4):23–31.

Pan, G., Z. Li. 2006. Investigation on incomplete condensation of non-azeotropic working fluids in high temperature heat pumps. *Energy Conversion and Management* 47(13–14):1884–1893.

Patel, K., A. Kar. April 2012. Heat pump assisted drying of agricultural produce—An overview. *Journal of Food Sciences and Technology* 49(2):142–160.

Pearson, A. 2007. Refrigeration with ammonia. *International Journal of Refrigeration* 31:545–551.

Pearson, A. February 30–34, 2008. Ammonia's future. *ASHRAE Journal* 50(2):30–34,36.

Peng, H., G. Ding, W. Jiang, H. Hu, Y. Gao. September 2009. Heat transfer characteristics of refrigerant-based nanofluid flow boiling inside a horizontal smooth tube. *International Journal of Refrigeration* 32(6):1259–1270.

Perera, C.O., M.S. Rahman. 1997. Heat pump demuhidifier drying of food. *Trends in Food Sciences Technology* 8:75–79.

Perez-Blanco, H. 1984. Absorption heat pump performance for different types of solution. *International Journal of Refrigeration* 7(2):115–122.

Pop, M.G., F. Chiriac, M. Ghitulescu, V. Gardus, A. Negrea, V. Minea. 1983. 7.5 Gcal/h hot household water preparation facility with an absorption—Compression heat pump. In: *United Nations—Economic Commission for Europe, Symposium on Rational Utilisation of Secondary Forms of Energy in the Economy, Particularly in Industry*, October 10–14, Paris, France.

Pope, J., D. Hude, M. Jeffrey, P.S.I. Lurgi. 2015. Industrial performance of the mechanical vapor recompression and multiple effect evaporator system: Successful operation and significant reduction in steam usage. http://www.sugarsonline.com/oldsite/SugarsPapers/Lurgi%20PSI%20Evaporator%20Paper.pdf. Accessed March 2015.

Prasertsan, S., P. Sean-saby, G. Prateepchaikul. 1997. Heat pump dryer. Part 3: Experiment verification of the simulation. *International Journal of Energy Research* 21:1–20.

Quack, H. 1999. Arbeitsleistende Expansion in Kaltdampf-Kältekreisläufen. *DKV-Tagungsbericht* 26:109–123.

Quack, H. 2000. Cryogenic Expanders ICEC 18, Mumbai, pp. 33–40.

Raaphorst, M. 2005. Optimale teelt in de gesloten kas—Teeltkundig verslag van de gesloten kas bij Themato in 2004. http://www.hdc.org.uk/sites/default/files/research_papers/PC%20256%20final%20report%20 2007.pdf. Accessed June 5, 2015.

Raldow, W.M., W.E. Wentworth. 1979. Chemical heat pumps—A basic thermodynamic Analysis. *Solar Energy* 23:75.

Rama, V., E. Siivola, T. Colpitts, B. O'Quinn. 2001. Thin-film thermoelectric devices with high room-temperature figures of merit. *Nature* 413(6856):597–602. DOI: 10.1038/35098012.

Richard, M.A., R. Labrecque. 2014. Techno-economic evaluation of combining heat pump and mechanical steam compression for the production of low pressure steam from waste heat. In: *11th IEA Heat Pump Conference*, May 12–16, Montréal, Québec, Canada.

Rivera, C.O., W. Rivera. 2003. Modeling of an intermittent solar absorption refrigeration system operating with ammonia–lithium nitrate mixture. *Solar Energy Materials and Solar Cells* 76(3):417–427.

Robinson, D.M., E.A. Groll. 1998. Efficiency of transcritical CO_2 cycles with and without an expansion turbine. *International Journal of Refrigeration* 21(7):577–589.

Rogers, R.D., K.R. Seddon. 2003. Ionic liquids—Solvents of the future? *Science* 302(5646):792–793.

Rolf, R., R. Corp. 1990. Chemical heat pump for drying of bark. Annual meeting: Technical section, Canadian Pulp and Paper Association, pp. 307–311.

Rowe, D.M. 1995. *CRC Handbook of Thermoelectrics*, CRC Press, Boca Raton, FL.

Sabareesh, K., N. Gobinath, V. Sajithb, S. Das, C.B. Sobhan. 2012. Application of TiO_2 nanoparticles as a lubricant-additive for vapour compression refrigeration systems—An experimental investigation. *International Journal of Refrigeration* 35(7):1989–1996.

Sager, J. 2013. Results from testing at the Canadian Centre for Housing Technology. http://www.chba.ca.pdf. Accessed May 28, 2015.

Saito, Y., H. Kameyama, K. Yoshida. 2007. Catalyst-assisted chemical heat pump with reaction couple acetone hydrogenation/2-propanol dehydrogenation for upgrading low-level thermal energy: Proposal and evaluation. *International Journal of Energy Research* 11:549–558.

Sand, J.R., S.K. Fisher, V.D. Baxter. 1999. TEWI analysis: It's utility, its shortcomings, and its results. In: *Proceedings of the Taipei International Conference on Atmospheric Protection*, September 13–14, Taipei, Taiwan.

Sarkar, J. 2008. Optimization of ejector-expansion transcritical CO_2 heat pump cycle. *Energy* 33(9):1399–1406.

Sarkar, J. 2009. Performance characteristics of ejector expander transcritical CO_2 refrigeration cycle. In: *Proceedings of the Institution of Mechanical Engineers, Part A: J. Power and Energy*, August 1, 2012, vol. 226, pp. 623–635.

Sarkar, J., S. Bhattacharyya, M.R. Gopal. 2004. Optimization of a transcritical CO_2 heat pump cycle for simultaneous cooling and heating applications. *International Journal of Refrigeration* 27:830–838.

Sarkar, J., S. Bhattacharyya, M.R. Gopal. 2009. Irreversibility minimization of heat exchangers for transcritical CO_2 systems. *International Journal of Thermal Sciences* 48:146–153.

Schmidt, E.L., K. Klöcker, N. Flacke, F. Steimle. 1998. Applying the transcritical CO_2 process to a drying heat pump. *International Journal of Refrigeration* 21(3):202–211.

Shah, R.K. 1991. Compact heat exchanger technology and applications. In: Foumaeny, E.A., P.J. Heggs (eds.) *Heat Exchange Engineering*, vol. 2, Compact Heat Exchangers: Techniques of Size Reduction. Ellis Horwood, London.

Sharma, A., C. Chen, V.L. Nguyen. 2009. Solar-energy drying systems: A review. *Renewable and Sustainable Energy Reviews* 13:1185–1210.

She, X., Y. Yin, X. Zhang. 2014. A proposed sub-cooling method for vapor compression refrigeration cycle based on expansion power recovery. *International Journal of Refrigeration* 43:50–61.

Soponronnarit, S., A. Nathakaranakule, S. Wetchacama, T. Swasdisevi, P. Rukprang. 1998. Fruit drying using heat pump. *International Energy Journal* 20:39–53.

Spinner, B. 1996. Changes in research and development objectives for closed solid-sorption systems. In: *Proceedings of the International Absorption Heat Pump Conference*, September 17–20, Montreal, Québec, Canada, pp. 82–96.

Spoelstra, S., M.E.H. Tijani. 2005. Thermo-acoustic heat pumps for energy savings. Seminar "Boundary crossing acoustics" of the Acoustical Society of the Netherlands, November.

Sporn, P., E.R. Ambrose. 1955. The heat pump and solar energy. In: *Proceedings of the World Symposium on Applied Solar Energy*, Phoenix, AZ, pp. 159–170.

Srikhirin, P., S. Aphornratana, S. Chungpaibulpatana. 2001. A review of absorption refrigeration technologies. *Renewable and Sustainable Energy Review* 5:343–372.

Stene, J. 2005. Residential CO_2 heat pump system for combined space and heating, and hot water heating. *International Journal of Refrigeration* 28:1259–1265.

Stene, J. 2007. Integrated CO_2 heat pump systems for space heating and hot water heating in low-energy houses and passive houses. International Energy Agency (IEA) Heat Pump Programme—Annex 32. Workshop in Kyoto, Japan, December 6.

Stene, J. 2008. Design and application of ammonia heat pump systems for heating and cooling of non-residential buildings. In: *Proceedings of the 8th IIR Gustav Lorentzen Conference on Natural Working Fluids*, Copenhagen, Denmark.

Stošić, N. 2002. Screw compressors in refrigeration and air conditioning. Centre for Positive Displacement Compressor Technology, City University, London, England. http://lms.iknow.com/pluginfile.php/28915 .pdf. Accessed March 26, 2015.

Stošić, N., I.K. Smith, A. Kovacević. 2002. A twin screw combined compressor and expander for CO_2 refrigeration systems. Purdue University Purdue e-Pubs, http://docs.lib.purdue.edu/cgi/viewcontent .cgi?article=2590&context=icec. Accessed May 15, 2015.

Strommen, I., T.M. Eikevik, A.F. Odilio. 1999. Optimum design and enhanced performance of heat pump dryers. In: Abudullah, K., A.H. Tamaunan, A.S. Maujumdar (eds.) *Proceedings of the 1st Asian-Australian Drying Conference*, vol. 68.

Strommen, I., O. Jonassen. 1996. Performance tests of a new 2-stage counter-current heat pump fluidized bed dryer. In: *Proceedings of the Tenth International Drying Symposium*, October, Bali, Indonesia, pp. 563–568.

Subiantoro, A., K.T. Ooi. 2013. Economic analysis of the application of expanders in medium scale air-conditioners with conventional refrigerants, R1234yf and CO_2. *International Journal of Refrigeration* 36:1472–1482.

Sveine, T., S. Grandum, H. Baksaas. 1998. Design of high temperature absorption-compression heat pump. In: *Proceedings of IIR Gustav Lorentzen conference of Commission B2 with B1, E1 & E2*, June 2–5, Oslo, Norway.

Takushima, A., R. Inoshiri, T. Okada. 2000. Condensation performance of air-cycle dryer system. In: *Proceedings of the 4th IIR-Gustav Lorentzen Conference on Natural Working Fluids*, Purdue University, West Lafayette, IN, pp. 465–471.

Thomas, W.J. 1996. RF drying provides process savings: New systems optimize radio frequency drying for the ceramic and glass fiber industries. *Ceramic Industry Magazine* pp. 30–34.

Tian, C.Q., N. Liang. 2006. State of the art of air-source heat pump for cold regions. *Renewable Energy Resources and a Greener Future* 8.

Tian, C.Q., N. Liang, W.X. Shi, X.T. Li. 2006. Development and experimental investigation on two-stage compression variable frequency air source heat pump. In: *International Refrigeration and Air Conditioning
Conference at Purdue*, Purdue University, West Lafayette, IN http://docs.lib.purdue.edu/iracc/799
(accessed July 15, 2015), p. 799.

Tleimat, B. 2015. Water and wastewater treatment technologies—Mechanical vapor compression distillation,
Encyclopedia of Life Support Systems (EOLSS). http://www.eolss.net/sample-chapters/c07/e6-144-26
.pdf. Accessed June 19, 2015.

Todd, B.J., T. Douglas, T. Reindl. 2008. ROTREX turbo compressor for water vapor compression (IEA HPP
Annex 13/35 2015).

Tritt, T.M., M.A. Subramanian. 2011. Thermoelectric materials, phenomena, and applications: A Bird's Eye
View. *MRS Bulletin* 31(3):188. DOI: 10.1557/mrs2006.44.

Tuckerman, D.B., R.F.W. Pease. 1981. High-performance heat sinkingfor VLSI. *IEEE Electron Device Letters*
EDL-2(5):126–129.

Ueda, K., Y. Hasegawa, K. Wajima, M. Nitta, Y. Kamada, A. Yokoyama. 2012. Deployment of a new series of
eco turbo ETI chillers. *Mitsubishi Heavy Industries Technical Review* 49:56–62.

UNEP Ozone Secretariat. http://ozone.unep.org/new_site/fr/montreal_protocol.php. Accessed May 14, 2015.

van de Bor, D.M., C.A. Infante Ferreira, A.A. Kiss. 2014. Optimal performance of compression–resorption heat
pump systems. *Applied Thermal Engineering* 65:219–225.

van Valkenburg, M.E., R.L. Vaughn, M. Williams, J.S. Wilkes. 2005. Thermochemistry of ionic liquid heat-
transfer fluids. *Thermochimica Acta* 425(1–2):181–188.

Vining, C.B. 2007. ZT ≈ 3.5: Fifteen years of progress and things to come. In: Paper presented at the *European
Conference on Thermoelectrics*, Odessa, Ukraine, September 10–12.

Voneshen, D.J., K. Refson, E. Borissenko, M. Krisch, A. Bosak, A. Piovano et al. 2013. Suppression of thermal conductivity by rattling modes in thermoelectric sodium cobaltate. *Nature Materials* DOI: 10.1038/nmat3739.

Wang, C., P. Zhang, R. Wang. 2008. Review of recent patents on chemical heat pump. *Recent Patents on
Engineering* 2(3):208–216.

Wang, L.W., R.Z. Wang, R.G. Oliveira. 2009. A review on adsorption working pairs for refrigeration. *Renewable
and Sustainable Energy Reviews* 13(3):518–534.

Wang, X., Y. Hwang, R. Radermacher. 2009. Two-stage heat pump system with vapor-injected scroll compressor using R410A as a refrigerant. *International Journal of Refrigeration* 32:1442–1451.

Werle, R., E. Bush, B. Josephy, J. Nipkow. 2011. Granda, Energy-efficient heat pump driers—European experiences and efforts in the USA and Canada, June. http://www.topten.info/uploads/File/040_Rita_Werle_
final_driers.pdf. Accessed March 2015.

Winandy, E.L., J. Lebrun. 2002. Scroll compressors using gas and liquid injection: Experimental analysis and
modelling. *International Journal of Refrigeration* 25(8):1143–1156.

Wongsuwan, W., S. Kumar, P. Neveu, F. Meunier. 2001. A review of chemical heat pump technology and applications. *Applied Thermal Engineering* 21:1489–1519.

Xu, X. 2012. Investigation of vapor injection heat pump system with a flash tank utilizing R419A and low-
GWP refrigerant R32. Dissertation submitted to the Faculty of the Graduate School of the University
of Maryland, College Park, MD (in partial fulfillment of the requirements for the degree of Doctor of
Philosophy).

Yang, J.L., Y.T. Ma, M.X. Li, H.Q. Guan. 2005. Exergy analysis of transcritical carbon dioxide cycles with an
expander. *Energy* 30:1162–1175.

Yang, J.L., Y.T. Ma, S.C. Liu. 2007. Performance investigation of transcritical carbon dioxide two-stage compression cycle with expander. *Energy* 32:237–245.

Yari, M. 2009. Performance analysis and optimization of a new two-stage ejector-expansion transcritical
CO_2 refrigeration cycle. *International Journal of Thermal Science* 48(10):1997–2005.

Yu, Y.Q., P. Zhang, J.Y. Wu, R.Z. Wang. 2008. Energy upgrading by solid–gas reaction heat transformer: A critical review. *Renewable and Sustainable Energy Reviews* 12:1302–1324.

Yun, R., Y. Kim, C. Park. 2007. Numerical analysis on a microchannel evaporator designed for CO_2 air-
conditioning systems. *Applied Thermal Engineering* 27:1320–1326.

Zamora, M., M. Bourouis, A. Coronas, M. Valles. 2014. Pre-industrial development and experimental characterization of new air-cooled and water-cooled ammonia/lithium nitrate absorption chillers. *International
Journal of Refrigeration* 45:189–197.

Zbicinski, I., A. Jakobsen, J.L. Driscoll. 1992. Application of infrared radiation for drying of particulate material. In: Mujumdar, A.S. (ed.) *Drying 92*. Elsevier Science Publisher B.V., Amsterdam, the Netherlands,
pp. 704–711.

Zhang, H., D.V. Talapin. 2014. Thermoelectric tin selenide: The beauty of simplicity. *Angewandte Chemie International Edition* 53:9126–9127. DOI:10.1002/anie.201405683.

Zhao, L.-D., S.-H. Lo, Y. Zhang, H. Sun, G. Tan, C. Uher, C. Wolverton, P. Dravid, G. Vinayak. 2014. Ultralow thermal conductivity and high thermoelectric figure of merit in SnSe crystals. *Nature* 508(7496): 373–377. DOI: 10.1038/nature13184.

Zhu, L., J. Yu. 2015. Theoretical study of a thermoelectric-assisted vapor compression cycle for air-source heat pump applications. *International Journal of Refrigeration* 51:33–40.

Ziegler, F., P. Riesch. 1993. Absorption cycles. A review with regard to energetic efficiency. *Heat Recovery Systems and CHP* 13(2):147–159.

Zilio, C., J.S. Brown, G. Schiochet, A. Cavallini. 2011. The refrigerant R-1234yf in air conditioning systems. *Energy* 36:6110–6120.

Zyhowski, G.J., M.M. Spatz, S.Y. Motta. 2002. An overview of properties applications of HFC-245fa. In: *International Refrigeration and Air Conditioning Conference*, School of Mechanical Engineering, Purdue e-Pubs, Purdue University, West Lafayette, IN.

2 Modeling, Simulation, and Optimization of Heat Pump Drying Systems

Lu Aye

CONTENTS

2.1 INTRODUCTION

Heat pumps have enormous potential for saving energy and greenhouse gas emissions, as they provide a very efficient mean of recovering both sensible and latent heat. A traditional method of drying high moisture content products has been to heat ambient air and pass it over the product to be dried. It has been known that heating of the air for product drying with the aid of heat pumps is more economical than using direct electric heaters or diesel burners. For the application to be more economically attractive, the system configuration and the operating parameters should be carefully optimized whenever possible. The literature has established the existence of local optima parameters, such as airflow rates through evaporator, condenser, and dryer; inlet air temperatures and compressor speed, for specific types of heat pump drying systems. By applying an appropriate numerical optimization technique, the global optima on design or operating parameters for heat pump drying systems can be found. This chapter provides principles and practices for optimization of heat pump drying systems.

Optimization is the process of finding the conditions that give maximum or minimum values of a function. Stoecker (1989) demonstrated that an optimal thermal system can be designed by mathematically simulating the performance of a particular concept of a thermal system and then subjecting the simulation to a procedure of optimization. The mathematical models for the heat pump components and the dryer are required to assemble a simulation program for the optimization study. In the following section, the optimization problems of heat pump drying systems are identified and described.

2.2   OPTIMIZATION OF DESIGN PARAMETERS AND OPERATING VARIABLES

The performance parameters of heat pump drying systems can be described as function subprograms using the mathematical models. Most equations encountered in the heat pump drying system models are nonlinear. The design parameters and operating variables of the heat pump drying systems are generally subjected to physical constraints. Therefore, the problems in the design and operation of heat pump drying systems can be mathematically cast in the form of a constrained nonlinear function minimization problem as follows:

$$\text{Minimize: } f(x) \tag{2.1}$$

$$\text{Subject to: } g_j(x) \le x_j \le h_j(x); \quad (j = 1, 2, \ldots, m) \tag{2.2}$$

where:

$f(x)$ is an objective function
x is an n-vector of design or operating variables $= (x_1, x_2, \ldots, x_n)^T$
$g_j(x)$ is the jth lower constraint function
$h_j(x)$ is the jth upper constraint function
m is the total number of inequality constraints

The implicit variables $x_{n+1}, \ldots, x_m$ are dependent functions of the explicit independent variables $x_1, x_2, \ldots, x_n$. In constraint set (Equation 2.2), the constraints for the explicit independent variables $(j = 1, 2, \ldots, n)$ are called *explicit constraints*, and the constraints for the implicit variables $(j = n + 1, n + 2, \ldots, m)$ are called *implicit constraints*.

One way of dealing the problems that involve more than one objective function to be optimized simultaneously is the use of weighting factors to deduce single objective function. It should be noted that some methods (Miettinen, 1999; Branke et al., 2008) are available if deducing to a single objective function is not possible. Using the notation in Equation 2.1, if a particular optimization problem requires maximization, we simply minimize the negative of the function. If equality constraints are encountered in a problem, they can be eliminated by solving each for one variable and substituting the resulting expression in the problem formulation.

The optimization subroutine subprogram solves the above-mentioned general nonlinear constrained optimization problem. In other words, the optimization subroutine subprogram searches a set of design or operating variables x that allow optimization of the objective function $f(x)$. On the other hand, the performance simulation function subprogram computes the objective function $f(x)$ for a known set of design or operating variables. The heat pump drying system simulation function subprogram and the optimization subroutine subprogram are coupled together to form the heat pump drying system optimization program (Figure 2.1). The requirements of the optimization subroutine subprogram to be coupled are discussed in Section 2.2.1.

The objective function for a heat pump drying system optimization could be to maximize specific moisture extraction rate and to minimize drying time, energy consumption, life cycle

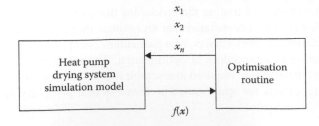

FIGURE 2.1   Heat pump drying system optimization program structure.

cost, or life cycle greenhouse gas emission. If the main focus is energy, the total energy consumption of an electric heat pump drying system for a batch of drying can be estimated as a function of power consumption of the compressor motor and fan motors, and the drying time required to achieve the desired moisture level of the product.

$$f(x) = E_{\text{tot}} = \int_{0}^{t_{\text{drying}}} (P_{\text{comp}} + P_{\text{fan}})dt \qquad (2.3)$$

where:

E_{tot} is the total energy consumption of the heat pump drying system for a batch (kJ)
t_{drying} is the drying time (s)
P_{comp} is the compressor motor power (kW)
P_{fan} is the total fan motor power (kW)

The heat pump drying system simulation function subprogram allows us to predict the instantaneous power requirements of the compressor motor and fan motors with respect to time and the drying time required; then the program is used to compute the total energy requirement of the system for a known set of design parameters and operating variables. Generally, the operating variables are constrained to lie in a given range. In general, the uncontrollable operating variables are the ambient air temperature and humidity ratio, and the controllable operating variables are the compressor speed, the condenser air mass flow rate, and the evaporator air mass flow rate for the fixed design parameters.

2.2.1 Coupling with an Optimization Algorithm

No single optimization algorithm has emerged as generally successful for heat pump drying system optimization, and the best method in any one instance is likely to be highly problem dependent. Generally, for the heat pump drying system simulation, the objective function cannot be computed outside the limit of the constraints. Therefore, the numerical optimization methods that require calculation of the objective function values outside the boundaries of the constraints cannot be applied.

In the heat pump drying systems optimization problems, the objective function is not analytically differentiable and the partial derivatives are impossible to find directly. One possibility of finding the partial derivatives is computing them numerically. The difficulty with this approach is in balancing the influence of rounding errors and truncation errors when using finite differences to estimate derivatives (Brent, 1973; Press et al., 2007). Hence, an algorithm that uses only computed values of the objective function is desirable.

These derivative-free types of algorithms are commonly called *direct search methods of optimization*. Himmelblau (1972) pointed out that in solving nonlinear programming problems, gradient and second-derivative algorithms converge faster than direct search methods. However, slow, safe algorithms are preferred in practice to fast algorithms, which may occasionally fail.

Random search methods are computationally less efficient than deterministic methods, but they are the most easily implemented of the direct search methods (Vanderplaats, 1984). Another advantage of random search methods is that very little computer storage is required for coding. As discussed by Edgar and Himmelblau (1988), random search methods do give the possibility of locating a global minimum rather than just a local minimum, and a number of investigations of random type algorithms to optimized unconstrained functions indicate high probabilities of success.

As shown in Figure 2.1, the heat pump drying system simulation model computes the objective function for a set of design or operating variables that is determined by the optimization routine. Based on the results for the previous known sets, the optimization routine iteratively searches for the set of design or operating variables that produce minimum objective function value until convergence criterion is met.

2.3 SIMULATION PROGRAM DEVELOPMENT

A heat pump drying system simulation program may be used in the design stage to help achieve an improved design. It may be applied to an existing system to explore prospective modifications. It may also be used for off-design (part-load or overload) performance predictions and operational optimization. There are several classes of computer simulation programs:

- Continuous versus discrete
- Deterministic versus stochastic
- Steady-state versus dynamic (transient)

In general, continuous, deterministic steady-state system simulations are preferable for optimization studies. A computer simulation model involves the calculation of operating variables (such as temperatures, pressures, flow rates of working fluid, and energy) for a set of design parameters in a heat pump drying system. Steady-state heat pump models have been applied for the simulations of heat pump drying systems, because the transient duration for the heat pump is in the order of minutes and the duration of drying is in the order of days. Quasi-steady-state models for drying have been applied for this purpose.

Knowledge of the performance characteristics of all heat pump dryer components as well as equations for thermophysical and transport properties of working substances, including the drying characteristic of the product, are required. The components of a typical heat pump drying system simulation model are shown in Figure 2.2.

2.3.1 AIR-TO-AIR HEAT PUMP MODEL

A heat pump model applied in drying simulation predicts the performance parameters and the state point of the moist air entering the drying chamber for a set of operating conditions. Depending on the nature of the investigation, the heat pump model could be empirical, semi-empirical, or phenomenological one. For system operational optimization of a known heat pump, semi-empirical model is sufficient. On the other hand, a phenomenological model is required for system design optimization. A schematic diagram of the vapor compression heat pump system is shown in Figure 2.3.

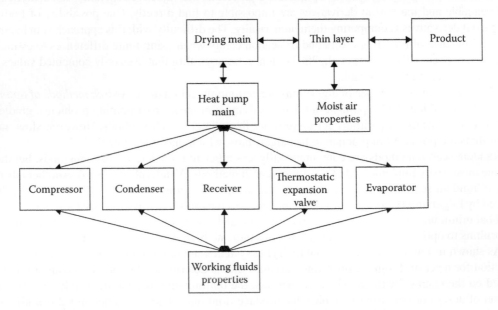

FIGURE 2.2   Example structure of a heat pump drying system simulation model.

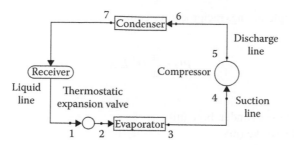

FIGURE 2.3 Schematic diagram of a vapor compression heat pump. (From Aye, L., Optimisation of heat pump grain drying system, PhD thesis, Department of Mechanical and Manufacturing Engineering, The University of Melbourne, Parkville, Victoria, Australia, 1995. With permission.)

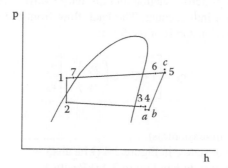

FIGURE 2.4 Pressure–enthalpy diagram of a practical heat pump cycle. (From Aye, L., Optimisation of heat pump grain drying system, PhD thesis, Department of Mechanical and Manufacturing Engineering, The University of Melbourne, Parkville, Victoria, Australia, 1995. With permission.)

The skeleton pressure–enthalpy diagram representing a single-stage vapor compression heat pump cycle using a reciprocating compressor is shown in Figure 2.4.

The phenomenological heat pump model includes component submodels for compressor, condenser, receiver, thermostatic expansion valve, and evaporator. It may also include heat transfer models of the suction line, the discharge line, and the liquid line. Pressure losses in various parts of the system may also be incorporated. A set of example submodels are presented in the following sections. Simulation and optimization of heat pump drying systems using a phenomenological model requires thousands of property evaluations of refrigerant. Fast computing time algorithms for accurate properties calculations are desirable.

2.3.1.1 Compressor

The refrigerant mass flow rate of a reciprocating compressor can be calculated by the following equation:

$$\dot{m}_r = \frac{\eta_v PD \cdot \text{rps}}{v_4} \tag{2.4}$$

where:
$\dot{m}_r$ is the refrigerant mass flow rate (kg s^{-1})
η_v is the volumetric efficiency (decimal)
PD is the piston displacement (m^3)
rps is the revolutions of the crankshaft per second (s^{-1})
v_4 is the refrigerant specific volume at entrance to the compressor (m^3 kg^{-1})

For a multicylinder, single acting, reciprocating compressor, the piston displacement is computed as follows:

$$PD = \frac{\pi}{4} D^2 L N_c \tag{2.5}$$

where:
    D is the diameter of the cylinder, bore (m)
    L is the length of the stroke (m)
    N_c is the number of cylinders

Kuehn et al. (1998) derived a relationship for the volumetric efficiency of a reciprocating compressor. In the derivation, the compression process is assumed to be a polytropic process. Re-expansion of clearance vapor, pressure drops across suction and discharge valves, and cylinder heating of vapor on the intake stroke are taken into account. The back flow from the valve and the gas pulsation effects are neglected. The relationship is of the form:

$$\eta_v = \left[1 + C - C \left(\frac{p_c}{p_b} \right)^{1/n} \right] \frac{v_4}{v_b} \tag{2.6}$$

where:
    C is the clearance factor (dimensionless)
    p_c is the absolute pressure at state c in Figure 2.4 (kPa abs.)
    p_b is the absolute pressure at state b in Figure 2.4 (kPa abs.)
    v_b is the refrigerant specific volume at state b in Figure 2.4 (m^3 kg^{-1})
    n is the polytropic compression index (dimensionless)

For polytropic compression, the theoretical work input per unit mass of refrigerant, W (kJ kg^{-1}), can be calculated as follows:

$$W = \frac{n}{n-1} p_b v_b \left[\left(\frac{p_c}{p_b} \right)^{(n-1)/n} - 1 \right] \tag{2.7}$$

By introducing the motor efficiency, η_{mot}, and compressor mechanical efficiency, η_{mech}, the actual power required by the compressor, P (kW), may be estimated as

$$P = \frac{\dot{m}_r W}{\eta_{mot} \eta_{mech}} \tag{2.8}$$

2.3.1.2 Condenser

The condenser can be simulated as a finned-tube cross-flow heat exchanger with a single pass circuit. It is assumed as a cross-flow heat exchanger with both fluids unmixed. The condenser can be divided into three zones: desuperheating, two-phase condensing, and subcooling regimes. We may assume that the condensation of refrigerant takes place in the two-phase zone at a constant saturation temperature corresponding to the condensing pressure. Heat transfer analyses of these zones are handled by the effectiveness or number of transfer units (NTUs) method. The effectiveness or NTU method has advantages for computer execution speed and convergence stability (Berg, 1993).

The effectiveness is defined as the ratio of the actual heat transfer rate for a heat exchanger to the maximum possible heat transfer rate.

$$\varepsilon = \frac{q}{q_{max}} \tag{2.9}$$

or

$$\varepsilon = \frac{C_c(t_{co} - t_{ci})}{C_{min}(t_{hi} - t_{ci})} = \frac{C_h(t_{hi} - t_{ho})}{C_{min}(t_{hi} - t_{ci})} \tag{2.10}$$

where:
  ε is the effectiveness (dimensionless)
  q is the actual heat transfer rate (kJ s^{-1})
  q_{max} is the maximum possible heat transfer rate (kJ s^{-1})
  C is the capacity rate (kW K^{-1}) $= \dot{m}c_p$
  $\dot{m}$ is the mass flow rate (kg s^{-1})
  c_p is the specific heat at constant pressure (kJ kg^{-1} K^{-1})
  t is the temperature (°C)
  $C_{min} = C_c$ or C_h, whichever is smaller
  The subscripts c and h refer to the cold and hot fluids, whereas i and o designate the fluid inlet
    and outlet conditions

For any heat exchanger, it can be shown that effectiveness can be expressed as a function of NTU and heat capacity ratio as in Equation 2.11 (Kays and London, 1984).

$$\varepsilon = f(N, C_r) \tag{2.11}$$

where:
  C_r is the heat capacity ratio $= C_{min}$ (smaller capacity rate [kWK^{-1}])/C_{max} (greater capacity rate
    [kWK^{-1}])
  N is the NTU and defined as:

$$N = \frac{UA}{C_{min}} \tag{2.12}$$

where:
  U is the overall heat transfer coefficient (kW m^{-2} K^{-1})
  A is the area (m^2)

If ε, t_{hi}, and t_{ci} are known, the actual heat transfer rate may be readily be determined from the expression

$$q = \varepsilon C_{min}(t_{hi} - t_{ci}) \tag{2.13}$$

For the desuperheating and subcooling zones, the effectiveness can be estimated by an equation given in Kreider and Kreith (1981). The relationship is

$$\varepsilon = 1 - \exp\left\{ \frac{N^{0.22}}{C_r}[\exp(-C_r N^{0.78}) - 1] \right\} \tag{2.14}$$

Equation 2.11 is valid for $N > 0.25$. If $N \leq 0.25$, the effectiveness is computed from Equation 2.15.

$$\varepsilon = 1 - \exp(-N) \tag{2.15}$$

It should be noted that for the range $N \leq 0.25$, all heat exchangers have the same effectiveness regardless of the value of C (Incropera and DeWitt, 1990).

For the two-phase zone, the heat capacity ratio is zero and hence, the relationship is (Kays and London, 1984):

$$\varepsilon_{tp} = 1 - \exp(-N_{tp}) \tag{2.16}$$

where N_{tp} is the NTU for two phase. It should be noted that for calculation of N_{tp}, $C_{min} = C_a$ in Equation 2.12, where C_a is the air side capacity rate (kW K^{-1}).

2.3.1.3 Receiver

To predict the temperature of refrigerant at the outlet of the receiver for a known refrigerant temperature at the inlet and the ambient air temperature, the following relationship is used:

$$\dot{m}_r c_{pr}(t_{rri} - t_{rro}) = h_{ro} A_o(\bar{t} - t_{am}) \tag{2.17}$$

where:

$\dot{m}_r$ is the mass flow rate of refrigerant (kg s^{-1})
c_{pr} is the specific heat of liquid refrigerant at $\bar{t}$ (kJ kg^{-1} K^{-1})
$\bar{t}$ is the mean temperature of the receiver (°C) $= 0.5(t_{rri} + t_{rro})$
t_{rri} is the temperature of liquid refrigerant at inlet of the receiver (°C)
t_{rro} is the temperature of liquid refrigerant at outlet of the receiver (°C)
h_{ro} is the combined convective and radiative heat transfer coefficient of the receiver (kW m^{-2} K^{-1})
     $= h_{roc} + h_{ror}$
A_o is the outside surface area of receiver (m^2)
t_{am} is the ambient air temperature (°C)

The receiver is assumed to be big enough to occur further subcooling and as a heated horizontal cylinder loosing heat to the ambient air. The natural convective heat transfer coefficient of the receiver, h_{roc}, is approximated by a simplified equation for free convection from horizontal cylinder to air at atmospheric pressure, given in Holman (2008).

$$h_{roc} = \begin{cases} 1.32 * 10^{-3} \left(\dfrac{\Delta t}{D_o} \right)^{1/4} & 10^4 < Gr_f Pr_f < 10^9 \\ 1.24 * 10^{-3} (\Delta t)^{1/3} & Gr_f Pr_f > 10^9 \end{cases} \tag{2.18}$$

where:

h_{roc} is the natural convective heat transfer coefficient (kW m^{-2} K^{-1})
$\Delta t = \bar{t} - t_{am}$ (°C)
D_o is the outside diameter (m)
Gr_f is the Grashof number evaluated at film temperature t_f
Pr_f is the Prandtl number evaluated at film temperature t_f
$t_f = 0.5(\bar{t} + t_{am})$ (°C)

The radiative heat transfer coefficient of the receiver, h_{ror}, is approximated by an equation given in Holman (2008).

$$h_{ror} = 5.669 * 10^{-11} \varepsilon_r \left(\bar{T}^2 + T_{am}^2 \right)(\bar{T} + T_{am}) \tag{2.19}$$

where:

h_{ror} is the radiative heat transfer coefficient (kW m^{-2} K^{-1})
ε_r is the emissivity of the receiver

$\bar{T}$ is the mean absolute temperature of the receiver (K)

T_{am} is the absolute temperature of ambient air (K)

2.3.1.4 Thermostatic Expansion Valve

The operation of the thermostatic expansion valve (TXV) is based on maintaining a constant degree of suction superheat at the evaporator outlet. The valve regulates the refrigerant flow entering the evaporator. The TXV is treated like an orifice and the refrigerant entering the valve is assumed to be in liquid state. The steady-state TXV model can be derived from the equation of flow through an orifice, which is as follows:

$$\dot{m}_r = A_e c_v \sqrt{2\rho\Delta p} \tag{2.20}$$

where:

$\dot{m}_r$ = refrigerant mass flow rate (kg s^{-1})

A_e is the effective expansion valve flow area (m^2)

c_v is the expansion valve flow coefficient (dimensionless)

ρ is the density of liquid refrigerant entering the expansion valve (kg m^{-3})

Δp is the pressure differential across the expansion valve (Pa abs.)

2.3.1.5 Evaporator

The evaporator is divided into two-phase refrigerant with dry air side, two-phase refrigerant with wet air side, and refrigerant superheat zones. Heat transfer analyses are similar to the ones used for the condenser. Equation 2.15 is used for the two-phase zones and Equations 2.13 and 2.14 are used for the superheat zone. The major tasks of the evaporator model in the heat pump drying model are the prediction of the evaporating pressure, the air condition and the refrigerant temperature at the outlet of the evaporator, and the refrigerant state at the inlet of the evaporator for a known set of evaporator air mass flow rate ($\dot{m}_{ae}$), refrigerant mass flow rate ($\dot{m}_r$), inlet air temperature (t_{aei}), and degree of superheat at the outlet of the evaporator (DSH). It is assumed that dehumidification occurs only on the two-phase surface. At the beginning of the computation, the refrigerant exit pressure from the evaporator (p_3, Figure 2.4) is assumed, and it is corrected at the end. Methods for determining the extent of the two-phase refrigerant with dry air-side zone and two-phase surface with dehumidification (wet air-side) may be found in Aye (1995).

2.3.2 A THIN-LAYER DRYING MODEL

Drying models for various products (e.g., grains, fruits, and vegetables) generally based on the one-dimensional thin-layer drying model. Figure 2.5 shows an elemental thin layer of the product. Drying air at initial temperature (T_o) and initial absolute humidity (W_o) is passed through a thin layer of the product of thickness (Δx) at initial moisture content (M_o) and initial temperature (θ_o) for a time interval (Δt). After the drying time interval, the product moisture content decreases to M_f and the absolute humidity of drying air increases to W_f. At the same time, the temperature of the drying air decreases to T_f and the temperature of the product increases to θ_f.

The energy balance equation is derived from the conservation of energy for a control volume.

$$E_{in} + E_g - E_{out} = \Delta E_{st} \tag{2.21}$$

where:

E_{in} is the energy entering through the control surface

E_g is the energy generation

E_{out} is the energy leaving through the control surface

ΔE_{st} is the change of energy stored within the control volume

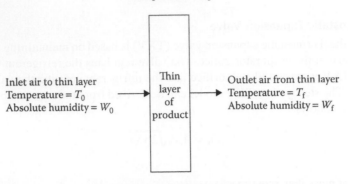

FIGURE 2.5   An elemental thin layer of a product. (From Aye, L., Optimisation of heat pump grain drying system, PhD thesis, Department of Mechanical and Manufacturing Engineering, The University of Melbourne, Parkville, Victoria, Australia, 1995. With permission.)

The following energy balance may be written for the process taking place in any product layer during the time interval (Δt).

$$m_a h_{ao} + m_a W_0 h_{vo} + m_{go} h_{go} + m_w h_{wo} = m_a h_{af} + m_a W_f h_{vf} + m_{gf} h_{gf} + m_w \Delta L \qquad (2.22)$$

where:

m_a is the mass of dry air passing during the time step (kg) = $\dot{m}_a \Delta t$

$\dot{m}_a$ is the mass flow rate of dry air (kg s^{-1})

Δt is the time step (s)

m_g is the mass of wet product in the layer (kg)

m_w is the mass of water evaporated (kg)

h is the specific enthalpy (kJ kg^{-1})

W is the absolute humidity (kg of water per kg of dry air)

ΔL is the latent heat of bound water in excess of that of free water (kJ kg^{-1}) = $h_s - h_v$

h_s is the enthalpy of vaporization of bound water of product (kJ kg^{-1})

h_v is the evaporation enthalpy of free water at temperature T (kJ kg^{-1})

The subscripts are as follows:

a, dry air

g, wet product

o, initial condition

f, final condition

v, water vapor

w, liquid water

Dividing both sides of Equation 2.22 by m_a yields

$$h_{ao} + W_0 h_{vo} + \frac{m_{go}}{m_a} h_{go} + \frac{m_w}{m_a} h_{wo} = h_{af} + W_f h_{vf} + \frac{m_{gf}}{m_a} h_{gf} + \frac{m_w}{m_a} \Delta L \qquad (2.23)$$

Assuming $m_{gf} = m_{go}$ for a time step (Δt) and substituting the enthalpy values in Equation 2.23 gives

$$c_a T_o + W_o(h_{fg} + c_v T_o) + c\theta_o + c_w \Delta W \theta_o = c_a T_f + W_f \left(h_{fg} + c_v T_f\right) + c\theta_f + \Delta L \Delta W \quad (2.24)$$

where:

c_a is the specific heat of dry air (kJ kg^{-1} K^{-1})
c_v is the specific heat of water vapor (kJ kg^{-1} K^{-1})
c_w is the specific heat of liquid water (kJ kg^{-1} K^{-1})
$c = (m_{go}/m_a)c_g$
c_g is the specific heat of wet product (kJ kg^{-1} K^{-1})
T_o is the air temperature into the layer (°C)
T_f is the air temperature out of the layer (°C)
W_o is the absolute humidity into the layer (kg of water per kg of dry air)
W_f is the absolute humidity out of the layer (kg of water per kg of dry air)
h_{fg} is the enthalpy of vaporization of water (kJ kg^{-1})
θ_o is the initial temperature of product in the layer before time step (°C)
$\theta_f = T_f$, the final temperature of product in the layer after time step (°C)
$\Delta W = W_f - W_o$

The equation of mass conversation for water may be written as follows:

$$m_a W_o + m_{dg} M_o = m_a W_f + m_{dg} M_f \quad (2.25)$$

where:

m_{dg} is the mass of dry matter of product (kg)
M_o is the initial moisture content of product before Δt (decimal dry basis)
M_f is the final moisture content of product after Δt (decimal dry basis)

Equation 2.25 can be rearranged as:

$$W_f = W_o + R\left(M_o - M_f\right) \quad (2.26)$$

where R is the dry matter of product to dry air ratio (kg of dry matter per kg of dry air)

$$= \frac{m_{dg}}{m_a} = \frac{m_{go}}{m_a\left(1 + M_o\right)} \quad (2.27)$$

The final moisture content of product after time interval, M_f, may be estimated by using an appropriate thin-layer equation such as

$$\frac{M_f - M_e}{M_i - M_e} = MR \quad (2.28)$$

where:

M_e is the equilibrium moisture content of product (decimal dry basis)
M_i is the initial moisture content of product before drying (decimal dry basis)
MR is the instantaneous moisture ratio of the product for the time interval (dimensionless)

For a deep-bed drying or a tray drying by solving the set of equations for each thin layer sequentially, the outlet air conditions for each time step can be estimated. Repeating the computations for each time step, the final product's moisture content can be predicted.

2.4 CONCLUSIONS

Computer simulation models of heat pump drying systems have been applied for performance prediction, system design optimization, operational, and control optimization. In this chapter, generalized computer program structures for heat pump drying system simulation and optimization have been presented. Algorithms of example phenomenological models of a vapor compression heat pump and a thin-layer drying model are provided. The type and complexity of the submodels incorporated depend on the aims and objectives of the system simulation and optimization.

REFERENCES

Aye, L. (1995). Optimisation of heat pump grain drying system, PhD thesis, Department of Mechanical and Manufacturing Engineering, The University of Melbourne, Parkville, Victoria, Australia.

Berg, J.L. (1993). Estimating the size of air cooled heat exchangers with the NTU method, in *Heat Transfer-Atlanta*, ed B. G. Volintine, AIChE Symposium Series Number 295, vol. 89, American Institute of Chemical Engineers, New York, pp. 274–284.

Branke, J., Deb, K., Miettinen, K., and Slowinski, R. (2008). *Multiobjective Optimization: Interactive and Evolutionary Approaches*, Springer, Berlin, Germany.

Brent, R.P. (1973). *Algorithms for Minimization without Derivatives*, Prentice-Hall, Englewood Cliffs, NJ.

Edgar, T.F. and Himmelblau, D.M. (1988). *Optimization of Chemical Processes*, McGraw-Hill, New York.

Himmelblau, D.M. (1972). *Applied Nonlinear Programming*, McGraw-Hill, New York.

Holman, J.P. (2008). *Heat Transfer (SI Units)*, 9th edn., McGraw-Hill, New York.

Incropera, F.P. and DeWitt, D.P. (1990). *Fundamentals of Heat and Mass Transfer*, 3rd edn., Wiley, New York.

Kays, W.M. and London, A.L. (1984). *Compact Heat Exchangers*, 3rd edn., McGraw-Hill, New York.

Kreider, J.F. and Kreith, F. (1981). *Solar Energy Handbook*, McGraw-Hill, New York.

Kuehn, T.H., Ramsey, J.W., and Threlkeld, J.L. (1998). *Thermal Environmental Engineering*, 3rd edn., Prentice Hall, Upper Saddle River, NJ.

Miettinen, K. (1999). *Nonlinear Multiobjective Optimization*, Kluwer Academic Publishers, Boston, MA.

Press, W.H., Flannery, B.P., Teukolsky, S.A., and Vetterling, W.T. (2007). *Numerical Recipes: The Art of Scientific Computing*, 3rd edn., Cambridge University Press, England, U.K.

Stoecker, W.F. (1989). *Design of Thermal Systems*, 3rd edn., McGraw-Hill, New York.

Vanderplaats, G.N. (1984). *Numerical Optimization Techniques for Engineering Design with Applications*, McGraw-Hill, New York.

3 Advances in Heat Pump-Assisted Agro-Food Drying Technologies

Conrad Oswald Perera

CONTENTS

3.1 INTRODUCTION

During the past few decades, environmental issues, such as climate change, have arisen because of pollution and depletion of natural resources. Fossil fuel consumption contributes to environmental changes. From this point of view, development of sustainable and energy-efficient systems has become important. This has led to the need to develop new methods and equipment for energy-intensive processes such as drying, which must comply with the recent environmental and energy policies, especially in the United States (NRDC 2014).

Drying usually refers to the removal of water from a substance using some energy source, and it is one of the most common ways to preserve food products. Many drying technologies have been developed over the years, such as hot air drying (AD), vacuum drying, and freeze drying. It is well known that the quality of dried products is strongly influenced by drying methods and the drying process employed (Antal et al. 2011, Bonazzi and Dumoulin 2011). In order to identify the advantages and disadvantages of each method, the drying characteristics and product quality must be evaluated as accurately as possible. Recently, there has been a great interest in the use of heat pump dryers (HPDs) for drying fruits, vegetables, and biological materials (Chua and Chou 2014). An HPD is usually a closed system, which was reported to consume less energy compared with other drying methods.

Heat pumps are widely used in water and space heating applications throughout the world. They are able to transfer heat in the opposite direction of spontaneous heat flow by absorbing heat from an area of low temperature and releasing it to a warmer area. They generally work via mechanical vapor compression, absorption, or combined (hybrid) compression–absorption cycles. Although vapor-compression cycle dates back to 1834, the first commercialized machine was produced in 1850. Heat pumps were not originally very popular because of their high installation costs (Calm 1997). In the 1970s, however, alternative applications of heat pumps for dehumidification and food drying were employed.

Food drying processes are energy intensive, and a large portion of energy required in the food industry is used to remove water. Much of the energy is lost as heat with the air discharged from dryers. Most drying industries use heat recovery devices to maximize energy usage. Heat pumps used in drying have the potential to save some of the primary energy used and thus improve the drying efficiency. Therefore, a sound knowledge of energy efficiency and optimum operating conditions is vital for the economic operation of dryers. In drying, a combination of heat transfer by conduction, convection, and radiation is used to heat the product to be dried. Much work has been done to increase the drying efficiency of convective AD, including spray drying. Recently, Minea (2013) reviewed the requirements for R&D needs and future developments for heat pump-assisted drying. He suggested more rigorous studies and increased interaction among drying, heat pump, and process control researchers and specialists at both the academia and industry levels.

In order to determine the suitable type of dryers and drying regimes for a particular product, the drying mechanism and physical properties of the drying material have to be well understood. Such information is required not only for the design and control of dryers but also for setting standards for different categories of dried products.

A number of reviews have been written about heat pumps and heat pump dehumidifier drying systems and their improvements (Chua et al. 2010, Colak and Hepbasli 2009a,b, Li et al. 2011, Minea 2013, Kivevele and Huan 2014). Recently, Minea (2010, 2013) reported that design features and information are missing to accurately operate laboratory and industrial-scale HPDs. Many of the works in the literature neither specify the HPD configuration nor the quantity of material used when reporting high-energy performance. They include general statements, like "the drying conditions are controlled by adjusting the capacity of the heat pump components," without indicating how capacity control was performed in any particular case (Minea 2010). Few of them explained how the temperature and relative humidity (RH) control were achieved, in order to provide the relatively high dehumidifying performance being reported.

Many drying technologies have been developed to obtain stable food products and such drying processes are highly energy intensive. The optimal operation of a dryer is one of the most cost-effective methods for energy saving. Product quality is another important factor to be considered simultaneously with energy conservation. Heat pump dehumidifier dryers offer several advantages over conventional hot air dryers for the drying of food products, such as higher energy efficiency, better product quality, and the ability to operate independently outside ambient weather conditions. In addition, this technology is environmental friendly due to low energy requirement and no release of gases and fumes into the atmosphere (Perera and Rahman 1997).

Sosle et al. (2000) found that the final quality of the HP-dried apple slices in terms of consumer preference was very high. Perera (2001) observed that modified atmosphere HPD-dried apples showed excellent color and retention of Vitamin C. Sosle et al. (2003) demonstrated that an HPD could be successfully used for drying of apple slices, and they also determined the performance of HP for apple drying.

HP dryers have been successfully employed to dry many agro-food products. Perera and Owen (2008) showed that mechanically bruised vanilla beans when dried intermittently (4 h drying and 4 h resting) in an HPD at 28°C and 80% relative humidity over 20 days (to dry slowly but adequately to prevent mold growth) and allowed to recondition to a final moisture content of 25% on a wet-weight basis under ambient conditions produced the highest level of vanillin (over 5.25% on a wet-weight basis).

Rahman et al. (2007) used a rectangular-shaped potato and apple slices to form model composite samples to verify a theoretical model of drying using heat pump convective drying and found that the predictions of the numerical model are in good agreement with the experimental results. Namsanguan et al. (2004) studied the effect of superheated steam drying (SSD) followed by drying using HPD or AD on shrimp. The results showed that SS/HP dried shrimp had a much lower degree of shrinkage, higher degree of rehydration, better color, less tough and softer, and more porous than single-stage SS dried shrimp. They also found that SSD/HPD gave redder shrimp compared to those dried in a single-stage superheated steam dryer.

Shi et al. (2008) studied drying of horse mackerel using HPD and found that the specific moisture extraction rate (SMER) was maximum when the bypass air ratio was 0.6–0.8. The optimum air velocity for drying was 2.0–3.0 m/s, and HPD provided an alternative way to process intermediate moisture fillets of fish.

Claussen et al. (2012) used HPD to study the drying characteristics of salted cod and found that an increase in air velocity (from 0.8 to 3.0 m/s) and temperature (from 15°C to 22°C) and a decrease of relative humidity (from 60% to 30%) increased the water removal rate in the first 12 h. After 12 h, the difference was small. They also showed that by lowering the relative humidity in the drying air and utilization of the condenser, when the amount of *wet* salted fish is increased in the drying tunnel, the dryer capacity could be increased by 50%. Hossain et al. (2013) developed mathematical models for evaluating the performance of HPDs for aromatic plants. They found that the average coefficient of performance (COP), moisture extraction rate (MER), SMER, and drying efficiency were 5.45 kg/h, 140.03 kg/h, 0.038 kg/kWh, and 78.23%, respectively. Their model may be used for designing HPDs to dry aromatic plants as well as other heat sensitive crops. Fatouh et al. (2006) compared the drying characteristics of different herbs using an HPD and found that parsley required the lowest specific energy consumption (3684 kJ/kgH$_2$O) followed by spearmint (3982 kJ/kgH$_2$O) and Jew's mallow (4029 kJ/kgH$_2$O).

Alves-Filho and Eikevik (2008) tested the use of heat pump dehumidifiers for the drying and ripening of Spanish cheeses (Manchego and Mallorquin) and found that this technology fits well with this application, and that it promoted similar or higher removal rates of moisture than the traditional methods.

Flores (2012) studied the use of HPD to dry mixtures of protein matrices and found that protein samples incorporated with a carbohydrate matrix dried at 30°C had lower shrinkage values for materials with the same size. The smallest particles presented higher porosity and less shrinkage, leading to higher drying rates.

Drying of cod fish (clipfish) in Norway has been exclusively transformed from the traditional sun drying to employing HPDs in the past two decades, which has resulted in uniformly high-quality finished product (Strommen, Ingvald. 1998. Personal communication).

3.2 BASIC HEAT PUMP DRYER

A basic HPD consists of an air-to-air heat pump that functions in a manner similar to a domestic refrigerator. It consists of a condenser (hot heat exchanger), a compressor, and an evaporator (cold heat exchanger). The heat pump is located in a well-insulated chamber that can be hermetically sealed. The product is taken on trays and hot air from the condenser is blown over it as in a normal convective air dryer. The warm, moisture laden air is blown over the evaporator coil of the heat pump, where it is rapidly cooled to a temperature below its dew point, resulting in water condensing out. The latent heat recovered in the process (−2255 kJ/kg of water condensed) is released at the condenser of the refrigeration circuit as shown in Figure 3.1. The latent heat given up by the moist air is taken up by the refrigerant and is recirculated within the refrigeration circuit. The absorbed energy (latent heat) is given out at the condenser coil, and the cool dry air from the evaporator is again heated by the heat energy that is given out. The process is repeated until the product is dried to the desired moisture level. The air within the enclosed system is recirculated, leading to a thermal efficiency, approaching 100% (Oliver 1982) at the optimum humidity in the circulating air and temperature. Because the water is condensed and removed in a liquid state rather than its vapor state, it allows the latent heat of vaporization to be captured, and only a small amount of sensible heat is

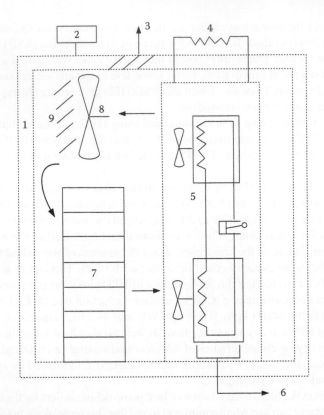

FIGURE 3.1 A schematic diagram of the operation of a typical heat pump dryer. 1, Vapor-sealed and insulated structure; 2, humidifier; 3, overheat vent; 4, external condenser; 5, heat pump dehumidifier; 6, condensate; 7, product tray; 8, primary air circulation fan; 9, air distributor. (Produced with permission *Trends Food Sci. Technol.*, 8, Perera, C.O. and Rahman, M.S., Heat pump dehumidifier drying of food, 75–79, Copyright 1997, with permission from Elsevier.)

lost (Oliver 1982). In practice, design modifications, such as partial evaporator bypass systems and additional heat exchangers, are used to maximize efficiency (Jia et al. 1990). Although most of the currently available HP dryers recirculate all the air, a few nonrecirculatory units are also now available (Hesse 1994, Prasertsan et al. 1996).

The energy efficiency of the HPD is strongly influenced by the relative humidity of the air passing over the evaporator coil of the HPD system shown in Figure 3.1. When the relative humidity of the air is low, the evaporator temperature needs to be lowered to remove the moisture in the air, which leads to a large temperature difference between the evaporator and the condenser, resulting in low energy efficiency in the heat pump and the HPD.

Drying efficiency is a measure of the quantity of energy used in removing a unit mass of water from a product. Normally, it is measured in terms of kJ/kg, although when considering electrically operated HPD dryers, units of kWh/kg are used. The efficiency of a heat pump can be expressed as its COP:

$$\text{COP} = \frac{\dot{Q}_\text{h}}{W_\text{c}} \tag{3.1}$$

where:

$\dot{Q}_\text{h} = W_\text{c} + \dot{m}_\text{w} * h_\text{fg}$ is the thermal power supplied by the heat pump's condenser (kW)

W_c is the compressor electrical power input (kW)

$\dot{m}_\text{w}$ is the water extraction rate (kg/s)

h_fg is the water latent heat of vaporization (kJ/kg) (Perera and Rahman 1997)

Although it is usual to determine the efficiency of a heat pump by its COP value, in the case of a dehumidifier, a more useful measure is the amount of water condensed per unit of electricity consumed. This is termed as the SMER, which can be defined as the kilograms of water extracted per kilowatt–hour of energy input into the system; it defines the effectiveness of the energy used in the drying process.

$$COP = 1 + SMER * h_{fg} \qquad (3.2)$$

where:
   SMER is given in kg/kWh
   h_{fg} is the latent heat of vaporization

The SMER for a well-designed dehumidifier is in the range of 1–4 kg/kWh, with an average value of approximately 2.5 kg/kWh. It is useful to compare this figure with the latent heat of vaporization of water, which is 2255 kJ/kg at 100°C or 1.596 kg/kWh. The low regeneration temperature can help achieve a high COP of the heat pump (COP_{HP}) and a high SMER (Perera and Rahman 1997).

3.3 PRINCIPLE OF HEAT PUMP DRYING

3.3.1 MECHANISM OF HEAT PUMP DRYING

The drying of most biological materials, including food, follows a falling rate profile, and the period during which this occurs is controlled by the mechanism of liquid and/or vapor diffusion. Theoretical models for predicting the drying rates are based on the simplified solutions of Fick's second law. They are valid only within the temperature, relative humidity, airflow rates, and the range of moisture contents that they were developed for. The key advantages of such models are that the geometry of the food to be dried, mass diffusivity, and conductivity were not involved. Thin-layer drying models that describe the drying of these materials mainly fall into three categories, namely theoretical, semi-theoretical, and empirical, which are generally based on mass transfer, neglecting the effect of heat transfer. Recently, Kivevele and Huan (2014) described the development of drying rate prediction models for food drying. They used Fick's second law of diffusion, assuming that the resistance to moisture flow was uniform throughout the material, the diffusion coefficient was constant, and the volume shrinkage was negligible. Fick's second law of diffusion can be stated as

$$\frac{dM}{dt} = D \frac{d^2 M}{dr^2} \qquad (3.3)$$

where:
   M is the local moisture content (kg water/kg dry solids)
   r is the diffusion path (m)
   t is the time (s)
   D is the moisture-dependent diffusivity (m²/s)

The analytical solution of Equation 3.3 was given by Crank (1975) for various regularly shaped bodies, such as rectangular, cylindrical, and spherical. The drying kinetics of many foodstuffs, such as tomatoes (Hawlader et al. 1991), carrots (Reyes et al. 2002), pine nut seeds (Karatas and Pinarli 2001), and pineapple (Nicoleti et al. 2001), has been predicted using the analytical solution of Equation 3.3.

Most agricultural food materials such as cereals (e.g., rice, corn, and wheat) change little in volume during the drying process. Therefore, the analytical solution of Equation 3.3 applies satisfactorily to the study of these materials (Kivevele and Huan 2014). However, for foods with high moisture content, such as fruits and vegetables, the analytical solution of Equation 3.3 obtained for constant diffusivity and volume is not always applicable, because shrinkage and diffusivity as functions of moisture content often need to be taken into account (Madamba et al. 1996).

3.3.2 Modeling of Heat Pump Drying

A simple HPD is a convective air dryer, and the theories derived for classical convective AD systems are applicable. For the purpose of developing mathematical models, it is often sufficient to use simple semiempirical equations, which can adequately describe the drying kinetics when the external resistance to heat and mass transfer is eliminated or minimized. A common way to achieve this is to carry out experiments using a thin layer of the material being dried. Thin-layer drying kinetics is needed to design full-scale HP drying systems. Rahman et al. (1997) have developed drying kinetic models for the desorption isotherm of peas using two-component models in an HP dryer.

Numerous experimental and modeling efforts on thin-layer drying have been proposed in the literature for predicting the drying kinetics of mushrooms, pollen, and pistachio (Midilli et al. 2002); red bell pepper (Vega et al. 2007); apple pomace (Wang et al. 2007); canola seeds (Mohsenimanesh and Gazor 2010); pepper (Darvishi et al. 2014); and many other agricultural and biological products.

Henderson and Pabis (1961) developed a semi-theoretical model for thin-layer drying, known as the *two-term model*, which relates the moisture of the food being dried to time by an exponential relationship. They used the model successfully to predict the drying rate of corn. This model was later used by Watson and Bhargava (1974), Wang and Singh (1978), and Karathanos (1999) for wheat, rice, and dried fruits, respectively. However, it requires a constant product temperature and assumes a constant diffusivity.

The Lewis model (1921) is a special case of the Henderson and Pabis (1961) model and was used to describe the drying rates of barley (Bruce 1985), wheat (O'Callaghan et al. 1971), and cashew nuts (Chakraverty 1984). The two-term model was shown to give the highest correlation coefficients between experimental and predicted data values. Recently, the Lewis model was also proved to be the best for drying of jackfruit by Soares et al. (2014).

The Lewis model assumes that the internal resistance to moisture movement and the moisture gradients within the material is negligible (Lewis 1921). This model considers only the surface resistance and is given by

$$MR = \frac{M - M_E}{M_O - M_E} = \exp(-k_L t) \tag{3.4}$$

where:
The average moisture content was expressed as nondimensional moisture ratio MR
M_O and M_E are the initial and equilibrium moisture contents, respectively
M is the moisture content at time t
k_L is the Lewis drying coefficient

This model was used primarily because it is simple. However, one of the drawbacks of this model is that it tends to overpredict the early stages of the drying curve (Parti 1990, Rahman et al. 1997). Therefore, the Page (1949) model was introduced as a modification of Equation 3.4 to overcome this shortcoming. This model has produced good fits in predicting drying times of food and agricultural materials:

$$MR = \frac{M - M_E}{M_O - M_E} = \exp(-k_p t^n) \tag{3.5}$$

where k_p and n are the Page drying coefficients that determine the precise shape of the drying curve. Although neither of them has a direct physical significance, empirical regression equations have been developed, which relate both the parameters to drying conditions (Cronin and Kearney 1998, Hossain and Bala 2002, Jayas et al. 1991).

Therefore, the drying rate for the Page equation could be given by

$$\frac{dM}{dt} = (-k_p n t^{n-1})(M - M_E) \tag{3.6}$$

A modified Page drying coefficient can be defined as

$$k^* = k_p n t^{n-1} \tag{3.7}$$

If $n < 1$, k^* decreases during the drying process. Thus, higher values of k_p can be used to more closely approximate the diffusion equation in the initial stages of drying without overpredicting drying in the later stages. If $n = 1$, Equations 3.6 and 3.7 reduce to the Lewis model, approaching the solution of diffusivity equation. Several authors have compared different drying models and have obtained better results for the Page equation than for other existing drying models (Cronin and Kearney 1998, Hossain and Bala 2002, Panchariya et al. 2002).

HPD dryers are more suited for batch operations than for continuous operation, because batch operations give total circulation without much air leakage rates and results in high thermal efficiencies. However, continuous HPD systems have been built for the drying of vegetables and gelatin (Anon 1988). Strommen and Jonassen (1996) and Alves-Filho and Strommen (1996) described the development of countercurrent HP-fluidized bed dryers with high SMERs for the drying of heat-sensitive products.

Some of the earlier failures of HP dryers were because of inadequate thermal insulation and gas tightness of the seals of the chamber structure, resulting in loss of energy efficiency. In addition to the electrical energy required to drive the compressor, energy is also required to preheat the product and chamber structure, to drive the fan for primary airflow over the product that is to be dried, and to replace any heat loss through conduction and air leakages. The motors driving the fan and the compressor can be located within the chamber, so that the residual heat produced by them is absorbed within the drying chamber instead of being lost to the atmosphere.

Increasing the humidity in the drying air slows down the drying process but improves energy efficiency (Perera and Rahman 1997). In general, the dehumidifier efficiency and capacity are proportional to increases in temperature and humidity. However, at high temperatures (e.g., 60°C), the increase in SMER with increase in relative humidity is compromised (Figure 3.2).

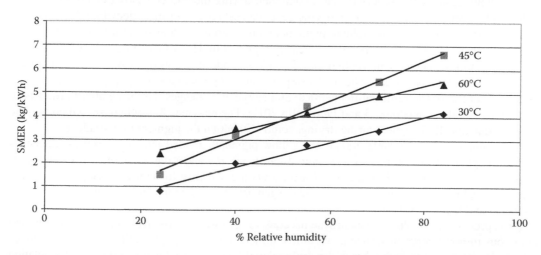

FIGURE 3.2 Specific moisture extraction rate (SMER) as a function of relative humidity and dry bulb temperature. (Adapted from *Trends Food Sci. Technol.*, 8, Perera, C.O. and Rahman, M.S., Heat pump dehumidifier drying of food, 75–79, Copyright 1997, with permission from Elsevier.)

3.4 QUALITY CHANGES OF CHEMICAL, BIOLOGICAL, AND BIOCHEMICAL PRODUCTS DURING HEAT PUMP DRYING

Major quality parameters associated with dried food products are the change in color and visual appeal, flavor/odor, microbial load, nutrients, porosity-bulk density, shape, texture, rehydration properties, water activity, and chemical stability. These aspects of dried food quality were discussed by Perera (2005).

The basic HP drying system is a convective AD system. Many of the chemical and biochemical changes are the same as those that happen during convective AD. Many chemical and biochemical reactions can be induced by temperature increase in foods, for example, Maillard reactions, vitamin degradation, fat oxidation, and denaturation of thermally unstable proteins, resulting in the variation of the rehydration properties of dried products and enzyme reactions (which can be either promoted or inhibited) (Bonazzi and Dumoulin 2011). Some of these chemical and biochemical reactions are beneficial, for example, flavor and color development during baking. However, others may be undesirable, for example, undesirable color changes, such as brown color development and loss of vitamins and nutrients during drying. Vitamin C (ascorbic acid [AA]) is an important nutrient, and it is often taken as an index of the nutrient quality of processes. AA can undergo oxidation to dehydroascorbic acid under aerobic conditions. Dehydroascorbic acid could further undergo oxidation to AA degradation products. Recently, Kurozawa et al. (2014) showed that the degradation of vitamin C was higher in papaya dried at 70°C than that dried at 40°C, and they concluded that this was because at 70°C, papaya sample remained in the rubbery state until the end of drying, because product temperature (T_p) was above the glass transition temperature (T_g) along the process. This state is characterized by great molecular mobility inside the food, which facilitates the degradation. At 40°C, the rate of nutrient degradation was very slow as T_g was close to T_p, and papaya has suffered phase transition from rubbery to glassy state.

Color is one of the most important attributes with respect to the quality of dried foods, because most of the time, one important criterion taken into account by consumers when choosing a new product is its visual appearance. Color can change during drying because of chemical and biochemical reactions. The rates of such reactions depend strongly on time, temperature, and other drying parameters. Most fruits and vegetables contain pigments such as carotenoids, chlorophylls, anthocyanins, and betalains, which are susceptible to degradation by enzymatic or nonenzymatic reactions induced by drying, and these reactions continue during storage (Bonazzi and Dumoulin 2011). Chong et al. (2013) reported that a combined drying method consisting of HP pre-drying followed by vacuum microwave finish drying gave the best results for most dehydrated fruits.

Chlorophylls are green lipid-soluble pigments with a porphyrin ring structure and an Mg^{++} ion at the center of its molecule, which can be easily replaced by H^+ in food products during heating and drying to pheophytin. Chlorophyll can degrade to undesirable gray–brown compounds such as pheophytin or pheophorbide. The initial step of chlorophyll degradation in processed foods is a disruption of the tissue, resulting in chemical and enzymatic changes that bring about products of chlorophyll catabolism (Bonazzi and Dumoulin 2011). The enzyme chlorophyllase, which may be released due to cell disruption during drying, could also act on the pigment to degrade it.

Carotenoids are lipid-soluble orange and yellow pigments divided into two families: carotenes and xanthophylls. Carotenes exists as α- and β-carotenes, the major pigments in carrots, and lycopene is the major pigment in tomatoes. Carotenoids are sensitive to light and temperature; high temperature, long-processing time, light, and oxygen have been shown to enhance their degradation (Bonazzi and Dumoulin 2011).

Food products are multicomponent systems consisting of water, carbohydrate, protein, lipids, and numerous minor components, such as vitamins, minerals, acids, and phenolics. These compounds may easily interact with each other at the drying conditions employed and result in the degradation of food quality (Sokhansanj and Jayas 1987). Consequently, the products desirability for consumption is affected. Thus, appropriate temperature plays an important role during drying. Browning of foods can occur by enzymatic (polyphenol oxidase [PPO]) and nonenzymatic (Maillard, AA browning) reactions, which usually impairs the sensory properties of products because of the associated changes in color, flavor, texture, and nutritional properties (Martinez and Whitaker 1995). Enzymatic browning requires four components: oxygen, enzyme, copper, and a phenolic substrate (Laurila et al. 1998). The PPO group of enzymes catalyzes the oxidation of phenolic compounds in plants to ortho-quinones. Immediately, the quinones condense and react nonenzymatically with other phenolic compounds and amino acids to produce dark brown, black, or red pigments of indeterminate structures (Whitaker and Lee 1995). Martinez and Whitaker (1995) have found that low pH values help to decrease the activity of PPO because of loose binding of Cu, which activates the enzyme, thus allowing the acid molecules to sequester the Cu, and hence reduce browning. Consequently, exclusion of oxygen and/or application of low pH environments can ease browning. To date, sulfiting agents are widely used in the drying industry, but they are considered harmful for asthmatics and are therefore unsuitable to such consumers. Many studies have shown that AA is able to reduce browning (Gill et al. 1998, Pizzocaro et al. 1993, Sapers and Miller 1992). However, AA itself may undergo oxidation during drying and therefore, it is unsuitable for preventing browning in dried products.

3.5 PHYSICAL TRANSFORMATIONS DURING HEAT PUMP DRYING

Several physical phenomena associated with temperature change, time, and water loss have been known. They include decrease in water activity, glass transition and crystallization, melting (of fat), evaporation (of volatile components), and migration or retention of components. The consequences of water loss on product characteristics are complex and are interconnected. For example, a decrease in a_w corresponds to a reduction of water availability for microbial growth. In addition, molecular mobility within the food matrix is reduced; chemical and biochemical reactions are delayed, thus achieving both chemical and biological stabilities. On the other hand, a decrease in water activity slows down the water transfer and therefore, the drying rate is reduced, thus increasing the time that the product must spend at a relatively high temperature during which reactions may develop. Most of the aroma compounds of foods are more volatile than water because of a combination of high vapor pressure and low solubility of the aroma compounds in aqueous solutions (Saravacos and Krodika 2014).

However, it has been widely observed that the release of aroma compounds during drying is much less than expected when considering only the volatility or vapor pressure of the molecules. Three mechanisms have been proposed, namely, selective diffusion, entrapment within microregions, and interactions with the substrate (Saravacos and Krodika 2014). Retention of volatiles is a process mainly controlled by diffusion, and diffusivity of aroma molecules decreases much faster than that of water when the moisture content decreases. As a result, the product changes during drying in such a manner that its surface becomes virtually impermeable to aroma components. The retention of volatile compounds increased with the increase in dry matter content. The concept of micro-regions has been used to explain the retention of volatile compounds during freeze-drying (Flink and Karel 1970). During drying, sugars might create amorphous micro-regions, entrapping the volatile molecules by hydrogen bonding. This is a description of the retention phenomenon at the microscopic level. It is also possible that retention is influenced by specific interactions with the substrate, such as covalent bonds (aldehydes with $-NH_2$ and $-SH$ groups), steric entrapment, or sorption (on proteins, lipids). However, the selective diffusion is often presented as the main and most general mechanism. Coumans et al. (1994) outlined that the concept of selective diffusion can be exploited to create high aroma retention conditions, by promoting a rapid decrease in water content at the surface of the product. The retention or loss of aroma compounds is also influenced by the structure (amorphous or crystallized) of the dried product. Structural changes can be used to explain the entrapment of aroma compounds in spray drying encapsulation processes (Ré 1998).

Crystallization tends to increase the loss of aroma, because it rejects impurities, including volatiles. Senoussi et al. (1995) measured the loss of diacetyl as a function of the rate of crystallization of lactose during storage. They found that when the lactose was stored at 20°C above the glass transition temperature, T_g, the amorphous product immediately crystallized and practically, all diacetyl was lost after 6 days. Levi and Karel (1995) also found increased rates of loss of volatile (1-n-propanol) as a result of crystallization in an initially amorphous sucrose system.

3.6   DEVELOPMENTS IN HEAT PUMP DRYING SYSTEMS

3.6.1   MICROWAVE-ASSISTED HEAT PUMP DRYERS

In the early stages of most food dying operations, when the material is still very wet, water easily reaches the surface by capillarity. Under fixed conditions of air temperature and relative humidity, the surface attains the wet-bulb temperature after a warming-up period and proceeds to dry at a constant rate, which is a function of the air speed. Eventually, however, the moisture content of the product falls to levels at which migration to the surface becomes controlled by heat transfer and diffusion within the solid. This leads to a progressive fall in the rate of evaporation. The final moisture content or the equilibrium moisture content required from a drying process is usually such that the final product will neither lose nor gain moisture when in contact with the atmosphere over extended periods. Consequently, the final moisture content required for the overwhelming bulk of dried food products falls well within the falling rate period, which means that nearly all industrial convective driers have problems over throughput, size of plant, uniformity of final moisture content, and energy consumption.

During the constant rate period, hot AD is relatively fast (because the vapor pressure at the surface is high) and efficient (because the moisture burden of the air leaving can be high, while still providing a driving force for drying). However, a large amount of latent heat is rejected. It takes six times as much energy to evaporate water as it does to raise it from 20°C to 100°C. If there is some way of recycling this energy, heat pumps may have a role here. During the falling rate period, hot air undergoes every disadvantage, in that it is slow, uneven, and inefficient. If the manufacturers were satisfied with very slow rates of drying (over a period), then the problems of excessive energy consumption at the dry end could be overcome. This is just not practical in most cases. What is required is a heating technique that liberates heat throughout the volume of the product at rates that are greatest where the moisture is found, thus avoiding the surface heat transfer limitations inherent in hot air systems (Zhang et al. 2006). This requirement clearly indicates that electromagnetic fields, such as microwaves, could have an important part to play in view of their ability to penetrate into dielectric materials and their strong absorption by water at appropriate frequencies (Anon 1988, Lawton 1978).

Jia et al. (1993) were the first to report on heat pump drying assisted by microwave energy. They reported the overall performance of a microwave-assisted HP drying system. A prototype dryer with 5 kW of heat pump compressor and 10 kW microwave power was constructed in the experiment. The results of the study indicated that with careful design, heat pump-assisted microwave drying is superior to conventional AD in terms of energy consumption.

3.6.2   MODIFIED ATMOSPHERE HEAT PUMP DRYING

Because the heat pump unit in a basic HPD is enclosed in an insulated drying chamber, which is hermetically sealed, it is possible to modify the atmosphere within the chamber. The concept of modifying the atmosphere in heat pump drying was first conceived by O'Neill et al. (1998). They compared the apple cubes dried in air and in a modified atmosphere, and showed that those dried in a modified atmosphere (N_2) were less brown and more porous than those dried in air (see Table 3.1).

A prototype of modified atmosphere heat pump dryer (MAHPD) was built by Hawlader et al. (2006a) and they dried apple, guava, and potato under an inert environment (nitrogen or carbon dioxide). The impact on color, surface porosity, and rehydration abilities were investigated. Lemon juice and

TABLE 3.1

Density and Porosity of Apple Cubes Dried under HPD, MAHPD, and Vacuum

Density/Porosity	HPD Dried in Air	MAHPD Dried in N_2	Dried in Vacuum
Apparent density (kg/m³)	670 ± 10	569 ± 16	517 ± 20
Particle density (kg/m³)	772 ± 15	638 ± 18	841 ± 10
Material density (kg/m³)	1578 ± 23	1506 ± 25	1659 ± 40
Open pore porosity	0.511	0.577	0.493
Total porosity	0.576	0.622	0.689

peel were used as natural inhibitors to prevent browning in air-dried apple slices. Comparisons were made between MAHP-dried samples and those dried by freeze and vacuum drying. Results showed that MAHP operated at a relatively low temperature of about 45°C and relative humidity around 10% led to better physical properties, such as reduced shrinkage, decreased firmness, and more porous structure of the materials, which resulted in quicker rehydration. They also found that when using inert gas, the color of heat pump-dried food was similar to those obtained by vacuum or freeze drying.

Hawlader et al. (2006b) compared the drying characteristics and retention of volatile gingerol under different methods of drying. They found that MAHP drying had an improved effective diffusivity. The concentrations of the main pungent principle (6-gingerol) of ginger, extracted from the samples by different methods, were determined by high-pressure liquid chromatography. The retention of 6-gingerol increased in the order of normal AD, freeze drying, nitrogen MAHP drying, carbon dioxide drying MAHPD, and vacuum drying. From this point of view, inert gas MAHPD showed a better retention of flavor compared to AD or freeze drying.

Hawlader et al. (2006c) found that when using CO_2 as the inert gas, the effective diffusivity during the drying process was 44% and 16.34% higher in guava and papaya, respectively, when compared to normal HPD drying. There was less browning, faster rehydration, and more vitamin C retention in the final products. All of these reveal a great potential for MAHPD in the food drying industry.

Substituting O_2 with an inert gas during the drying process is a promising and an effective method to control physicochemical reactions, protect nutrient components, and improve product quality. Flushing inert gases such as N_2 and CO_2 to replace O_2 has been reported to reduce browning, increase porosity, and shorten the drying period (O'Neill et al. 1998). Ramesh et al. (1999, 2001) found that inert gas drying with N_2 increases the drying rate, mass and heat transfer, and nutrient component retention. However, the N_2 drying method requires continuous N_2 gas usage, which causes substantial consumption of inert gas. Hawlader et al. (2006a) combined heat pump drying technology with inert environmental conditions to develop a modified atmosphere heat pump drying process. Several fruits and vegetables have been dried using this method, and the results showed that the color protection and nutrient retention of the products dried with inert gas (N_2 and CO_2) were significantly higher than AD and similar to products dried by vacuum or freeze drying; thus, the modified atmosphere heat pump drying could protect the overall quality of products (Hawlader et al. 2006a,b,c).

Liu et al. (2014) used an MAHPD system to dry *Flos Lonicerae* (flower bud of *Lonicera japonica* Thunb.), a common herb in traditional Chinese medicine and showed that the increase in drying temperature and the decrease in oxygen content could improve the drying rate and effective moisture diffusivity. Increased retention of chlorogenic acid and chlorophyll contents as well as decrease in color differences were observed with decreasing drying temperature and oxygen content. High product quality could be achieved at all drying temperature levels when oxygen content was as low as 5%. Therefore, they concluded that reducing the oxygen content in the drying atmosphere is an effective method to improve product quality.

3.6.3 Microwave-Assisted Modified Atmosphere Heat Pump Drying

In 1997, an experimental modified atmosphere HPD with microwave heating facilities was developed at the Mount Albert Research Centre in Auckland, New Zealand. A schematic diagram of the drying rig is shown in Figure 3.3. It consists of a cylindrical drying chamber with a rotating drying rack along a vertical axis at the center of the cylinder. The drying rack is connected to a load cell, so that the weight is constantly monitored and recorded during drying, and the drying curves were plotted by the computer. A 5 kW microwave generator delivered microwave power vertically down the wall of the drying chamber. The drying chamber is connected to an external heat pump unit. The heat exchanger unit and the refrigeration system were situated externally as shown in Figure 3.3.

The drying chamber is connected to a nitrogen gas cylinder and to a vacuum pump. The operation of the dryer is controlled by programmable logic control system attached to a computer. The drying chamber after loading the product is partially evacuated to remove oxygen and then flushed with nitrogen. The process was repeated around four to five times until the oxygen level in the drying chamber was less than 0.1%, and the whole chamber was filled with nitrogen gas at normal atmospheric pressure. The dryer was operated with a full load under MAHPD conditions until the moisture content dropped down to about 30%–40%, when the rate of drying begins to drop off significantly. When the moisture content drops to this level, microwave generator was operated to heat up the product and improve the drying rate. The products produced by this dryer were comparable to vacuum dried products in terms of color, rehydration properties, and retention of nutrients, such as vitamin C. Apple cubes dried by microwave-assisted MAHPD process had high porosity and low bulk density (Table 3.1).

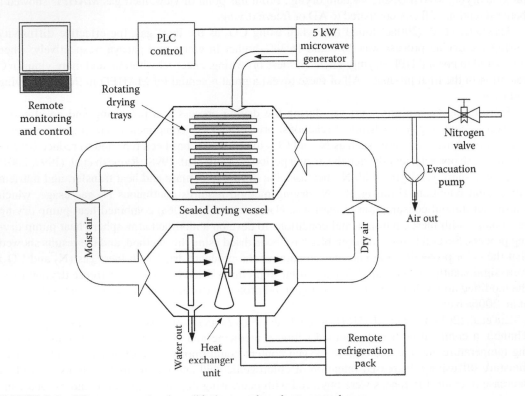

FIGURE 3.3 Microwave-assisted modified atmosphere heat pump dryer.

From Table 3.1, the apparent or bulk density of the dried apple cubes was lower than those dried in air and higher than those dried by vacuum. The particle density, however, was the lowest, showing that the particles had more pores. The open pore porosity was the highest in microwave-assisted MAHPD-dried apple cubes. However, the total porosity was in between air- and vacuum-dried samples. The high degree of open pore porosity gives rise to rapid rehydration rate of the MAHPD samples. This property is very important when considering dried products that need rapid rehydration rates; for example, dried vegetables used in Maggi 2 Minute Noodles, not only need to float onto the surface when hot water is added to the noodle mix for esthetic appeal, but they also need to be rehydrated within 2 min to have the desired texture.

The drawback of this system was that when the moisture content falls below 10% on a wet-weight basis, the magnetron fuse tends to blow frequently, hindering the operation.

3.6.4 DESICCANT-ASSISTED HEAT PUMP DRYING

Kivevele and Huan (2014) described an alternate way to solve the problem of low energy efficiency, as drying progresses in a normal HPD by incorporating a desiccant unit parallel to the evaporator to share part of the moisture-removing load, as illustrated in Figure 3.4. There are two parallel air ducts in this system. One is the heat pump air duct, which includes air valve 1, evaporator, and condenser. The other is the desiccant duct, which includes air valve 2 and the desiccant unit. The heat pump and the desiccant unit work alternately. The heat pump in this system is different from that of the basic HPD shown in Figure 3.1, in that there are two parallel refrigerant circuits. The first circuit consists of the compressor, valve 1, the heat sink, condenser, expansion valve, and evaporator. The second circuit consists of the compressor, valve 2, heating tubes, expansion valve, and the heat sink. The heat pump refrigerant cycles through different circuits in different drying stages. The working period is divided into three stages for the batch drying of biological materials. In the first stage, when the material contains high moisture content, the heat and mass transfer between the hot air and material is extensive, so the relative humidity of the exhausted air is high. In this stage, the heat pump works with valve 1 open and valve 2 closed, and the heat pump refrigerant flows in the first

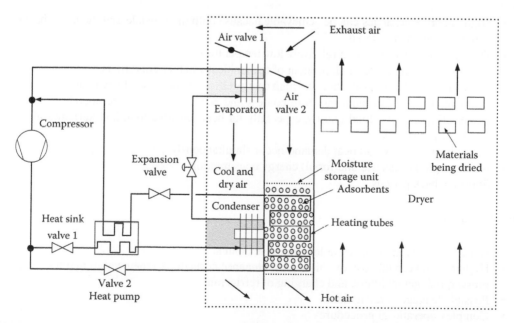

FIGURE 3.4 Desiccant-assisted heat pump dryer. (Produced with permission from Kivevele, T. and Huan, Z., *S. Afr. J. Sci.*, 110(5/6), Art. #2013-0236, 11 pages, 2014.)

circuit. Simultaneously, air valve 1 is open and air valve 2 is closed. The exhausted airflows through the heat pump duct. The moisture in the exhausted air is condensed and removed in the evaporator, and then heated in the condenser in the usual manner. After most of the moisture in the materials has been removed, the hot air cannot extract enough moisture from the materials in the dryer, and so the relative humidity of the exhausted air becomes low, and the drying process enters the second stage.

In the second stage, the heat pump stops, air valve 1 is closed and air valve 2 is opened, the exhausted airflows through the desiccant duct, and the desiccant unit begins to work. The moisture in the exhausted air is absorbed by the desiccant, and the dry air is heated simultaneously by the heat of absorption. When the moisture content in the materials reduces to the required level, the second stage ends and the drying process using the desiccant also ends. After the dried materials have been taken out of the dryer, the third stage begins. In this stage, the heat pump works to regenerate the desiccant. As valve 1 is closed and valve 2 is opened, the heat pump refrigerant flows in the second circuit, and the heat pump receives heat from the heat sink, which in turn heats the desiccant in the desiccant unit. At the same time, air valves 1 and 2 and the dryer door are all open. As fresh airflows into the desiccant unit, it is heated and carries water vapor discharged by the desiccant out of the desiccant unit. When the water content in the desiccant has been reduced to the required level, the third stage ends and the process is complete. Desiccant-assisted HPDs have great energy-saving potential for batch drying of thermally sensitive biological materials. The energy consumed by this heat pump can be 30–50% lower than that of a basic HPD (Figure 3.1). However, its energy efficiency is mainly affected by steam capacity and the regeneration temperature of the desiccant.

3.7   ADVANTAGES AND DISADVANTAGES OF HEAT PUMP DRYERS

The advantages and disadvantages of HPD dryers compared to basic hot-air convective driers are listed by Minea (2013) as follows:

Advantages:

- High energy efficiencies can be achievable because both the sensible and the latent heat of vaporization are recovered.
- Drying can be carried out at relatively low temperatures.
- Drying can be conducted independent of the ambient weather conditions.
- High annual factors of utilization and high energy efficiency because of low temperature lifts.
- Efficient control of product moisture content and wet- and dry bulb temperatures of the air resulting in better final quality, especially for heat-sensitive food and agricultural and biomedical products.
- Waste heat recovery and heat demand occur simultaneously.
- Reduced global (electric and fossil) energy consumption.
- Short payback periods.

Disadvantages:

- Relative energy inefficiency at low moisture content
- Higher initial capital cost and higher maintenance costs due to the need to maintain compressor, refrigerant filters, and charging of refrigerant
- Possible leakage of refrigerant
- Complex operational procedures
- Additional floor space required
- Requirements for competent design engineers and operating technicians

One of the major disadvantages of standard HPD drying systems is the energy inefficiency at low moisture contents. However, this could be overcome with proper design modifications and adequate control strategies.

Although heat pump-assisted drying technology has been in operation for a number of decades and is considered as a mature technology, its industrial application has not been as widespread as it should be, particularly in the field of agricultural food products drying. HPD dryers have not so far achieved the success of being installed for drying of such food products, even when the payback period has been demonstrated to be relatively short. The following are among the reasons for their neglect:

- Uncertainty of reliability of HPD dryer by potential users.
- Lack of good hardware for some types of potential applications.
- Lack of experimental evidence and demonstration of installations for different types of industries.
- Lack of required knowledge of chemical engineering and heat pump technology in target industries.
- Commercial viability of heat pump drying may be affected by the relative cost of electricity and fossil fuels.

Optimal integration of driers and heat pumps remains a challenging R&D task, and wider dissemination of existing and future knowledge may also contribute to future applications (Minea 2013).

Knowledge of drying mechanism is important in designing HP dryers. The vapor pressure in the drying atmosphere must be reduced, in order to bring about drying, which can be achieved by increasing the temperature, decreasing the relative humidity together with an increase in the airflow rate over the product.

3.8   FUTURE OF HEAT PUMP DRYING SYSTEMS

According to Minea (2013), a number of factors need to be addressed before HPD drying systems could be used more widely for commercial drying of food and agro-based products. He pointed out that the knowledge of the drying mechanisms, selection of the appropriate drying systems and dryer designs, relevant drying data, relevant drying models and controls, workable integrated systems, final product quality, and ecological issues are paramount in getting HP drying systems more widely accepted in the food industry. These are all complex issues that require systematic approaches to find the appropriate solutions. Minea (2013) also stated that without substantial structural modifications and adjustments of the system integration and/or drying schedules and control strategies, a given HPD system cannot be efficiently used:

- For drying a wide number of products, in different physical states, such as solids, liquids, semiliquids, agro-food, heat-sensitive, and biological products
- In any drying mode (batch, conveyor, continuous, and intermittent)
- Over large temperature ranges (e.g., from −30°C to 110°C)
- With any drying medium (air, oxygen, inert gases, CO_2, etc.)
- Using the same refrigerant, without modifying its refrigerating configuration, heat exchanger design, and/or controls

Most of the improvements in HP drying systems have to focus on the simplification of the concepts in order to increase the reliability and dehumidification efficiency of industrial drying applications.

Although improving the heat pump design will improve the dryer operating conditions, compressor performance is also very important, because the compressor is the main component in any heat pump system. Improving compressor performance may reduce power input that is required. One of the most efficient ways to improve compressor performance is by cooling.

The performance of the heat pump system can also be improved by using different refrigerants. Rakhesh et al. (2003) have studied the performance of HP drying systems using different refrigerants such as HFC227 and CFC114 at different condensing and evaporating temperatures. They compared the heating capacity and COP of the heat pump with the two different refrigerants. The results showed that HFC227 was the better working refrigerant for evaporating temperatures higher than 30°C.

The performance degradation due to switching from CFC-12, a chlorofluorocarbon refrigerant, to hydrofluorocarbon refrigerant in a direct expansion solar-assisted heat pump was investigated by Gorozabel Chata et al. (2005). These results showed that CFC-12 produced the highest value of COP, followed by HFC-22 and R-134a. The system performance degradation was about 2%–4% in the 0°C–20°C collector temperature range when CFC-12 was replaced with HFC-134a. For a mixture of refrigerants, HFC-410A was shown to be more efficient than either HFC-407C or HFC-404A, but not as good as HFC-134a. The refrigerant HFC-410A produced COP values that were 15%–20% lower than those obtained with HFC-134a.

In the refrigerant cycle of the heat pump, the refrigerant gas could become superheated at the evaporator and compressor. During the evaporation process, the refrigerant is completely vaporized part way through the evaporator. As the cool refrigerant vapor continues through the evaporator, additional heat is absorbed, which superheats the vapor. Pressure losses caused by friction further heats the vapor. When superheating occurs at the evaporator, the enthalpy of the refrigerant is raised, extracting additional heat and increasing the refrigeration effect of the evaporator. When superheating occurs at the compressor, no useful cooling occurs (Energy Management Series 11). Alves-Filho and Eikevik (2007) used refrigerant desuperheaters to reject the system's excess heat to outdoors, thereby improving the efficiency of the HPD system.

3.9 CONCLUSIONS

The ability of HPDs to convert the latent heat of condensation into sensible heat at the hot condenser makes them unique heat-recovering devices for drying applications. It is well recognized that drying using HPD is an energy-efficient way of drying heat- and oxygen-sensitive products such as bioactive food and pharmaceutical products. However, the simplest models are only efficient in drawing out free moisture and loosely bound moisture from the product. Efficiency drops dramatically once free moisture and loosely bound moisture are removed. In most cases, the efficiency of HPD used for drying falls off with increase in temperature above 45°C. Modifications to HP dryers, such as desiccant- and microwave-assisted drying, are used to overcome this limitation. Because dehumidification takes place inside a closed chamber, it is possible to change the atmosphere within the drying chamber with a suitable inert gas such as nitrogen or carbon dioxide. An HP dryer in which the atmosphere is modified using an inert gas is known as an MAHPD. Such dryers will have the combined benefits of low energy usage, relatively low temperature of drying, and the possibility of employing an inert gas to minimize oxidation of sensitive products. Over the past few years, it was shown that the dried products obtained from MAHPD dryers had better flavor retention, less color degradation, lower oxidation, higher porosity, and higher rehydration rate of dried products compared to normal HP-dried products. Such physical and chemical characteristics are important in certain dried products, such as dried vegetables for incorporation into 2-minute noodles and dried fruits for incorporation into breakfast cereals. With improved efficiency of drying, the MAHPD will have novel applications in the food and pharmaceutical industries. Improving the COP of the heat pump system is important for improving the SMER and dryer performance. In the development of an energy-efficient drying system, the economic factors such as cost, system efficiency, and performance are important. Hybrid drying technology using HPD with increased performance of the system may also increase the capital cost greatly, and there is a need to balance the two.

Drying using HPD has been commercially used in Norway for over two decades to dry uniformly high-quality dried cod (bacalao or clipfish). There are numerous applications of HPD technology cited in the literature to produce high-quality dried fish, meat, protein products, fruits, vegetables, spices, and herbs.

Most of the world's dried spices such as black pepper, cinnamon, clove, and cardamom come from the developing countries where they are dried out in the open sun. Thus, they are prone to get contaminated by dust, grime, and microorganisms. Most of these spices are used in foods at the end of cooking or after cooking, which results in contamination of the food products. However, drying using HPD is done at low temperature, in a low relative humidity atmosphere and in a closed container; thus, this technology is especially suitable for the drying of high-quality spices and herbs that retain a high level of volatile aromatic constituents free of dust, grime, and microbial contamination.

Therefore, the opportunities for using heat pump drying technologies in the dehydration of a number of agro-food industries are high. It is now up to the scientists and technologists to disseminate the information about energy efficiency and high-quality finished products that can be obtained by employing HPD and to work with manufacturers to develop cheap and affordable equipment.

REFERENCES

Alves-Filho, O., T. Eikevik. 2007. Atmospheric freeze drying with heat pumps—New possibilities in drying of biomedical materials. Paper presented at the *5th Asia-Pacific Drying Conference*, August 13–15, Hong Kong, China, pp. 461–467.

Alves-Filho, O., T.M. Eikevik. 2008. Heat pump drying kinetics of Spanish cheese. *International Journal of Food Engineering* 4(6): Article 6. http://www.degruyter.com/view/j/ijfe.2008.4.6/ijfe.2008.4.6.1481/ijfe.2008.4.6.1481.xml (accessed July 24, 2015).

Alves-Filho, O., I. Strommen. 1996. Performance and improvements in heat pump dryers. In *Drying'96—Proceedings of the 10th International Drying Symposium*, eds. Strumillo, C. and Z. Pakowski, pp. 405–416. Technical University, Lodz, Poland.

Anon. 1988. Drying of vegetables. In *Industrial Drying by Electricity*, ed. Working Group 'Heat Recovery', pp. 52–53, 58–59. International Union for Electroheat, Paris, France.

Antal, T., A. Figiel, B. Kerekes et al. 2011. Effect of drying methods on the quality of the essential oil of spearmint leaves (*Mentha spicata* L.). *Drying Technology* 29(15): 1836–1844.

Bonazzi, C., E. Dumoulin. 2011. Quality changes in food materials as influenced by drying processes. In *Modern Drying Technology, Vol. 3: Product Quality and Formulation*, 1st edn., eds. Tsotsas, E. and A.S. Mujumdar, pp. 1–20. Wiley-VCH Verlag GmbH & Co. KGaA, Weinheim, Germany. http://onlinelibrary.wiley.com/doi/10.1002/9783527631667.ch1/pdf (accessed June 01, 2015).

Bruce, D.M. 1985. Exposed-layer barley drying, three models fitted to new data up to 150 °C. *Journal of Agricultural Engineering Research* 32: 337–348.

Calm, J. M. 1997. Heat pumps in USA. *International Journal of Refrigeration* 10: 190–196.

Chakraverty, A. 1984. Thin-layer characteristics of cashew nuts and cashew kernels. In *Drying'84*, ed. A.S. Mujumdar, pp. 396–400. Hemisphere, Washington, DC.

Chong, C.H., C.L. Law, A. Figiel et al. 2013. Colour, phenolic content and antioxidant capacity of some fruits dehydrated by a combination of different methods. *Food Chemistry* 141: 3889–3896.

Chua, K.J., S.K. Chou. 2014. Recent advances in hybrid drying technologies. In *Emerging Technologies for Food Processing*, ed. D.W. Sun, pp. 447–460. Academic Press, Salt Lake City, UT. http://www.science-direct.com/science/article/pii/B9780124114791000243 (accessed June 01, 2015).

Chua, K.J, S.K. Chou, W.M. Yang. 2010. Advances in heat pump systems: A review. *Applied Energy* 87: 3611–3624.

Claussen, I.C., E. Indergård, Magnussen et al. 2012. Factors, influencing the drying process of salted fish (cod-Gadus microcephalus and Saithe–Pollachius virens) Part B: Water removal rate from salted fish based on weight and drying air conditions. file:///C:/Users/Prof%20Conrad/Downloads/Claussen,%20Inderg%C3%A5rd,%20Magnussen,%20Gullsv%C3%A5g.pdf (accessed June 1, 2015).

Colak, N., A. Hepbasli. 2009a. A review of heat-pump drying (HPD): Part 1—Systems, models and studies. *Energy Conversion and Management* 50(9): 2180–2186.

Colak, N., A. Hepbasli. 2009b. A review of heat-pump drying (HPD): Part 2—Applications and performance assessments. *Energy Conversion and Management* 50(9): 2187–2199.

Coumans, W.J., P.J.A.M. Kerkhof, S. Bruin. 1994. Theoretical and practical aspects of aroma retention in spray-drying and freeze drying. *Drying Technology* 12(1 & 2): 99–149.

Crank, J. 1975. *The Mathematics of Diffusion*. Pergamon Press, Oxford, UK.

Cronin, K., S. Kearney. 1998. Monte Carlo modelling of a vegetable tray dryer. *Journal of Food Engineering* 35: 233–250.

Darvishi, H., A.R. Asl, A. Asghari et al. 2014. Study of the drying kinetics of pepper. *Journal of Saudi Society of Agricultural Sciences* 13(2): 130–138.

Energy Management Series 11, Canada. Refrigeration and heat pumps. http://www.nrcan.gc.ca/sites/oee.nrcan.gc.ca/files/pdf/commercial/password/downloads/EMS_11_refrigeration_and_heat_pump.pdf (accessed June 1, 2015).

Fatouh, M., M.N. Metwally, A.B. Helali et al. 2006. Herbs drying using a heat pump dryer. *Energy Conversion and Management* 47: 2629–2643.

Flink, J.M., M. Karel. 1970. Retention of organic volatiles in freeze-dried solution of carbohydrates. *Journal of Agricultural and Food Chemistry* 18: 259–297.

Flores, M.R.F. 2012. New and energy efficient drying for protein mixtures. MSc thesis. http://www.diva-portal.org/smash/get/diva2:722311/FULLTEXT01.pdf (accessed June 1, 2015).

Gill, M.I., J.R. Gorny, A.A. Kader. 1998. Responses of Fuji apple slices to ascorbic acid treatment and low oxygen atmospheres. *HortScience* 33(2): 305–309.

Gorozabel Chata, F.B., S.K. Chaturvedi, A. Almogbel. 2005. Analysis of a direct expansion solar-assisted heat pump using different refrigerants. *Energy Conversion and Management* 46:2614–2624.

Hawlader, M.N.A., C.O. Perera, M. Tian. 2006a. Properties of modified atmosphere heat pump dried foods. *Journal of Food Engineering* 74: 392–401.

Hawlader, M.N.A., C.O. Perera, M. Tian. 2006b. Comparison of the retention of 6-gingerol in drying of ginger under modified atmosphere heat pump drying and other drying methods. *Drying Technology* 24: 51–56.

Hawlader, M.N.A., C.O. Perera, M. Tian., K.L. Yeo. 2006c. Drying of guava and papaya: Impact of different drying methods. *Drying Technology* 24: 77–87.

Hawlader, M.N.A., M.S. Uddin, A.B. Ho et al. 1991. Drying characteristics of tomatoes. *Journal of Food Engineering* 14: 259–268.

Henderson, S.M., S. Pabis, 1961. Grain drying theory. I. Temperature effect on drying coefficient. *Journal of Agricultural Engineering Research* 6(3): 169–174.

Hesse, B. 1994. Energy efficient electric drying systems for industry. In *Drying'94—Proceedings of the 9th International Drying Symposium*, eds. Rudolph, V., R.B. Keey, and A.S. Mujumdar, pp. 591–598. University of Queensland, Brisbane, Queensland, Australia.

Hossain, M.A., B.K. Bala. 2002. Thin-layer drying characteristics for green chilli. *Drying Technology* 20(2): 489–505.

Hossain, M.A., K. Gottschalkb, M.S. Hassanc. 2013. Mathematical model for a heat pump dryer for aromatic plants. *Procedia Engineering* 56: 510–520.

Jayas, D.S., S. Cenkowski, S. Pabis et al. 1991. Review of thin-layer drying and wetting equations. *Drying Technology* 9(3): 551–588.

Jia, X., S. Clements, P. Jolly. 1993. Study of heat pump assisted microwave drying. *Drying Technology* 11(7): 1583–1616.

Jia, X., P. Jolly, S. Clements. 1990. Heat pump assisted continuous drying: Part 2. Simulation results. *International Journal of Energy Research* 14: 771–782.

Karatas, S., I. Pinarli. 2001. Determination of moisture diffusivity of pine nut seeds. *Drying Technology* 19(3–4): 701–708.

Karathanos, V.T. 1999. Determination of water content of dried fruits by drying kinetics. *Journal of Food Engineering* 39(4): 337–344.

Kivevele, T., Z. Huan. 2014. A review on opportunities for the development of heat pump drying systems in South Africa. *South African Journal of Science* 110(5/6): 2013-0236.

Kurozawa, L.E., L. Terng, M.D. Hubinger et al. 2014. Ascorbic acid degradation of papaya during drying: Effect of process conditions and glass transition phenomenon. *Journal of Food Engineering* 123: 157–164.

Laurila, E., R. Kervinen, R. Ahvenainen. 1998. Inhibition of browning in minimally processed vegetables and fruits. *Postharvest News and Information*, 9(4), 53N–66N. http://ucanr.edu/datastoreFiles/608-361.pdf (accessed May 19, 2015).

Lawton, J. 1978. Drying: The role of heat pumps and electromagnetic fields. *Physics in Technology* 9: 214–220.

Levi, G., M. Karel. 1995. The effect of phase transitions on release of *n*-propanol entrapped in carbohydrate glasses. *Journal of Food Engineering* 24(1): 1–13.

Lewis, W.K. 1921. The rate of drying of solids materials. *The Journal of Industrial and Engineering Chemistry* 13(5): 427–432.

Li, J.G., M.Y. Othman, S. Mat et al. 2011. Review of heat pump systems for drying application. *Renewable and Sustainable Energy Reviews* 15: 4788–4796.

Liu, Y.-H., S. Miao, J.-Y. Wu et al. 2014. Drying and quality characteristics of *Flos lonicerae* in modified atmosphere with heat pump system. *Journal of Food Process Engineering* 37: 37–45.

Madamba, P.S., R.H. Driscoll, K.A. Buckle. 1996. The thin-layer drying characteristics of garlic slices. *Journal of Food Engineering* 29: 75–97.

Martinez, M.V., J.R. Whitaker. 1995. The biochemistry and control of enzymatic browning. *Trends in Food Science & Technology* 6: 195–200.

Midilli, A., H. Kucuk, Z. Yapar. 2002. A new model for single layer drying. *Drying Technology* 20(7): 1503–1513.

Minea, V. 2010. Improvements of high-temperature drying heat pumps. *International Refrigeration and Refrigeration Conference*. School of Mechanical Engineering, Purdue University, West Lafayette, IN. http://docs.lib.purdue.edu/cgi/viewcontent.cgi?article=2102&context=iracc (accessed June 1, 2015).

Minea, V. 2013. Heat-pump-assisted drying: Recent technological advances and R&D needs. *Drying Technology* 31(10): 1177–1189.

Mohsenimanesh, A., H.R. Gazor. 2010. Modelling the drying kinetics of canola in fluidised bed dryer. *Czech Journal of Food Science* 28(6): 531–537.

Namsanguan, Y., W. Tia, S. Devahastin et al. 2004. Drying kinetics and quality of shrimp undergoing different two-stage drying processes. *Drying Technology* 22(4): 759–778.

Nicoleti, J.F., J. Telis-Romero, V.R.N. Telis. 2001. Air-drying of fresh and osmotically pre-treated pineapple slices: Fixed air temperature versus fixed slice temperature drying kinetics. *Drying Technology* 19(9): 2175–2191.

NRDC. 2014. Energy savings and industrial competitiveness Act of 2014. http://www.nrdc.org/energy/files/energy-savings-industrial-competitiveness-act-IB.pdf (accessed June 1, 2015).

O'Callaghan, D.G., D.J. Menzies, P.H. Bailey. 1971. Digital simulation of agricultural dryer performance. *Journal of Agricultural Engineering Research* 16: 223–244.

Oliver, T.N. 1982. Process drying with dehumidifying heat pump. In *International Symposium on the Industrial Application of Heat Pumps*, pp. 73–88. BHRA Fluid Engineering, Cranfield, Bedford, UK.

O'Neill, M., M.S. Rahman, C.O. Perera, L.D. Melton, B. Smith. 1998. Colour and density of apple cubes dried in air and in modified atmospheres. *International Journal of Food Properties* 1(3): 197–205.

Page, C. 1949. Factors influencing the maximum rate of drying shelled corn in layers. MS thesis, Purdue University, West Lafayette, IN.

Panchariya, P.C., D. Popovic, A.L. Sharma. 2002. Thin-layer modeling of black tea drying process. *Journal of Food Engineering* 52: 349–357.

Parti, M. 1990. A theoretical model for thin-layer grain drying. *Drying Technology* 8(1): 101–122.

Perera, C.O. 2001. Modified atmosphere heat pump drying of food products. In *Proceedings of the Second Asia-Oceania Drying Conference*, pp. 469–476. Penang, Malaysia.

Perera, C.O. 2005. Selected quality aspects of dried foods. *Drying Technology* 23: 717–730.

Perera, C.O., E. Owen. 2008. Effect of tissue disruption by different methods followed by incubation with hydrolysing enzymes on the production of vanillin in Tongan vanilla beans. *Food Bioprocess Technology* 3(1): 49–54.

Perera, C.O., M.S. Rahman. 1997. Heat pump dehumidifier drying of food. *Trends in Food Science & Technology* 8(3): 75–79.

Pizzocaro, F., D. Torreggiani, G. Gilardi. 1993. Inhibition of apple polyphenoloxidase (PPO) by ascorbic acid, citric acid and sodium chloride. *Journal of Food Processing and Preservation* 17: 21–30.

Prasertsan, S., P. Saen-Saby, G. Prateepchaikul et al. 1996. Effects of drying rate and ambient conditions on the operating modes of heat pump dryer. In *Drying'96—Proceedings of the 10th International Drying Symposium*, eds. Strumillo, C. and Z. Pakowski, pp. 529–534. Lodz Technical University, Lodz, Poland.

Rahman, M.S., C.O. Perera, C. Thebaud. 1997. Desorption isotherm and heat pump drying kinetics of peas. *Food Research International* 30(7): 485–491.

Rahman, S.M.A., M.R. Islam, A.S. Mujumdar. 2007. A study of coupled heat and mass transfer in composite food products during convective drying. *Drying Technology* 25(7–8): 1359–1368.

Rakhesh, B., G. Venkatarathnam, S.S. Murthy. 2003. Performance comparison of HFC227 and CFC114 in compression heat pumps. *Applied Thermal Engineering* 23: 1559–1566.

Ramesh, M.N., W. Wolf, D. Tevini, G. Jung. 1999. Studies on inert gas processing of vegetables. *Journal of Food Engineering* 40: 199–205.

Ramesh, M.N., W. Wolf, D. Tevini, G. Jung. 2001. Influence of processing parameters on the drying of the spice paprika. *Journal of Food Engineering* 49: 63–72.

Ré, M.I. 1998. Microencapsulation by spray drying. *Drying Technology* 16(6): 1195–1236.

Reyes, A., P.I. Alvarez, F.H. Marquardt. 2002. Drying of carrots in a fluidized bed: Effects of drying conditions and modelling. *Drying Technology* 20(7): 1463–1483.

Sapers, G.M., R.L. Miller. 1992. Enzymatic browning control in potatoes with ascorbic acid-2-phosphate. *Journal of Food Science* 57: 1132–1135.

Saravacos, G.D., M. Krodika. 2014. Mass transfer properties of food. In *Engineering Properties of Food*, 4th edn., eds. Rao, M.A., S.S.H. Rizvi, A. Datta, and J. Ahmed, pp. 311–358. CRC Press, Boca Raton, FL.

Senoussi, A., E.D. Dumoulin, Z. Berk. 1995. Retention of diacetyl in milk during spray drying and storage. *Journal of Food Science* 60(5): 894–905.

Shi, Q.L., C.H. Xue, Y. Zhao, Z.J. Li, X.Y. Wang. 2008. Drying characteristics of horse mackerel (*Trachurus japonicus*) dried in a heat pump dehumidifier. *Journal of Food Engineering* 84(1): 12–20.

Soares, D.S.C., D.G. Costa, A.K.S. Abud et al. 2014. The use of performance indicators for evaluating models of drying jackfruit (*Artocarpus heterophyllus* L.): Page, Midilli, and Lewis. *International Journal of Biological, Food, Veterinary and Agricultural Engineering* 8(2): 132–135.

Sokhansanj, S., D.S. Jayas. 1987. Drying of foodstuffs. In *Handbook of Industrial Drying*, Vol. 2. ed. Mujumdar, A.S., pp. 517–554. Marcel Dekker, New York.

Sosle, V.R., G.S.V. Raghavan, R. Kittler. 2000. Low temperature drying process for heat sensitive agro-food products. In *ASAE Annual International Meeting*, pp. 1–9. Milwaukee, WI.

Sosle, V.R., G.S.V. Raghavan, R. Kittler. 2003. Low-temperature drying using a versatile heat pump dehumidifier. *Drying Technology* 21(3): 539–554.

Strommen, I., O. Jonassen. 1996. Performance tests of a new 2-stage counter-current heat pump fluidized bed dryer. In *Drying'96—Proceedings of the 70th International Drying Symposium*, eds. Strumilo, C. and Z. Pakowski, pp. 563–568. Lodz Technical University, Lodz, Poland.

Vega, A., P. Fito, A. Andres, R. Lemus. 2007. Mathematical modeling of hot-air drying kinetics of red bell pepper (var. Lamuyo). *Journal of Food Engineering* 79: 1460–1466.

Wang, C.Y., R.P. Singh. 1978. A single layer drying equation for rough rice. ASAE Paper No. 78-3001, St. Joseph, MI.

Wang, Z., J. Sun, F. Chen, X. Liao, X. Hu. 2007. Mathematical modelling on thin-layer microwave drying of apple pomace with and without hot air pre-drying. *Journal of Food Engineering* 80: 536–544.

Watson, E.L., V.K. Bhargava. 1974. Thin-layer drying studies on wheat. *Canadian Agricultural Engineering* 16(1): 18–22.

Whitaker, J.R., C.Y. Lee. 1995. Recent advances in the chemistry of enzymatic browning. In *Enzymatic Browning and Its Prevention*, eds. Lee, C.Y. and J.R. Whitaker, pp. 2–7. American Chemical Society Symposium Series 600, Washington, DC.

Zhang, M., J. Tang, A.S. Mujumdar, S. Wang. 2006. Trends in microwave related drying of fruits and vegetables. *Trends in Food Science & Technology* 17: 524–534.

4 Advances in Heat Pump-Assisted Drying of Fruits

Sze Pheng Ong

CONTENTS

4.1 INTRODUCTION

Fruits are important sources of simple sugars (e.g., fructose, glucose, and sucrose), minerals (e.g., calcium, copper, iron, magnesium, manganese, phosphorus, and potassium), water-soluble vitamins (e.g., betaine, choline, folate, niacin, pantothenic acid, pyridoxine, riboflavin, thiamine, and vitamin C), fat-soluble vitamins (e.g., vitamin A, E, and K), as well as dietary fiber (Lim 2012; Shahidi and Tan 2012; Siddiq et al. 2012). Moreover, some fruits are rich in functional phytochemicals that possess high pharmacological values and medicinal effects. These bioactive compounds are vital and essential to maintain good health particularly in preventing cardiovascular diseases, inactivating carcinogens, inhibiting oxidation of low-density lipoprotein (LDL), reducing blood pressure, and inhibiting type-II diabetes (Rice-Evans et al. 1997; Evans 2002; Leong and Shui 2002; Shui and Leong 2005; Shetty et al. 2006; Skinner and Hunter 2013).

However, dried fruits that serve as the concentrated form of their fresh counterparts contain much lower moisture content and have longer shelf life. Large proportion of their water is removed (greater than 80% wet basis), either by thermal drying or by osmotic dehydration, in order to prevent postharvest deterioration, to minimize packaging requirements, to reduce freight weight, and most importantly, to enhance the storage stability (Stefan et al. 2003; Baker 2005; Raghavan et al. 2005; Islam and Mujumdar 2008). Furthermore, dried fruits provide higher total energy, fiber content, dietary antioxidants, and nutrients per serving than the corresponding fresh fruits as a consequence of the removal of water from the fruit matrix (Bennett et al. 2011; Shahidi and Tan 2012). Some studies have revealed that dried fruits have a superior quality of antioxidant properties due to the concentration of polyphenol and also due to the formation of its derivatives in monomer forms attributed to the breakdown or hydrolysis of the complex polymeric molecules during drying (Al-Farsi et al. 2005; Chang et al. 2006; Shahidi and Tan 2012; Lutz et al. 2015). Also, a rise in the consumption of dried fruits has been observed over the past decade.

Global statistical review reports indicate that the global production of dried fruits in 2013–2014 had increased by about 10% as compared to 2009, with a total production recorded at 9.2 million of metric tonnes (MT). In the year 2014, dates were the highest produced dried fruit on a global scale with a total production of 7.5 million MT, followed by grapes (1.3 million MT), prunes (230 thousand MT), figs (123.7 thousand MT), and apricots (68.8 thousand MT) (INC 2015).

Fruits and their products that are sensitive to heat must be carefully dried in order to retain their best qualities, especially in terms of color, flavor, texture, aroma, and nutrition contents (Mujumdar 2006; Chen 2008). Convective drying is a simple yet common method in the preservation of fruits. Moisture removal from fruits may deactivate enzymes or microorganisms that often cause undesired biochemical reactions which subsequently lead to quality deterioration (Jayaraman and Das Gupta 2006). However, application of a relatively high temperature during convective drying often degrades the heat labile substances in foodstuffs and also impairs their physical as well as sensory properties (Zhang et al. 2002; Sablani 2006; Xu et al. 2006). Hence, drying at a low temperature, for instance temperature below the water triple point, is preferred in producing dried fruits with premium quality. Freeze drying could be the best option in this case, but it is not economical especially when the final products fetch low margins and the time required for drying is quite long (Krokida and Philippopoulos 2005; Marques et al. 2006).

Heat pump-assisted dryers come into play as they are well recognized as a cost-competitive drying method, and they are immensely helpful in drying such heat-sensitive materials at a comparatively low temperature and humidity. Almost any type of convective dryer can be fitted with a heat pump system. Generally, a heat pump-assisted dryer is versatile and can be operated in different configurations. A number of operating parameters can be adjusted by manipulating settings of a few components in the basic heat pump-assisted dryer (Ong and Law 2009). Basic drying principles and fundamentals of a heat pump-assisted dryer have been reviewed extensively elsewhere (Perera and Rahman 1997; Islam and Mujumdar 2008; Colak and Hepbasli 2009; Fadhel et al. 2011; Jangam and Mujumdar 2011; Patel and Kar 2012; Minea 2013a,b, 2014). Interested readers can refer to the related literature. This chapter will review and discuss the applications of heat pump-assisted dryers exclusively in fruit drying. First, the types and categories of dried fruit will be reviewed and analyzed. Then, configurations and drying modes used in heat pump-assisted dryers will be compiled and examined based on the aspects of engineering design and energy performance of the drying unit as well as the quality of the dried fruit products.

4.2 CLASSIFICATION AND CHARACTERISTICS OF DRIED FRUITS

Before discussing further about fruit drying, it is essential to know the fundamental classification and characteristics of fruits. From a biological and morphological point of view, a fruit refers to the mature ovary of a plant that contains its seeds. Generally, the edible portion of most fruits is the fleshy part that surrounds the seeds. There are three different layers of tissue in the ovary: the exocarp, mesocarp, and endocarp. These layers may develop into distinct parts of the fruit, whereby the exocarp (outermost) becomes the peel or skin, the mesocarp (middle) becomes the flesh, and the endocarp (innermost) becomes the deepest part of the flesh or a specialized tissue surrounding the seeds. Nevertheless, not all fruits have such clearly demarcated tissues, and the three layers are just indistinguishable. Then, the term *pericarp* is used to denote all the ovarian tissues surrounding the seeds. Fruits can be classified based on botanical structure, climatic requirements, chemical composition, floral morphology, ovary position, and so on. Figure 4.1 illustrates the classification of fruit types based on floral morphology. The fruits are categorized into three major groups, namely the simple fruits, aggregate fruits, and multiple fruits, depending on the number of ovaries and the number of flowers involved in their formation. Detailed descriptions about the fruit types and categories can be found elsewhere (Harris and Harris 2001; Bendre and Kumar 2009). Besides, fruits can be categorized based on their sugar and acid content as well. The characteristic

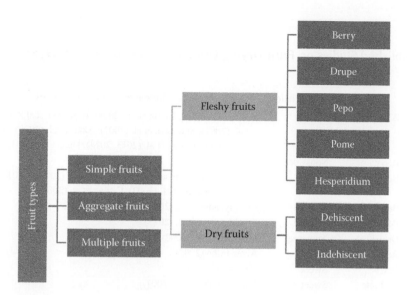

FIGURE 4.1 Fruit types.

flavor of a fruit is often associated closely with its sugar and acid content ratio. Therefore, the fruit will taste sweet when the sugar and acid ratio is high (high sugar but low acid content), whereas it will taste sour at the other end of the scale (high acid but low sugar content). There are four major categories, covering sweet, nonsweet, acid, and subacid fruits.

Considering the abovementioned fruit classification, Table 4.1 summarizes the studies on heat pump-assisted fruit drying that have been performed in the past two decades according to the fruit types and characteristics. Overall, there were about 26 different fruits being used as the fruit models in heat pump-assisted drying experiments. Figure 4.2 shows the fruit models studied by the relevant researchers using heat pump-assisted dryers. Analysis of the statistics obtained from the published studies and literature reveals that apple was the most popular fruit model, followed by banana, guava, and papaya. Meanwhile, some fruits like salak fruit, mango, coconut, pear, and plum were moderately studied in the heat pump-assisted drying. To the best of my knowledge and literature review, little information is available about the other dried fruits such as avocado, blueberry, chokeberry, cranberry, dragon fruit, fig, grapes, honeydew, kiwi, longan, morello, nectarines, pineapple, raspberry, sapota, and strawberry. Figure 4.3a elucidates the corresponding fruit types of the fruit models. Further examination of the statistics discloses that a majority of the fruit models are from the multiple fruits category (25%), followed by pome (23%), berry (21%), drupe (16%), pepo (12%), and aggregate fruits (3%). No open access literature is available on hesperidium category (e.g., orange, lemon, and grapefruit) thus far. Meanwhile, Figure 4.3b shows the characteristics of the fruit models. It is observed that a majority of the fruit models are from subacid category (58%), followed by sweet (30%), acid (7%), and nonsweet (5%). This suggests that heat pump-assisted dryer can be a good technique in handling fruits with moderate to high sugar content.

There are some fruits that are generally treated as vegetables, for instance the pumpkin, tomatoes, peas, eggplant, cucumber, and beans. There are also fruits (particularly those under the dry fruits category) that are better known as grains, nuts, or seeds, for example the almonds, macadamias, walnuts, rice, wheat, and corns. Therefore, these culinary fruits, grains, nuts, and seeds are not included and discussed in this chapter. Interested readers may refer to the related literature elsewhere (Queiroz et al. 2004; Pal et al. 2008; Borompichaichartkul et al. 2009; Gaware et al. 2010; Zielinska et al. 2013).

TABLE 4.1

Selected Published Studies on Fruit Drying Using Heat Pump-Assisted Dryer

Fruit			
Name	Type	Characteristic	Researchers and References
Apple	Pome	Subacid	Sosle et al. (2003); Uddin et al. (2004); Hawlader et al. (2006a); Rahman et al. (2007); Stawczyk et al. (2007); Aktas et al. (2009); Brushlyanova et al. (2013); Chong et al. (2013; 2014); Duan et al. (2013)
Avocado	Berry	Nonsweet	Ceylan et al. (2007)
Banana	Berry	Sweet	Prasertsan and Saen-saby (1998); Chua et al. (2000b, 2001; 2002); Dandamrongrak et al. (2003); Ceylan et al. (2007)
Blueberry	Berry	Subacid	Brushlyanova et al. (2013)
Chokeberry	Pome	Subacid	Brushlyanova et al. (2013)
Coconut	Drupe	Nonsweet	Mohanraj et al. (2008); Mohanraj (2014)
Cranberry	Berry	Acid	Alves-Filho (2002)
Dragon fruit	Berry	Sweet	Yong et al. (2006)
Fig	Multiple	Sweet	Xanthopoulos et al. (2007)
Grapes	Berry	Subacid	Vázquez et al. (1997)
Guava	Multiple	Subacid	Chua et al. (2000a,b; 2002); Uddin et al. (2004); Hawlader et al. (2006a,b)
Honeydew	Pepo	Sweet	Uddin et al. (2004)
Kiwi	Multiple	Acid	Ceylan et al. (2007)
Longan	Multiple	Sweet	Nathakaranakule et al. (2010)
Mango	Drupe	Subacid	Teeboonma et al. (2003); Uddin et al. (2004); Chong et al. (2013)
Morello	Drupe	Subacid	Brushlyanova et al. (2013)
Nectarines	Drupe	Subacid	Sunthonvit et al. (2007)
Papaya	Pepo	Sweet	Soponronnarit et al. (1998); Achariyaviriya et al. (2000); Teeboonma et al. (2003); Uddin et al. (2004); Hawlader et al. (2006b); Chong et al. (2013)
Pear	Pome	Subacid	Uddin et al. (2004); Chong et al. (2013)
Pineapple	Multiple	Acid	Tan et al. (2001)
Plums	Drupe	Subacid	Hepbasli et al. (2010); Erbay and Hepbasli (2013)
Raspberry	Aggregate	Subacid	Brushlyanova et al. (2013)
Salak fruit	Multiple	Subacid	Ong and Law (2011a,b; 2012a,b)
Sapota	Berry	Sweet	Jangam et al. (2008)
Strawberry	Aggregate	Acid	Şevik (2014)

4.3 CONFIGURATIONS AND DRYING MODES OF HEAT PUMP-ASSISTED DRYER

In fruit drying, it is essential to have a good control on the temperature, humidity, air velocity, and sometimes the composition (e.g., oxygen level) of the drying medium in order to guarantee good quality of the dried fruit products. A heat pump-assisted dryer is a combination of a heat pump system and a drying system. Conventional convective dryers such as a hot air oven, fluidized bed dryer, spray dryer, superheated steam dryer, and others that are commonly used in fruit drying can be coupled with a heat pump system to enjoy the benefits of drying at a higher energy efficiency as well as better product quality. A basic heat pump system consists of an evaporator, a condenser, an expansion valve, and a compressor. Proper arrangement and integration of these basic components in a forced air ventilated drying chamber will allow the heat pump-assisted dryer to operate in a similar way to a refrigeration unit, where the evaporator in the refrigerant loop acts as a heat sink (to dehumidify the drying air), while the condenser serves as a heat source (to elevate the temperature of the drying air) (Perera and Rahman 1997; Chou and Chua 2007; Islam and Mujumdar 2008;

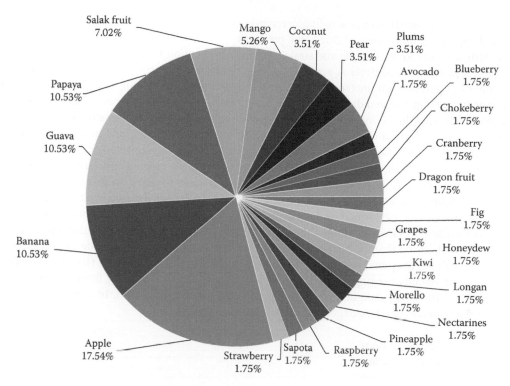

FIGURE 4.2 Fruit models used in heat pump-assisted drying.

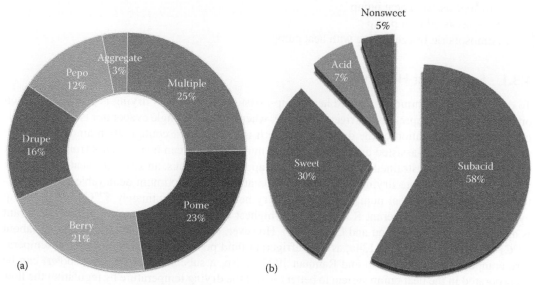

FIGURE 4.3 Analysis on fruit models: (a) fruit types and (b) fruit characteristics.

Jangam and Mujumdar 2011). During the lengthy drying, especially in the initial stage of many fruit drying processes, a useful amount of sensible and latent heat condensation from the exhausted moisture-laden drying air (from the dryer unit) is absorbed by the refrigerant fluid and recovered later in the condenser to heat up the drying air. As the dehumidified drying air regains the necessary drying potential at the evaporator, drying at a relatively low temperature becomes possible without compromising the drying kinetics and subsequently the drying time. A low temperature and short drying time are essential to ensure optimum retention of important heat sensitive bioactive ingredients and appealing appearance of dried fruit products.

Nonetheless, unlike the simple and straightforward application of heat pump in a refrigeration unit, a heat pump-assisted dryer is versatile and can be operated in many different configurations. A number of operating parameters can be adjusted by manipulating settings of a few components in a basic heat pump-assisted dryer such as the evaporator, condenser, compressor, airflow passageway, and refrigerant fluid (Ong and Law 2009). In addition, various external heating sources such as infrared, microwave, solar energy, and the like can be incorporated into a heat pump-assisted dryer to further improve its efficiency. A good heat pump and dryer integration scheme, by considering thermodynamics of both the drying air and refrigerant liquid, is crucial in determining the optimum operating and energy performance of the heat pump-assisted dryer. In this section, different possible configurations and settings of heat pump-assisted dryers are reviewed and compiled based on studies specifically performed on fruit drying in the recent years. Table 4.2 shows the selected works as well as the relevant configurations and drying conditions of the heat pump-assisted dryers. It can be observed that the heat pump-assisted dryers can operate over a wide range of drying conditions with the temperature ranging from −16°C to 80°C, relative humidity (RH) ranging from 3% to 85%, air velocity ranging from 0.03 to 5.4 m/s, and drying time ranging from 16 min to 80 h, depending on the dryer configuration, drying mode, and the dried material. In a nutshell, the heat pump-assisted dryers can be categorized in the following groups:

1. Single-stage heat pump
   a. Open-loop system
   b. Closed-loop system
   c. Partially closed system and miscellaneous configurations
2. Multistage heat pump
3. Infrared-assisted heat pump
4. Solar-assisted heat pump
5. Atmospheric freeze drying with heat pump

4.3.1 SINGLE-STAGE HEAT PUMP

To date, the most commonly applied heat pump-assisted dryer in fruit drying is principally based on a single-stage compression refrigeration cycle, whereby only a single evaporator is used for cooling and dehumidification of the drying medium. It is the simplest configuration among the other modes of heat pump-assisted dryers, but the maximum heat that can be recovered from the drying medium is rather constrained by the size of the evaporator, and thus, an auxiliary heater bank may be required to facilitate drying at a higher temperature. The maximum achievable air temperature in a single-stage heat pump without auxiliary heating is approximately 57°C for heat pump cycle charged with refrigerant R22, where the highest condensing temperature of the refrigerant is approximately 60°C (Chou and Chua 2007). However, operation at a higher temperature (about 75°C) is possible by using R-134a, as this refrigerant fluid possesses a higher condensing temperature compared to R22 (Perera and Rahman 1997). Also, a subcondenser (or a subcooler) can be incorporated in the heat pump system to better control the drying temperature by regulating the flow of refrigerant vapor to the main condenser and subcondenser. A fraction of the refrigerant vapor can be diverted to the subcondenser so that the excess heat can be rejected through the external air-cooled or water-cooled condensers. On the other hand, humidity of the drying air can be adjusted by regulating the airflow ratio passing through the bypass damper and the evaporator. Dehumidified air obtained at the downstream of the evaporator can be mixed with the bypassed air to maintain the desired humidity. Besides, the components of the heat pump system can be configured and arranged in different positions along the drying air passageway to create either an open-loop or closed-loop drying process.

TABLE 4.2
Selected Literatures on Heat Pump-Assisted Drying of Fruits and the Relevant Drying Conditions

Fruits and References	Heat Pump Configurations and Drying Conditions
Grapes Vázquez et al. (1997)	Single-stage heat pump with external condenser, closed-loop system Convective drying with dehumidified air 50°C–60°C 3 m/s ~24 h
Banana Prasertsan and Saen-saby (1998)	Single-stage heat pump, open and partially closed system Convective drying with dehumidified air 50°C–60°C Absolute humidity = 0.017–0.025 kg vapor/kg dry air 46–67.5 h
Glace of papaya Soponronnarit et al. (1998); Achariyaviriya et al. (2000)	Single-stage heat pump with external condenser, closed-loop system Convective drying with dehumidified air 50°C Air mass flow rate = 0.45 kg/s ~80 h
Banana and guava Chua et al. (2000a,b, 2001, 2002)	Multistage heat pump, partially closed system Convective drying with dehumidified air Time-varying drying conditions 20°C–40°C ~20%–100% RH 0.5–3.0 m/s
Pineapple Tan et al. (2001)	Infrared-assisted heat pump, partially closed system Convective drying with dehumidified air and intermittent irradiation with infrared 40°C 21% RH 2.0 m/s ~350–400 min
Cranberry Alves-Filho (2002)	Single-stage heat pump with external condenser, fluidized bed dryer, closed-loop system Atmospheric freeze-drying and followed by convective drying with dehumidified air −10°C–30°C 15%–85% RH 0.5–1.5 m/s ~140 min
Glace of papaya and mango Teeboonma et al. (2003)	Single-stage heat pump with subcondenser, partially closed system Convective drying with dehumidified air 55°C Air mass flow rate = 1400–1950 kg/h 40–48 h
Apple Sosle et al. (2003)	Single-stage heat pump with subcondenser, open and closed-loop system Convective drying with dehumidified air 17°C–27°C ~21 h
Banana Dandamrongrak et al. (2003)	Convective drying with dehumidified air 50°C 3.1 m/s 22.4–44.7 h

(Continued)

TABLE 4.2 (*Continued*)

Selected Literatures on Heat Pump-Assisted Drying of Fruits and the Relevant Drying Conditions

Fruits and References	Heat Pump Configurations and Drying Conditions
Apple, guava, papaya, pear, mango, and honeydew Uddin et al. (2004)	Convective drying with dehumidified air 30°C–50°C 21%–43% RH 1.2–3.3 m/s
Apple, guava, and papaya Hawlader et al. (2006a,b)	Single-stage heat pump with subcondenser, closed-loop system Convective drying with dehumidified and modified air (N_2 or CO_2) 45°C 10% RH 0.7 m/s
Dragon fruit Yong et al. (2006)	Infrared-assisted, single-stage heat pump with subcondenser, partially closed system Convective drying with dehumidified air 44°C 18.5% RH 0.85 m/s
Fig Xanthopoulos et al. (2007)	Single-stage heat pump, partially closed system Convective drying with dehumidified air 46.1°C–49.6°C; 12.60%–17.65% RH (without heater) 50°C–60°C; 9%–10.65% RH (with auxiliary heater) 1–5 m/s 23–53.3 h
Apple and potato composite Rahman et al. (2007)	Single-stage heat pump, closed-loop system Convective drying with dehumidified air 48°C 18% RH 1.48 m/s ~300 min
Kiwi, avocado, and banana Ceylan et al. (2007)	Single-stage heat pump, open-loop system Convective drying with heated air 40°C ± 0.2°C 0.03–0.39 m/s ~360 min
Nectarines Sunthonvit et al. (2007)	Convective drying 25°C 10% RH 1.6 m/s 19 h
Apple Stawczyk et al. (2007)	Single-stage heat pump, closed-loop system Atmospheric freeze-drying −16°C to 22°C
Plums Chegini et al. (2007)	Partially closed system Convective drying with dehumidified air 70°C–80°C ~3% RH ~16 min
Sapota Jangam et al. (2008)	Single-stage heat pump with sub-condensers, open-loop system Convective drying with dehumidified air 40°C 15% RH 1 m/s ~11 h

(Continued)

TABLE 4.2 (*Continued*)

Selected Literatures on Heat Pump-Assisted Drying of Fruits and the Relevant Drying Conditions

Fruits and References	Heat Pump Configurations and Drying Conditions
Apple Aktas et al. (2009)	Single-stage heat pump, partially closed system Convective drying with heated air 39°C–41°C 2.5 m/s 3.5 h
Longan Nathakaranakule et al. (2010)	Infrared-assisted, single-stage heat pump, partially closed loop Convective drying with dehumidified air and Far infrared (FIR) 55°C 9.5–15.5 h
Salak fruit Ong and Law (2011a,b; 2012a,b)	Single-stage heat pump with subcondenser, closed-loop system Convective drying with dehumidified air 26°C; 27% RH (without heater) 37°C; 17% RH (with heater) 5.4 ± 0.5 m/s 10–51 h
Plums Hepbasli et al. (2010); Erbay and Hepbasli (2013)	Single-stage heat pump with subcondenser, closed-loop system Conveyor convective drying with dehumidified air 45°C–55°C 1.5 m/s
Pomaces of apple, blueberry, raspberry, chokeberry, and morello Brushlyanova et al. (2013)	Convective drying 45°C 1.5 m/s ~70 min to 250 min
Apple Duan et al. (2013)	Single-stage heat pump, closed-loop system Atmospheric freeze-drying −10°C–40°C ~30% RH 1.5 m/s 34–64 h
Apple, pear, mango, and papaya Chong et al. (2013, 2014)	Single-stage heat pump with subcondenser, closed-loop system Convective drying with dehumidified air and followed by vacuum-microwave drying 35°C 20% RH 4 m/s ~480 min
Strawberry Şevik (2014)	Solar-assisted, single-stage heat pump, open-loop Solar-assisted convective drying with heated air 50°C 0.29–0.64 m/s 160 min
Coconut Mohanraj et al. (2008); Mohanraj (2014)	Solar-assisted, single-stage heat pump, open-loop system Solar-assisted convective drying with heated air 41°C–48°C 1.5 m/s ~48 h

4.3.1.1 Open-Loop System

Ceylan et al. (2007) performed convective drying of banana, kiwi, and avocado with hot air (about 40°C) generated using a single-stage heat pump-assisted dryer. The design of the open-loop system was relatively simple and consists of an evaporator, a capillary tube, a condenser, a compressor, and an axial fan. Referring to the schematic diagram of the dryer (Figure 4.4), the evaporator was positioned at the exhaust point of the dryer with an aim to recover some heat from the discharged drying air before it was released to the ambient. Meanwhile, the condenser was placed at the inlet point of the dryer to heat the drying air (from the ambient) prior to entering the drying chamber. No auxiliary heater was installed, and the temperature of the drying air was mainly controlled by adjusting the speed of the axial fan using a proportional integral derivative (PID) controller. The airflow increased when the temperature was too high and vice versa. The humidity of the drying air was not reported. It is predicted to be similar to that observed in a conventional hot air dryer since such open-loop configuration does not permit the use of the dehumidified air that generates at the evaporator. It took about six hours to complete the drying of fruit slices of 5 mm thickness. The initial moisture contents of banana, kiwi, and avocado were reported at 4.71, 4.31, and 1.51 g water/g dry matter, whereas the final moisture contents were 0.5, 0.75, and 0.35 g water/g dry matter, respectively.

Jangam et al. (2008) investigated the drying of sapota fruit pulp using an open-loop single-stage heat pump-assisted dryer with a temperature of 40°C, RH of 15%, and air velocity at 1 m/s. It took about 8–16 h to dry the pretreated sapota pulp (4, 6, 8, and 10 mm on an aluminium tray) from an initial moisture content of 68%–73% w/w (wet basis) to the required final moisture content of 6%–7% w/w (dry basis) and water activity below 0.45. The drying rate in the heat pump-assisted drying was found higher as compared to the corresponding hot air drying due to the increased mass transfer rate with the use of dehumidified air as drying medium. Moreover, the heat pump-assisted dried sapota powder was more stable as compared to the freeze dried sapota powder due to its averagely higher rehydration ratio. In the study, the evaporator was used to dehumidify the ambient air first before heating it with condensers (see Figure 4.5). The heat pump system was charged with refrigerant R-134a and with a total heat load of 1.5 kWh. The design of the dryer was rather complex with a series of condensers (internal condenser, desuperheater, and external condenser) and heat

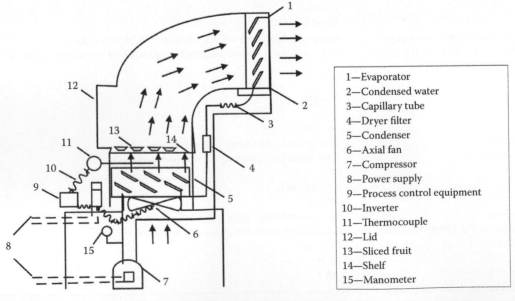

1—Evaporator
2—Condensed water
3—Capillary tube
4—Dryer filter
5—Condenser
6—Axial fan
7—Compressor
8—Power supply
9—Process control equipment
10—Inverter
11—Thermocouple
12—Lid
13—Sliced fruit
14—Shelf
15—Manometer

FIGURE 4.4 Schematic diagram of heat pump-assisted dryer used by Ceylan et al. (Redrawn and reproduced from *App. Therm. Eng.*, 27, Ceylan, I. et al., Mathematical modeling of drying characteristics of tropical fruits, 1931–1936, Copyright 2007, with permission from Elsevier.)

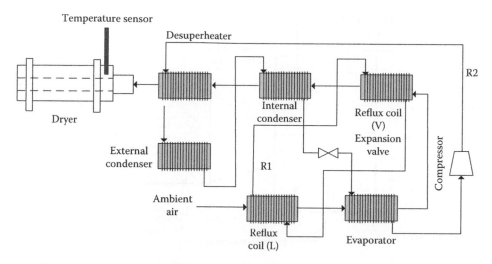

FIGURE 4.5 Heat pump-assisted dryer used by Jangam et al. in sapota fruit pulp drying. (Redrawn and reproduced with permission from Jangam, S.V. et al., *Dry. Technol.*, 26, 369–377, 2008.)

bypass coils that used to improve specific moisture extraction rate (SMER). However, the SMER value was not provided in the study. The energy performance of the dryer could not be evaluated and predicted especially when the energy in the hot and humid air at the dryer's exhaust was not recovered at the evaporator but was instead released to the ambient.

4.3.1.2 Closed-Loop System

Vázquez et al. (1997) reported that the drying of grapes at a high temperature (50°C–60°C) and low humidity (wet-bulb temperature of 30°C–36°C) in a closed-loop single-stage heat pump-assisted dryer could significantly reduce the drying time (about 24 h) as compared to conventional sun drying (40 days). The final moisture content was reported at about 15%–20% (wet basis), and the color as well as texture of dried products was qualitatively satisfactory. The pilot-scale heat pump-assisted dryer was equipped with a heat pump system (with an external condenser), an auxiliary heater, and a water atomizer that was used to humidify the drying air when necessary to maintain its RH. The heat pump system was charged with refrigerant HFC-134a and the velocity of drying air was reported to be 3 m/s.

Soponronnarit et al. (1998) conducted drying of papaya glace at a temperature of 50°C, air mass flow rate of 0.45 kg/s, and bypass recycled air of 63% using a closed-loop single-stage heat pump-assisted dryer. It took about 40 h to complete the first stage drying (sample dimension of 6.35 × 15 × 2.54 cm, initial moisture content of 74% on dry basis, and final moisture content of 38%) and another 40 h to further reduce the moisture content to 23% (cut to smaller sample cube of 0.98 cm³). Despite the long drying time, the color of the final product was reported as light reddish-orange (in code 34-C from R.H.S. color chart), which was lighter than that of a fruit dried in a hot air tunnel (0.45°C–75°C). The design of the dryer was not complex and the drying cabinet (0.90 × 0.80 × 0.75 cm) could accommodate 12 drying trays and about 70–132 kg papaya glace. The heat pump system composed of an evaporator (3.66 kW), an internal condenser (4.5 kW), an external condenser, a two-piston compressor (1.3 kW), and two cycles of capillary tube throttling valves (see Figure 4.6). A heater (3 kW) was fitted in the dryer as well. The type of refrigerant used in the heat pump cycle and RH of the drying air were not provided. However, the moisture extraction rate (MER) from the evaporator was estimated to be 0.78 kg water condensed/h, whereas the energy consumption was estimated at 9.93 MJ/kg water evaporated. The SMER and coefficient of performance for the heat pump (COP$_{hp}$) were reported at 0.63 kg water evaporated/kW and 3.71–3.85, respectively. Further a study by Achariyaviriya et al. (2000) on papaya glace drying revealed that ambient conditions

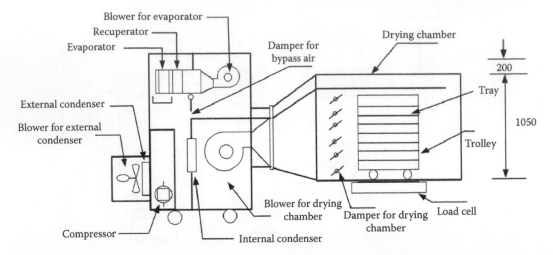

FIGURE 4.6 Heat pump-assisted dryer used by Soponronnarit et al. in papaya glace drying. (Redrawn and reproduced with permission from Soponronnarit, S. et al., *Int. Energy J.*, 20, 39–53, 1998.)

(28.9°C–29.3°C and 65.6%–77.5% RH) would significantly affect the performance of the dryer with either open-loop or partially closed configuration. Also, the fraction of evaporator bypass air, specific airflow rate, and drying air temperature were key factors that attributed to the performance of the heat pump-assisted dryer as well. The simulation results showed that the SMER would increase when the fraction of evaporator bypass air increased in the range of 0%–75%, but decrease when the bypass ratio was more than 80%.

Sosle et al. (2003) used a versatile heat pump-assisted dryer to perform drying of apple rings with 1 cm thickness at a temperature of 17°C–27°C (see Figure 4.7). The dryer was configured to closed-loop system during the drying by closing both dampers in the dryer. A commercial heat pump dehumidifier (2.3 kW) with an external water-cooled condenser and refrigerant R22 was incorporated in the dryer. The unit also consisted of a compressor and a centrifugal fan (1/8 hp). No further information is available on the evaporator capacity and the RH of drying air. However, it was observed that most of the heat exchanged at the evaporator had sensible heat and the temperature of the air entering the drying chamber kept increasing as the drying progressed after six hours of drying. Despite the drawbacks on high energy consumption due to the rejection of energy input at the secondary water-cooled condenser as excess heat, the closed-loop heat pump-assisted dryer could

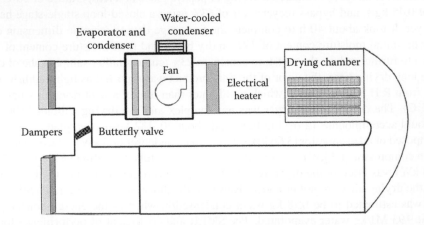

FIGURE 4.7 Schematic diagram of heat pump-assisted dryer used by Sosle et al. in drying of apple rings. (Redrawn and reproduced with permission from Sosle, V. et al., *Dry. Technol. Int. J.*, 21, 539–554, 2003.)

produce dried apples with lower water activity and better rehydration as compared to hot air drying (at 45°C and 65°C). Further examinations with nuclear magnetic resonance revealed that drying in mild air conditions would allow the water loss to occur mostly from the vacuole compartments in apple parenchyma cells and resulted in overall shrinkage of the cells. The apple tissue could retain the integrity of its cell walls much better in the heat pump-assisted drying as compared to the freeze dried product that is usually associated with cell wall damage and loss of turgor.

Hawlader et al. (2006a,b) had proposed a closed-loop airflow circuit in a heat pump-assisted dryer to enable drying of guava, papaya, and apple in inert atmosphere by replacing the air with inert gases such as nitrogen (N_2) and carbon dioxide (CO_2). All the drying experiments were conducted at 45°C, with RH of 10% RH and air velocity of 0.7 m/s. The single-stage heat pump consisted of an evaporator, an internal condenser, an external condenser, a compressor, and an expansion valve (see Figure 4.8). The details on the heat pump capacity and sizing were not available and hence the energy performance of the dryer could not be evaluated. Nonetheless, the studies showed that drying of guava and papaya in low concentration of oxygen (O_2) could yield more retention of vitamin C, less browning, and faster rehydration of the dried products. The quality of heat pump-assisted dried products was comparable with the freeze dried products. In addition, drying of apple cubes (about 1 cm^3) in the modified air resulted in less browning, more porous, and less firm dried products as compared to a normal heat pump-assisted dryer.

Rahman et al. (2007) studied the drying of apple and potato composite food products using a closed-loop single-stage heat pump-assisted dryer (see Figure 4.9). The drying of the composite slabs (20 × 10 × 20 mm and 20 × 10 × 10 mm) was performed at a temperature of 48°C, RH of 18%, air velocity of 1.48 m/s, and a total drying time of about 300 min. The initial moisture contents of apple and potato were reported at 6.4 kg/kg (dry basis) and 5.2 kg/kg (dry basis), respectively. However, the heat pump capacity and energy performance were not provided in this study. The main focus of the study was to investigate the variation of moisture content and temperature distribution

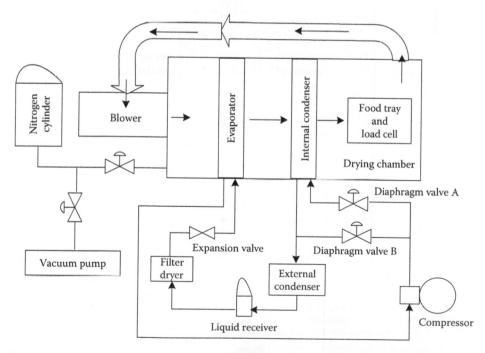

FIGURE 4.8 Schematic diagram of heat pump-assisted dryer with modified atmosphere proposed by Hawlader et al. (Redrawn and reproduced from *J. Food Eng.*, 74, Hawlader, M.N.A. et al., Properties of modified atmosphere heat pump dried foods, 392–401, Copyright 2006a, with permission from Elsevier.)

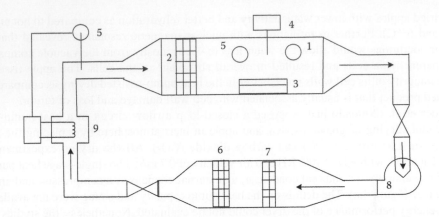

FIGURE 4.9 Heat pump-assisted dryer used in apple–potato composite drying performed by Rahman et al. (1, drying chamber; 2, honeycomb; 3, heating plate; 4, temperature/relative humidity transmitter; 5, T-type thermocouple; 6, condenser; 7, evaporator; 8, blower; 9, air heater). (Redrawn and reproduced with permission from Rahman, S.M.A. et al., *Dry. Technol.*, 25, 1359–1368, 2007.)

inside a composite product. It was observed that the drying rate for the composite material was relatively higher in comparison to single materials due to the higher apparent moisture diffusivity in the former. Color of the heat pump-assisted dried composite product was close to that of the original product.

Ong and Law (2011a,b, 2012a,b) performed the drying of indigenous salak fruit using a closed-loop single-stage heat pump-assisted dryer (see Figure 4.10). The heat pump system was charged

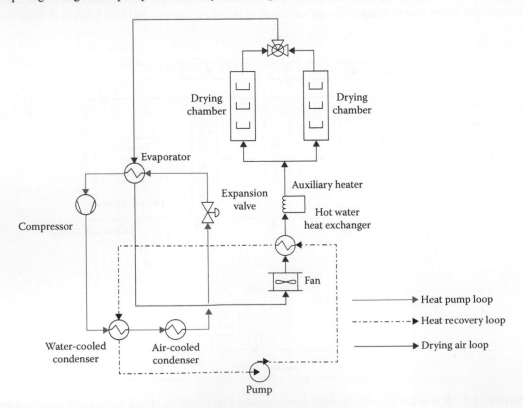

FIGURE 4.10 Schematic diagram of heat pump-assisted dryer used by Ong and Law in salak fruit drying. (Redrawn and reproduced with permission from Ong, S.P. and Law, C.L., *Dry. Technol.*, 29, 1954–1962, 2011b.)

with refrigerant R22 and composed of a primary condenser (water-cooled), a secondary condenser (air-cooled), an expansion valve, as well as a compressor with a capacity of 9.38 kW h. The temperature of the dehumidified air after the evaporator was elevated by the hot water heat exchanger and an auxiliary heater (when a higher temperature was required). Meanwhile, a blower was used to supply constant forced air (at about 5.4 m/s) in the drying chambers (33 × 98 × 33 cm). The drying temperatures and related humidity were recorded at 26°C, 27% RH (without heater) and 37°C, 17% RH (with heater), respectively. The fruits were cut into slices with a dimension of 40 × 20 × 3 mm and the initial moisture content was determined at about 81.3% (wet basis). It took about 10–51 h to achieve the equilibrium moisture content at 0.13–0.19 g water/g dry solid. The energy performance of the dryer was not determined in the studies. The reported work mainly focused on the effects of drying methods (heat pump-assisted drying, hot air drying and freeze drying), drying modes (isothermal and intermittent), and pretreatment (blanching) on the properties of salak fruit. Positive results were obtained in terms of higher ascorbic acid content, higher total phenolic content, better storage stability, and minimum total color change under the heat pump-assisted drying. The authors proposed that heat pump-assisted drying of salak fruit could be further optimized by dividing the dehydration process into three distinct phases, namely, the initial, intermittent, and final stages. Optimization was performed with the aim of determining the best combination of drying variables that could avoid an excessively long drying period and of improving the product quality. The optimized heat pump-assisted intermittent drying profile resulted in the reduction of drying time by 36%.

Chong et al. (2013, 2014) investigated the drying of apple, pear, mango, and papaya cubes (15 mm^3) in a closed-loop single-stage heat pump-assisted dryer that was similar to the one used by Ong and Law (2011b). Various drying profiles were tested and the study revealed that the best quality of the dried product in terms of color attributes, phenolic content, and antioxidant activity could be achieved by the use of heat pump-assisted dryer in the predrying (until a moisture content of 30%) stage, followed by vacuum-microwave oven to finish drying (until equilibrium moisture content). The heat pump-assisted drying was performed at a temperature of 35°C, RH of 20% and total drying period of 480 min, while the vacuum-microwave drying was conducted at a power level of 360 W for papaya and 240 W for apple, pear, and mango for about 4 min.

Hepbasli et al. (2010) and Erbay and Hepbasli (2013) investigated the drying of plums using a single-stage heat pump-assisted conveyor dryer (see Figure 4.11). The drying of plum slices with an average thickness of 4 mm was performed at 45°C–55°C and with an air velocity of 1.5 m/s. The moisture contents of the fresh and dried plums were determined at 84.49% and 15.70%, respectively. The air was heated by the heat pump system, which included a scroll compressor, two condensers (internal and external ones), an expansion valve, an evaporator, and a heat recovery unit (HRU). Refrigerant R407C was used in the heat pump cycle. Detailed information on the design of conveyor band system in the dryer is not provided. Nonetheless, the effect of drying temperature on the efficiencies of the systems was evaluated in terms of exergetic efficiency, improvement potential rate, and exergoeconomic analysis. It was observed that the most important system components were the evaporator and the HRU, followed by the compressor and the condenser. Inefficiencies within the evaporator and the HRU had a significant effect on the overall system efficiency, while inefficiencies in the condenser and compressor were mainly related to the internal operating conditions of the components. In addition, an increase in the drying air temperature caused a decrease in the proportion of avoidable endogenous exergy destructions to total exergy destructions within the HRU.

4.3.1.3 Partially Closed System and Miscellaneous Configurations

Prasertsan and Saen-saby (1998) proposed a partially closed single-stage heat pump-assisted dryer as shown in Figure 4.12 for the drying of banana at a temperature of 50°C–60°C. The unit was assembled from air conditioner parts that were charged with refrigerant R22 and were attached to a compressor with a capacity of 11 kW. The evaporator and condenser were made from 9 mm internal diameter copper tubes with total heat transfer areas of 61.30 and 87.44 m^2, respectively. The drying

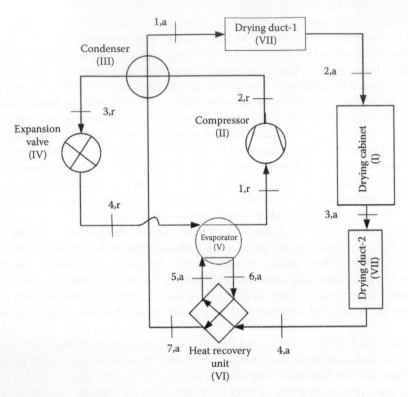

FIGURE 4.11 Schematic diagram of heat pump-assisted dryer used by Hepbasli et al. in drying of plum slices. (Redrawn and reproduced with permission from Hepbasli, A. et al., *Dry. Technol.*, 28, 1385–1395, 2010.)

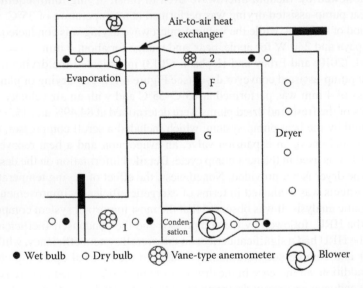

FIGURE 4.12 Heat pump-assisted dryer proposed by Prasertsan and Saen-saby. (Redrawn and reproduced with permission from Prasertsan, S. and Saen-saby, P., *Dry. Technol.*, 16, 235–250, 1998.)

chamber (1.1 × 1.1 × 2 m) was able to accommodate about 19 drying trays. It was observed that the moisture contents of the products were reduced by 150% during the first 24 h as compared to normal market practice that would usually take about 2 days with a moisture content less than 30%. The MER was reported at 1.184–2.710 kg/h, while the SMER was at 0.242–0.540 kg/kW h. The work provided few important notes about the performance of a heat pump-assisted dryer. The authors commented that MER decreased rapidly with drying time, and hence the heat pump-assisted system is more suitable for drying of high moisture content materials or operating during the early stage of batch drying. On the other hand, heat pump-assisted dryer had the lowest operating cost as compared to an electrically heated hot air dryer and a direct-fired dryer based on the economic analyses by the authors. Drying with an open system (vent all the hot drying air at dryer outlet and heat the fresh air at dryer inlet) was more efficient than the partially closed system, probably due to a higher air velocity and higher drying air temperature in the open system. Nevertheless, it would be slightly high in the total power consumption (additional 10–20 kW h) to run in open system mode. About 47%–48% of the hot drying air at the dryer outlet had to be purged in order to limit the heat recovered at the evaporator as well as to obtain a proper pressure ratio across the compressor, thus maintaining the stability of the heat pump system.

Xanthopoulos et al. (2007) designed a simple partially closed single-stage heat pump-assisted dryer (see Figure 4.13) for fig drying. The heat pump system was charged with refrigerant R22 and enabled drying at a temperature of 46.1°C–49.6°C (without heater) and 50°C–60°C (with auxiliary electric resistances). The corresponding RH was 12.60%–17.65% and 9%–10.65% RH, respectively. The air velocity could be adjusted from 1 to 5 m/s. The drying cabinet composed of two concentric circular tubes with metal frames at the bases to hold the figs. Drying air was directed to properly designed truncated cones and perforated metal sheets prior to entering the drying chamber in order to establish a homogenous air-stream profile. Meanwhile, an air vent system was installed in between the drying cabinet and evaporator to regulate the fresh air uptake as well as hot drying air vent when necessary. Drying time ranged from 23 h (60°C, 3 m/s) up to 53.3 h (47.2°C, 2 m/s) to reduce the moisture content from about 71.34% to below 30% (wet basis). No further data were available on the energy performance of the dryer as well as the quality of the dried product.

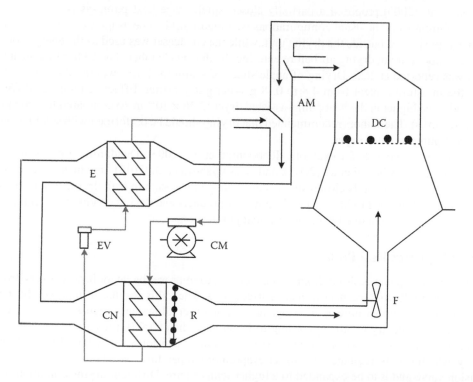

FIGURE 4.13 Heat pump-assisted dryer used by Xanthopoulos et al. in figs drying. DC, Drying cabinet; F, Fan; AM, Air mix; CM, Compressor; CN, Condenser; EV, Expansion valve; E, Evaporator; R, Resistance. (Redrawn and reproduced from *J. Food Eng.*, 81, Xanthopoulos, G. et al., Applicability of a single-layer drying model to predict the drying rate of whole figs, 553–559, Copyright 2007, with permission from Elsevier.)

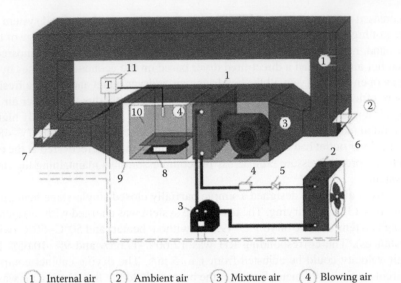

| ① Internal air | ② Ambient air | ③ Mixture air | ④ Blowing air |

FIGURE 4.14 Heat pump-assisted dryer used by Aktas et al. in drying of apple slices (1, condenser; 2, evaporator; 3, compressor; 4, dryer; 5, expansion valve; 6, fresh air inlet; 7, exhaust air; 8, digital balance; 9, drying chamber; 10, apple slices; 11, thermocouple and process control equipment). (Redrawn and reproduced from *Desalination*, 239, Aktas, M. et al., Determination of drying characteristics of apples in a heat pump and solar dryer, 266–275, Copyright 2009, with permission from Elsevier.)

Aktas et al. (2009) proposed a partially closed single-stage heat pump-assisted dryer that is slightly different from the usual configurations (see Figure 4.14). The evaporator of the heat pump system was positioned outside the drying unit, while the condenser was used as the energy source to raise the temperature of drying air before entering the drying chamber. The COP of the heat pump system was reported at 2.25. Drying of apple slices of 4 mm thickness would take about 3.5 h to reduce the moisture content from 4.8 to 0.18 g water/g dry matter. Effective moisture diffusivity was found to be higher in the heat pump-assisted dryer (2.36×10^{-8} m²/s) compared to a solar dryer (1.03×10^{-8} m²/s). The authors recommended that drying should be performed with solar heater first before heat pump-assisted dryer.

There is vast literature, for example, Teeboonma et al. (2003), Dandamrongrak et al. (2003), Uddin et al. (2004), Chegini et al. (2007), and Brushlyanova et al. (2013), in which the authors had reported the use of a partially closed single-stage heat pump system in fruit drying. Interested readers may refer to Table 4.2 to know the relevant fruit products and drying conditions. Details on the dryer design could be obtained from the original publication.

4.3.2 MULTISTAGE HEAT PUMP

A multistage heat pump-assisted dryer was developed to improve the latent heat recovery from the drying air by increasing the heat transfer surface area. Thus, this results in improved heat pump energy efficiency. Furthermore, a two-stage heat pump dryer allows easy regulation of temperature and humidity of the drying air. According to Chou and Chua (2007), refrigerant vapor in a two-stage heat pump is split into two streams; one enters an expansion valve with a higher discharge capacity, which is to be regulated to a lower evaporator temperature, while the other enters another expansion valve and is to be expanded to a higher temperature. Different drying conditions can be obtained by manipulating the flow ratio of the refrigerant vapor charged into the high and low pressure evaporators in the two-stage heat pump dryer.

Chua et al. (2000a,b; 2001; 2002) had designed a multistage partially closed heat pump-assisted dryer for the drying of banana and guava. The two-stage heat pump system composed of an economizer, high and low evaporators, a hot gas condenser, and two subcondensers. The dryer was also equipped with auxiliary heater and bypass dampers at the evaporators. Energy performance of the dryer was not reported but extensive works were performed on the intermittent drying modes with stepwise temperature and cyclic temperature profiles. In general, the drying experiments were conducted at a temperature of 20°C–40°C, RH of 20%–100%, and air velocity of 0.5–3.0 m/s. Chua et al. (2002) reported that step-up temperature profile could minimize color degradation, while stepwise temperature profile could reduce the drying time. Meanwhile, Chua et al. (2000a,b) observed that retention of ascorbic acid was about 20% higher in drying with cyclic temperature as compared to drying at isothermal, providing that proper temperature schedule was selected. The overall color change could be reduced while maintaining high drying rates when appropriate temperature–time variation was selected.

4.3.3 INFRARED-ASSISTED HEAT PUMP

Tan et al. (2001) investigated the performance of an infrared-assisted heat pump dryer by installing two infrared lamps (500 mm in length, 230 V, 1000 W) inside a two-stage heat pump-assisted dryer, which was similar to those used by Chua et al. (2000a,b; 2001; 2002). Drying of pineapple samples (20 × 20 × 5 mm) was conducted at a temperature of 40°C, RH of 21%, air velocity of 2 m/s, and intermittent infrared irradiation at intermittent ratio of 1, 1/2, and 1/3 with the cycle time of 30 min. It was observed that the drying rate increased by four to five times with intermittent infrared heating, while it minimized the overall color change of the pineapple samples.

Yong et al. (2006) had designed a multimode heat pump-assisted dryer where the unit composed of an evaporator, two condensers, a centrifugal blower (2.2 kW capacity), an auxiliary heater (6 kW), and a quartz-faced heater (3 kW capacity) to supply the radiative heat (see Figure 4.15). Nevertheless, drying of dragon fruit slices (30 × 30 × 5 mm) was conducted with the normal convective mode with dehumidified air at a temperature of 44°C, RH of 18.5%, and air velocity of 0.65 m/s, without the infrared radiation mode. The authors commented that the dragon fruit belongs to a distinct category of fruits, containing a highly viscous sugary liquid. Its stickiness could be due to the presence of sugar in the fruit. The dragon fruit samples were pretreated with drilled holes (12 or 24 holes on surface of 25 × 25 mm and diameter of 1 or 2.5 mm) to improve

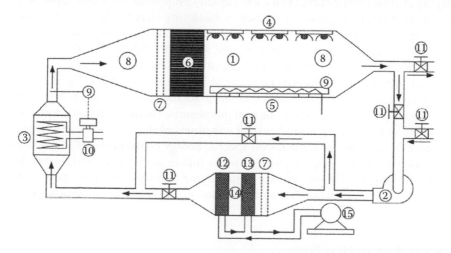

FIGURE 4.15 Heat pump-assisted dryer used by Yong et al. (1, drying chamber; 2, blower; 3, auxiliary air heater; 4, infrared lamps; 5, conduction heating plate; 6, honey comb; 7, perforated plate; 8, relative humidity sensor; 9, thermocouple; 10, temperature controller; 11, valve; 12, condenser; 13, evaporator; 14, heat pump; 15, compressor). (Redrawn and reproduced with permission from Yong, C.K. et al., *Dry. Technol. Int. J.*, 24, 397–404, 2006.)

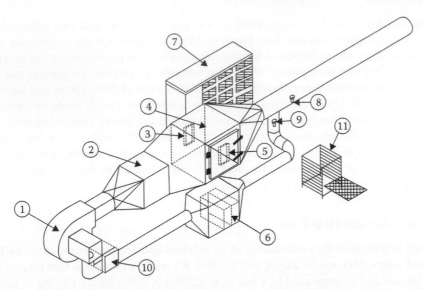

FIGURE 4.16 Schematic diagram of dual-function heat pump-assisted dryer proposed by Nathakaranakule et al. (1, fan; 2, heater; 3,5, FIR rods; 4, drying chamber; 6, evaporator; 7, condenser; 8,9, butterfly valves; 10, cover; 11, rack and tray). (Redrawn and reproduced from *J. Food Eng.*, 100, Nathakaranakule, A. et al., Far-infrared radiation assisted drying of longan fruit, 662–668, Copyright 2010, with permission from Elsevier.)

the drying rate. No further information was provided on the energy performance of the dryer or the product quality of the dried dragon fruit.

Nathakaranakule et al. (2010) studied the drying of longan by using a dual-function heat pump-assisted dryer that could be used either as a hot air or heat pump-assisted dryer (see Figure 4.16). The drying experiments were performed at a temperature of 55°C, and about 2 kg of longan were used in each drying experiment. The initial moisture content of longan was about 84%–86% (wet basis) and final moisture content was about 18% (wet basis). The drying unit was designed as a closed-loop system (could be partially closed as well when necessary) where the exhaust air leaving the drying chamber was directed to a 2 kW capacity evaporator and water vapor in the exhaust air was condensed before it was recycled back to the drying chamber by using a forward-curved blade centrifugal fan that was driven by a 2.2 kW motor. Drying air was heated by three 1 kW heaters prior to entering the drying chamber (0.52 × 0.52 × 0.52 m). Two 650 W ceramic heaters were installed in the drying chamber for generating the Far infrared (FIR) with wavelengths between 7 and 1000 μm and approximate infrared intensity of 0.6–1.2 W/cm². The condensing unit, which consisted of a 0.65 kW compressor and a 3.5 kW capacity condenser, was placed outside the drying air pathway and, hence, the heat that absorbed at the evaporator was not recovered through the external condenser. Nonetheless, it was observed that FIR-assisted heat pump drying at the highest FIR intensity gave the lowest specific energy consumption (SEC) as compared to solely heat pump-assisted drying, mainly due to the higher drying rates and shorter drying time in the former. Moreover, FIR-assisted heat pump drying also created a more porous structure in the dried longan and consequently less shrinkage and higher rehydration percentage in the dried products.

4.3.4 Solar-Assisted Heat Pump

Şevik (2014) dried strawberry slices (5 mm) in a solar-assisted heat pump dryer (see Figure 4.17) with a temperature of 50°C and air velocity ranging from 0.29 to 0.7 m/s depending on the temperature. The air-source heat pump system was charged with refrigerant R-134a and consisted of a compressor (1/3 HP), a condenser, and an evaporator (1/3 HP). The condenser was mounted under

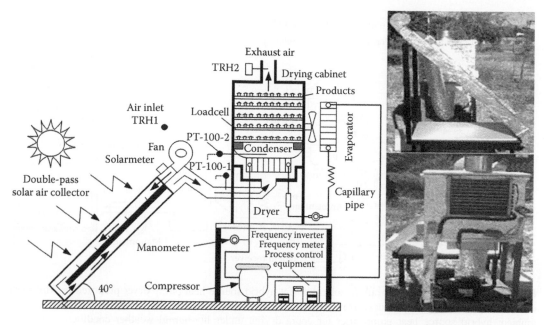

FIGURE 4.17 Heat pump-assisted dryer used by Şevik in drying of strawberry slices. (Reproduced from *Sol. Energy*, 105, Şevik, S., Experimental investigation of a new design solar-heat pump dryer under the different climatic conditions and drying behaviour of selected products, 190–205, Copyright 2014, with permission from Elsevier.)

the drying cabinet (at the drying air inlet point), while the evaporator was placed outside the drying cabinet (0.5 × 0.5 m). The drying unit was an open-loop system where it took in fresh ambient air and discharged the exhaust drying air to the atmosphere. The heat pump was supported by a double pass solar air collector (DPSAC) whereby the drying air was first processed in the DPSAC and then directed to the condenser of heat pump prior to entering the drying cabinet from the bottom. The system could run nonstop via the heat pump compressor when solar energy was insufficient, while power consumptions of fans and other instruments were provided with the PV unit. It took about 160 min to reduce the initial moisture content of 3.4 g water/g dry matter to a final moisture content of 0.08 g water/g dry matter. The ascorbic acid and polyphenols content of fresh strawberries were recorded as 230 mg and 14 µg/100 mg, while the dried ones were recorded as 150 mg and 5 µg/100 mg, respectively. It was observed that the collector efficiency was relatively low (16%–79%) due to insufficient solar energy in winter conditions. Similarly, the SMER values were fairly low (0.003–0.46 kg/kW h) due to low air velocity and a small amount of the products. The energy utilization ratio (EUR) was high (0.18–0.48) due to the high moisture content of strawberry.

Mohanraj et al. (2008) had performed the drying of copra using an open-loop single stage heat pump with an average COP of 3.5 and SMER of 0.85 kg/kW h. Further studies by Mohanraj (2014) had modified the unit to a solar-ambient hybrid source heat pump dryer by incorporating a glazed solar collector (2 × 1 m) with 0.8 mm thick copper fins as absorber surface into the heat pump-assisted dryer (see Figure 4.18). The average COP and SMER for this solar-assisted heat pump dryer were determined at 2.54 and 0.79 kg/kW h, respectively. The integration of solar energy with heat pump had enhanced the air temperature at the condenser outlet and also reduced the drying time by about 8 h. The copra drying was conducted under hot-humid weather conditions where the maximum solar intensity was measured at about 840 W/m², while ambient temperature and RH during the experiment was varied between 28.4°C and 33.4°C and 60%–73%, respectively. The drying unit was equipped with an air cooled condenser, a blower (735 W), a hermetically sealed reciprocating compressor of rated input 1020 W, and a drying chamber (1000 × 600 × 1000 mm) that can hold

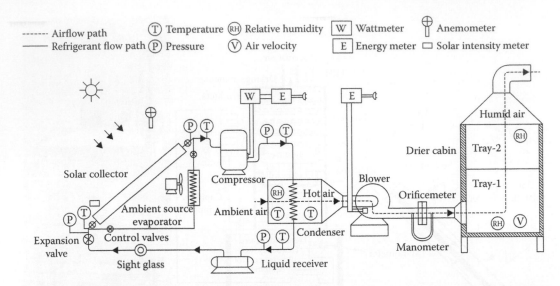

FIGURE 4.18 Schematic diagram of solar-ambient hybrid source heat pump dryer proposed by Mohanraj et al. in copra drying. (Reproduced from *Energy Sustain. Dev.*, 23, Mohanraj, M., Performance of a solar-ambient hybrid source heat pump drier for copra drying under hot-humid weather conditions, 165–169, Copyright 2014, with permission from Elsevier.)

500 coconuts per batch. The heat pump cycle was charged with refrigerant R22 and both the evaporators were designed to absorb a heat load of 2.8 kW from ambient sources (both solar and ambient). Ambient air was heated when it passed over the condenser coil and the air leaving the drying chamber was exhausted to the atmosphere considering the risk of harmful microorganisms to grow and accumulate at the wetted evaporator and adjacent surfaces. The processing cost of the copra obtained in heat pump-assisted dryer was observed to be higher as compared to other drying techniques. However, the better quality of copra obtained in a heat pump-assisted dryer made it more reasonable over other forced convection solar dryers and open sun drying.

4.3.5 Atmospheric Freeze Drying with Heat Pump

Freeze drying can be implemented at atmospheric pressure as the diffusion of water vapor to the surface through the matrix occurs mainly due to the vapor pressure gradient, rather than the difference in the absolute pressure (Claussen et al. 2007; Jangam and Mujumdar 2011).

Alves-Filho (2002) conducted the drying of granulated frozen cranberry using a closed-loop single-stage heat pump-assisted fluidized bed dryer (see Figure 4.19). The drying was performed by atmospheric freeze drying at −10°C in the first stage of drying, followed by drying with heated air at a temperature of 30°C. The air velocity was set at 0.5–1.5 m/s in accordance to the load of product. No further information was reported on the energy performance of the dryer, but the dried cranberry granule was found porous, free flowing, and possessed attractive natural color. The water activity of the final product was about 0.38–0.49 with moisture content at about 8%.

Stawczyk et al. (2007) designed a closed-loop single-stage heat pump-assisted packed bed dryer with vertical flow for the drying of apple cubes (1 cm³), applying atmospheric freeze drying method (see Figure 4.20). The heat pump system was used to cool and extract moisture from the exhaust drying air. The dehumidified air was then directed to an electric heater to elevate the temperature prior to entering the drying chamber. Various temperature profiles were tested and it was found that dewatering at temperatures around −10°C led to a highly porous product structure. The same process performed at temperatures around 0°C resulted in deterioration of product quality somehow. Furthermore, the dried products of atmospheric freeze drying at a lower temperature had similar characteristics of rehydration kinetics and hygroscopic properties as compared to the product

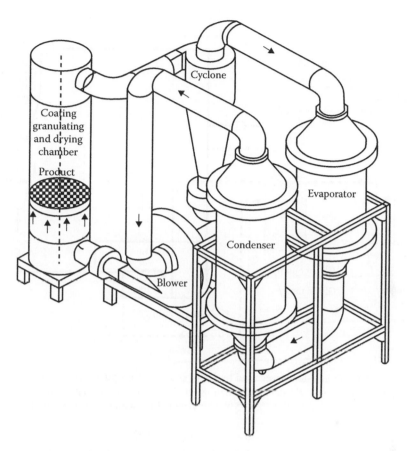

FIGURE 4.19 Heat pump-assisted dryer proposed by Alves-Filho for the atmospheric freeze drying of granulated cranberry. (Reproduced with permission from Alves-Filho, O., *Dry. Technol.*, 20, 1541–1557, 2002.)

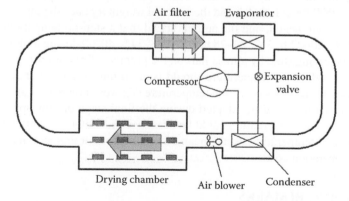

FIGURE 4.20 Heat pump-assisted dryer used by Stawczyk et al. in atmospheric freeze drying of apple cubes. (Reproduced with kind permission from Springer Science + Business Media: *Transp. Porous Media*, Kinetics of atmospheric freeze-drying of apple, 66, 2007, 159–172, Stawczyk, J. et al.)

obtained from vacuum freeze drying. The dried product also possessed a statistically higher value of antioxidant activity and polyphenol content as compared to conventional convective drying.

Duan et al. (2013) performed the atmospheric freeze drying apple cubes (1 cm^3) by using a closed-loop single stage heat pump-assisted dryer (see Figure 4.21). The apple cubes were placed on

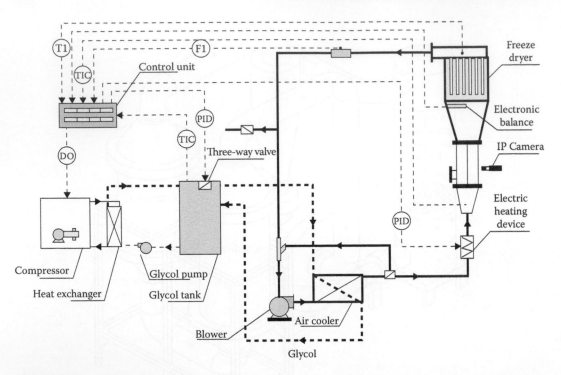

FIGURE 4.21 Heat pump-assisted dryer used by Duan et al. in atmospheric freeze drying of apple cubes. (Reproduced from *Food Bioprod. Process.*, 91, Duan, X. et al., The drying strategy of atmospheric freeze drying apple cubes based on glass transition, 534–538, Copyright 2013, with permission from Elsevier.)

perforated shelves in a cylindrical drying chamber which measured 0.4 m in diameter and 0.8 m in length. The drying experiments were performed with a temperature ranging from −10°C to 40°C, RH of about 30%, and air velocity of 1.5 m/s. The cubed samples were pretreated with citric acid before freezing at −25°C for at least 8 h and the material weight for each drying was 2 kg. Numerous temperature profiles were tested, and the authors found that a step-up temperature profile (0–6 h, −5°C; 6–20 h, −10°C; 20–24 h, 5°C; 24–28 h, 20°C; 28–34 h, 40°C) during atmospheric freeze drying could reduce the drying time by almost half on the premise of maintaining the product quality. It was proposed that −10°C air temperature should be used until the moisture content dropped to about 35% (wet basis) as the glass transition temperature (T_g) would increase quickly after that point and higher air temperature could be adopted during the final stage of atmospheric freeze drying. Although a much lower air temperature could be used to avoid water thawing in the early stage of drying, too low inlet air temperature would mean a lower evaporator temperature, which would lead to more expensive operation cost.

4.4 CONCLUDING REMARKS

Heat pump-assisted dryers offer many advantages in fruit drying, particularly in preserving the quality attributes of the dried fruit products, which are often assessed in terms of color, flavor, texture, aroma, and nutrition contents. Integration of a heat pump system into a fruit dryer has a big impact in the energy recovery and cost saving since the fruit drying process is usually lengthy with high power consumption. Overview of the applications and performance of heat pump-assisted dryers in fruit drying in the recent decades reveals that research and development in this area is moving relatively slow. There is tremendous potential for study and exploration in the dried fruit industry. Future studies could focus on development of more dried fruit products in terms of the variety and diversity. Currently, enough scientific studies are not conducted on fruits from the hesperidium group (e.g., orange, lemon, and grapefruit), acid category (e.g., kiwi, strawberry, and lime) and nonsweet category (e.g., avocado). In addition, a combination of heat pump systems and other types of dryers such as fluidized bed, microwave, spray dryer, and the like should be explored further, especially for their application in fruit drying. Last but not the least, incorporation of smart devices or control system in a heat pump-assisted dryer for automatic adjustment and optimum energy saving is a must for future research in order to maximize the benefits from the versatile dryer.

REFERENCES

Achariyaviriya, S., Soponronnarit, S., and Terdyothin, A. 2000. Mathematical model development and simulation of heat pump fruit dryer. *Drying Technology* 18(1–2): 479–491.

Aktas, M., Ceylan, I., and Yilmaz, S. 2009. Determination of drying characteristics of apples in a heat pump and solar dryer. *Desalination* 239(1–3): 266–275.

Al-Farsi, M., Alasalvar, C., Morris, A., Baron, M., and Shahidi, F. 2005. Comparison of antioxidant activity, anthocyanins, carotenoids, and phenolics of three native fresh and sun-dried date (*Phoenix dactylifera* l.) varieties grown in Oman. *Journal of Agricultural and Food Chemistry* 53(19): 7592–7599.

Alves-Filho, O. 2002. Combined innovative heat pump drying technologies and new cold extrusion techniques for production of instant foods. *Drying Technology* 20(8): 1541–1557.

Baker, C. G. J. 2005. Energy efficiency in drying. *Stewart Postharvest Review* 1(4): 1–11.

Bendre, A. M. and Kumar, A. (2009). *A Textbook of Practical Botany* II. New Delhi, India: Rastogi Publications.

Bennett, L. E., Jegasothy, H., Konczak, I. et al. 2011. Total polyphenolics and anti-oxidant properties of selected dried fruits and relationships to drying conditions. *Journal of Functional Foods* 3(2): 115–124.

Borompichaichartkul, C., Luengsode, K., Chinprahast, N., and Devahastin, S. 2009. Improving quality of macadamia nut (*Macadamia integrifolia*) through the use of hybrid drying process. *Journal of Food Engineering* 93(3): 348–353.

Brushlyanova, B., Petrova, T., Penov, N., Karabadzhov, O., and Katsharova, S. 2013. Drying kinetics of different fruit pomaces in a heat pump dryer. *Bulgarian Journal of Agricultural Science* 19(4): 780–782.

Ceylan, I., Aktas, M., and Dogan, H. 2007. Mathematical modeling of drying characteristics of tropical fruits. *Applied Thermal Engineering* 27(11–12): 1931–1936.

Chang, C.-H., Lin, H.-Y., Chang, C.-Y., and Liu, Y.-C. 2006. Comparisons on the antioxidant properties of fresh, freeze-dried and hot-air-dried tomatoes. *Journal of Food Engineering* 77(3): 478–485.

Chegini, G., Khayaei, J., Rostami, H. A., and Sanjari, A. R. 2007. Designing of a heat pump dryer for drying of plum. *Journal of Research and Applications in Agricultural Engineering* 52(2): 63.

Chen, X. D. 2008. Food drying fundamentals. In *Food Processing*, eds. X. D. Chen and A. S. Mujumdar, pp. 1–55. West Sussex, England: Blackwell Publishing.

Chong, C., Figiel, A., Law, C., and Wojdyło, A. 2014. Combined drying of apple cubes by using of heat pump, vacuum-microwave, and intermittent techniques. *Food and Bioprocess Technology* 7(4): 975–989.

Chong, C. H., Law, C. L., Figiel, A., Wojdyło, A., and Oziembłowski, M. 2013. Colour, phenolic content and antioxidant capacity of some fruits dehydrated by a combination of different methods. *Food Chemistry* 141(4): 3889–3896.

Chou, S. K. and Chua, K. J. 2007. Heat pump drying systems. In *Handbook of Industrial Drying*, ed. A. S. Mujumdar, pp. 1104–1130. Boca Raton, FL: Taylor & Francis Group.

Chua, K. J., Chou, S. K., Ho, J. C., Mujumdar, A. S., and Hawlader, M. N. A. 2000a. Cyclic air temperature drying of guava pieces: Effect on moisture and ascorbic acid contents. *Trans IChemE* 78: 72–78.

Chua, K. J., Hawlader, M. N. A., Chou, S. K., and Ho, J. C. 2002. On the study of time-varying temperature drying—Effect on drying kinetics and product quality. *Drying Technology: An International Journal* 20(8): 1559–1577.

Chua, K. J., Mujumdar, A. S., Chou, S. K., Hawlader, M. N. A., and Ho, J. C. 2000b. Convective drying of banana, guava and potato pieces: Effect of cyclical variations of air temperature on drying kinetics and color change. *Drying Technology: An International Journal* 18(4): 907–936.

Chua, K. J., Mujumdar, A. S., Hawlader, M. N. A., Chou, S. K., and Ho, J. C. 2001. Convective drying of agricultural products. Effect of continuous and stepwise change in drying air temperature. *Drying Technology: An International Journal* 19(8): 1949–1960.

Claussen, I. C., Ustad, T. S., Strømmen, I., and Walde, P. M. 2007. Atmospheric freeze drying—A review. *Drying Technology: An International Journal* 25(6): 947–957.

Colak, N. and Hepbasli, A. 2009. A review of heat pump drying: Part 1—Systems, models and studies. *Energy Conversion and Management* 50(9): 2180–2186.

Dandamrongrak, R., Mason, R., and Young, G. 2003. The effect of pretreatments on the drying rate and quality of dried bananas. *International Journal of Food Science & Technology* 38(8): 877–882.

Duan, X., Ding, L., Ren, G.-Y., Liu, L.-L., and Kong, Q.-Z. 2013. The drying strategy of atmospheric freeze drying apple cubes based on glass transition. *Food and Bioproducts Processing* 91(4): 534–538.

Erbay, Z. and Hepbasli, A. 2013. Advanced exergy analysis of a heat pump drying system used in food drying. *Drying Technology* 31(7): 802–810.

Evans, W. C. 2002. *Trease and Evans Pharmacognosy*. Edinburgh, Scottland: Saunders.

Fadhel, M. I., Sopian, K., Daud, W. R. W., and Alghoul, M. A. 2011. Review on advanced of solar assisted chemical heat pump dryer for agriculture produce. *Renewable and Sustainable Energy Reviews* 15(2): 1152–1168.

Gaware, T. J., Sutar, N., and Thorat, B. N. 2010. Drying of tomato using different methods: Comparison of dehydration and rehydration kinetics. *Drying Technology* 28(5): 651–658.

Harris, J. G. and Harris, M. W. 2001. *Plant Identification Terminology: An Illustrated Glossary*. Spring Lake, UT: Spring Lake Pub.

Hawlader, M. N. A., Perera, C. O., and Tian, M. 2006a. Properties of modified atmosphere heat pump dried foods. *Journal of Food Engineering* 74(3): 392–401.

Hawlader, M. N. A., Perera, C. O., Tian, M., and Yeo, K. L. 2006b. Drying of guava and papaya: Impact of different drying methods. *Drying Technology: An International Journal* 24(1): 77–87.

Hepbasli, A., Colak, N., Hancioglu, E., Icier, F., and Erbay, Z. 2010. Exergoeconomic analysis of plum drying in a heat pump conveyor dryer. *Drying Technology* 28(12): 1385–1395.

INC. 2015. Nuts and dried fruits global statistical review, International Nut and Dried Fruit Council, Tarragona, Spain.

Islam, M. R. and Mujumdar, A. S. 2008. Heat pump-assisted drying. In *Guide to Industrial Drying: Principles, Equipment and New Developments*, ed. A. S. Mujumdar, pp. 157–176. Mumbai, India: Colour Publications Pvt. Ltd.

Jangam, S. V., Joshi, V. S., Mujumdar, A. S., and Thorat, B. N. 2008. Studies on dehydration of sapota (*Achras zapota*). *Drying Technology* 26(3): 369–377.

Jangam, S. V. and Mujumdar, A. S. 2011. Heat pump assisted drying technology—Overview with focus on energy, environment and product quality. In *Modern Drying Technology, Vol. 4: Energy Savings*, eds. E. Tsotsas and A. S. Mujumdar, pp. 121–158. Berlin, Germany: Wiley-VCH.

Jayaraman, K. S. and Das Gupta, D. K. 2006. Drying of fruits and vegetables. In *Handbook of Industrial Drying*, ed. A. S. Mujumdar, pp. 606–631. Boca Raton, FL: Taylor & Francis Group.

Krokida, M. K. and Philippopoulos, C. 2005. Rehydration of dehydrated foods. *Drying Technology: An International Journal* 23(4): 799–830.

Leong, L. P. and Shui, G. 2002. An investigation of antioxidant capacity of fruits in Singapore markets. *Food Chemistry* 76(1): 69–75.

Lim, T. K. 2012. *Edible Medicinal and Non-Medicinal Plants: Fruits*. New York: Springer.

Lutz, M., Hernández, J., and Henríquez, C. 2015. Phenolic content and antioxidant capacity in fresh and dry fruits and vegetables grown in Chile. *CyTA—Journal of Food* 13(4): 541–547.

Marques, L. G., Silveira, A. M., and Freire, J. T. 2006. Freeze-drying characteristics of tropical fruits. *Drying Technology: An International Journal* 24(4): 457–463.

Minea, V. 2013a. Drying heat pumps—Part I: System integration. *International Journal of Refrigeration* 36(3): 643–658.

Minea, V. 2013b. Heat-pump–assisted drying: Recent technological advances and R&D needs. *Drying Technology* 31(10): 1177–1189.

Minea, V. 2014. Overview of heat-pump–assisted drying systems. Part I: Integration, control complexity, and applicability of new innovative concepts. *Drying Technology* 33(5): 515–526.

Mohanraj, M. 2014. Performance of a solar-ambient hybrid source heat pump drier for copra drying under hot-humid weather conditions. *Energy for Sustainable Development* 23: 165–169.

Mohanraj, M., Chandrasekar, P., and Sreenarayanan, V. V. 2008. Performance of a heat pump drier for copra drying. *Proceedings of the Institution of Mechanical Engineers, Part A: Journal of Power and Energy* 222(3): 283–287.

Mujumdar, A. S. ed. 2006. Principles, classification and selection of dryers. In *Handbook of Industrial Drying*, pp. 4–31. Boca Raton, FL: Taylor & Francis Group.

Nathakaranakule, A., Jaiboon, P., and Soponronnarit, S. 2010. Far-infrared radiation assisted drying of longan fruit. *Journal of Food Engineering* 100(4): 662–668.

Ong, S. P. and Law, C. L. 2009. Intermittent heat pump drying in fruit and vegetable processing. *The Sixth Asia-Pacific Drying Conference*, Bangkok, Thailand, pp. 316–322.

Ong, S. P. and Law, C. L. 2011a. Drying kinetics and antioxidant phytochemicals retention of salak fruit under different drying and pretreatment conditions. *Drying Technology* 29(4): 429–441.

Ong, S. P. and Law, C. L. 2011b. Microstructure and optical properties of salak fruit under different drying and pretreatment conditions. *Drying Technology* 29(16): 1954–1962.

Ong, S., Law, C., and Hii, C. 2012a. Effect of pre-treatment and drying method on colour degradation kinetics of dried salak fruit during storage. *Food and Bioprocess Technology* 5(6): 2331–2341.

Ong, S. P., Law, C. L., and Hii, C. L. 2012b. Optimization of heat pump–assisted intermittent drying. *Drying Technology* 30(15): 1676–1687.

Pal, U. S., Khan, M. K., and Mohanty, S. N. 2008. Heat pump drying of green sweet pepper. *Drying Technology: An International Journal* 26(12): 1584–1590.

Patel, K. K. and Kar, A. 2012. Heat pump assisted drying of agricultural produce—An overview. *Journal of Food Science and Technology* 49(2): 142–160.

Perera, C. O. and Rahman, M. S. 1997. Heat pump dehumidifier drying of food. *Trends in Food Science & Technology* 8(3): 75–79.

Prasertsan, S. and Saen-saby, P. 1998. Heat pump drying of agricultural materials. *Drying Technology* 16(1–2): 235–250.

Queiroz, R., Gabas, A. L., and Telis, V. R. N. 2004. Drying kinetics of tomato by using electric resistance and heat pump dryers. *Drying Technology* 22(7): 1603–1620.

Raghavan, G. S. V., Rennie, T. J., Sunjka, P. S. et al. 2005. Overview of new techniques for drying biological materials with emphasis on energy aspects. *Brazilian Journal of Chemical Engineering* 22: 195–201.

Rahman, S. M. A., Islam, M. R., and Mujumdar, A. S. 2007. A study of coupled heat and mass transfer in composite food products during convective drying. *Drying Technology* 25(7–8): 1359–1368.

Rice-Evans, C., Miller, N., and Paganga, G. 1997. Antioxidant properties of phenolic compounds. *Trends in Plant Science* 2(4): 152–159.

Sablani, S. S. 2006. Drying of fruits and vegetables: Retention of nutritional/functional quality. *Drying Technology: An International Journal* 24(2): 123–135.

Şevik, S. 2014. Experimental investigation of a new design solar-heat pump dryer under the different climatic conditions and drying behavior of selected products. *Solar Energy* 105: 190–205.

Shahidi, F. and Tan, Z. 2012. Raisins: Processing, phytochemicals, and health benefits. In *Dried Fruits: Phytochemicals and Health Effects*, eds. C. Alasalvar and F. Shahidi, pp. 372–388. Somerset, NJ: Wiley-Blackwell.

Shetty, K., Clydesdale, F. M., and Vattem, D. A. 2006. Clonal screening and sprout based bioprocessing of phenolic phytochemicals for functional foods. In *Functional Foods and Biotechnology*, eds. K. Shetty, G. Paliyath, A. L. Pometto, and R. E. Levin, pp. 1–16. Boca Raton, FL: CRC Press.

Shui, G. and Leong, L. P. 2005. Screening and identification of antioxidants in biological samples using high-performance liquid chromatography-mass spectrometry and its application on *Salacca edulis* reinw. *Journal of Agriculture and Food Chemistry* 53: 880–886.

Siddiq, M., Ahmed, J., and Lobo, M. G. 2012. *Tropical and Subtropical Fruits: Postharvest Physiology, Processing and Packaging*. Somerset, NJ: John Wiley & Sons.

Skinner, M. and Hunter, D. 2013. *Bioactives in Fruit: Health Benefits and Functional Foods*. Somerset, NJ: John Wiley & Sons.

Soponronnarit, S., Nathakaranakule, A., Wetchacama, S., Swasdisevi, T., and Rukprang, P. 1998. Fruit drying using heat pump. *International Energy Journal* 20(1): 39–53.

Sosle, V., Raghavan, G. S. V., and Kittler, R. 2003. Low-temperature drying using a versatile heat pump dehumidifier. *Drying Technology: An International Journal* 21(3): 539–554.

Stawczyk, J., Li, S., Witrowa-Rajchert, D., and Fabisiak, A. 2007. Kinetics of atmospheric freeze-drying of apple. *Transport in Porous Media* 66(1–2): 159–172.

Stefan, G., Hosahalli, S. R., and Michele, M. 2003. Drying of fruits, vegetables, and spices. In *Handbook of Postharvest Technology*, eds. H. S. Ramaswamy, G. S. Vijaya Raghavan, A. Chakraverty, and A. S. Mujumdar, pp. 653–695. CRC Press.

Sunthonvit, N., Srzednicki, G., and Craske, J. 2007. Effects of drying treatments on the composition of volatile compounds in dried nectarines. *Drying Technology: An International Journal* 25(5): 877–881.

Tan, M., Chua, K. J., Mujumdar, A. S., and Chou, S. K. 2001. Effect of osmotic pre-treatment and infrared radiation on drying rate and color changes during drying of potato and pineapple. *Drying Technology: An International Journal* 19(9): 2193–2207.

Teeboonma, U., Tiansuwan, J., and Soponronnarit, S. 2003. Optimization of heat pump fruit dryers. *Journal of Food Engineering* 59(4): 369–377.

Uddin, M. S., Hawlader, M. N. A., and Hui, X. 2004. A comparative study on heat pump, microwave and freeze drying of fresh fruits. *14th International Drying Symposium*. São Paulo, Brazil. Vol. C, pp. 2035–2042.

Vázquez, G., Chenlo, F., Moreira, R., and Cruz, E. 1997. Grape drying in a pilot plant with a heat pump. *Drying Technology: An International Journal* 15(3): 899–920.

Xanthopoulos, G., Oikonomou, N., and Lambrinos, G. 2007. Applicability of a single-layer drying model to predict the drying rate of whole figs. *Journal of Food Engineering* 81(3): 553–559.

Xu, Y., Zhang, M., Mujumdar, A. S., Duan, X., and Jin-cai, S. 2006. A two-stage vacuum freeze and convective air drying method for strawberries. *Drying Technology: An International Journal* 24(8): 1019–1023.

Yong, C. K., Islam, M. R., and Mujumdar, A. S. 2006. Mechanical means of enhancing drying rates: Effect on drying kinetics and quality. *Drying Technology: An International Journal* 24(3): 397–404.

Zhang, M., Li, C., Ding, X., and Cao, C. 2002. Thermal denaturation of some dried vegetables. *Drying Technology: An International Journal* 20(3): 711–717.

Zielinska, M., Zapotoczny, P., Alves-Filho, O., Eikevik, T. M., and Blaszczak, W. 2013. A multi-stage combined heat pump and microwave vacuum drying of green peas. *Journal of Food Engineering* 115(3): 347–356.

5 Advances in Heat Pump-Assisted Technologies for Drying of Vegetables

Chung Lim Law and Ho Hsien Chen

CONTENTS

5.1 INTRODUCTION

Vegetables are generally sold and consumed in fresh form; excessive vegetables that are not consumed within the short shelf life period have to be preserved for longer storage. In this regard, the vegetables require preservation. Drying is the most common and widely used preservation technique in reducing water activity of the perishable vegetables to prevent microbial activities, which degrade the product. Among the various dehydration techniques, hot air drying and sun drying are the most common techniques. Heat pump drying, on the other hand, is suitable for drying vegetables, and has been attracting interest in recent years because of its ability to produce dehydrated vegetables with better quality. In some places, coal and biomass are combusted to generate hot air; thus, hot air drying may cause food contamination. In many countries, direct contact of flue gas (hot air) and food is prohibited.

Chemical and physical changes occur during drying. The most common change that occurs during hot air drying and sun drying of vegetables is the conversion of chlorophyll to pheophytins (Rocha et al. 1993), which results in significant color change. Furthermore, the natural color pigment may be degraded due to high temperature exposure during the process. In addition, hot air drying can also cause changes to aromatic compounds, which results in loss of aroma. Furthermore, vegetables are also rich in phenolics compounds that are sensitive to high temperature. These compounds can be denatured and results in dehydrated vegetables with lower retention of bioactive ingredients.

In order to retain the color pigments, aromatic compounds, and bioactive ingredients that are sensitive to high temperature, researchers have reported that low temperature processing condition is the solution to the problem. In this regard, for the dehydration of vegetables, heat pump drying, vacuum drying, and freeze drying are alternatives that the vegetable processing industry may consider for the preservation of vegetables. Freeze drying and vacuum drying are relatively slow and require higher operating costs. These two drying techniques can be applied if the profit margin allows, else the heat pump drying is a suitable drying technique for preserving vegetables. Figure 5.1 shows the guides that lead to heat pump drying.

Heat pump drying is said to be an energy-efficient drying technique, as the heat pump can be used to recover energy from the exhaust air (Chua et al. 2002). In addition, the drying air temperature and humidity can be controlled (Prasertsan et al. 1996). Many researchers have acknowledged and reported the ability of this technique to provide better energy efficiency (Rossi et al. 1992) as

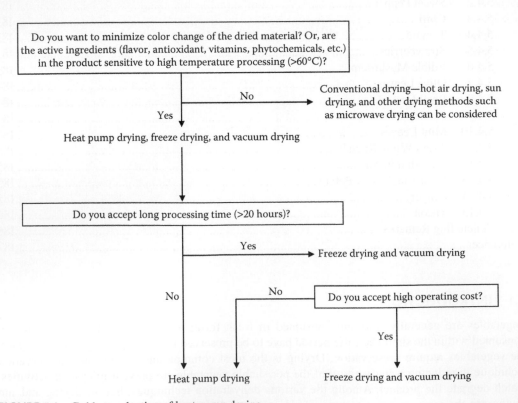

FIGURE 5.1   Guide to selection of heat pump drying.

well as to produce better quality of agricultural and bio-origin products (Prasertsan and Saen-saby 1998; Adapa et al. 2002; Alves-Filho 2002; Sosle et al. 2003).

Queiroz et al. (2004) reported that heat pump drying gave about 40%–50% reduction in the electric energy compared to hot air dryer using electrical heating elements. Similarly, Kohayakawa et al. (2004) reported that the savings were between 22% and 40%. Rossi et al. (1992) reported that the saving was 30%, and heat pump drying could produce better product quality. Lee and Kim (2009) compared the energy gain of heat pump dryer and hot air dryer, and found that heat pump drying has higher energy gain (three times more), but its drying time was about 1.5 times longer than hot air drying.

Readers (who read through Chapters 1 through 3) by now would have understood that the principal advantages of heat pump drying are the energy-saving potential, the controllability of drying operation parameters, its suitability in retaining heat-sensitive ingredients, its closed system, which avoids release of gases and fumes into atmosphere, as well as product contamination (Perera and Rahman 1997; Colak and Hepbasli 2009; Chua et al. 2010; Hii et al. 2012; Patel and Kar 2012). Therefore, the readers would appreciate that heat pump drying is a suitable drying technique for preserving vegetables.

5.2 VARIANTS OF HEAT PUMP DRYING

Before we discuss drying of vegetables using heat pump drying technique, let us briefly revisit the variants of heat pump drying techniques that researchers have applied to dry and dehydrate vegetables. Apart from the normal heat pump drying, which is carried out in continuous mode, it can be operated in intermittent mode, which gives us intermittent heat pump drying. It can be operated in vacuum condition, which gives us vacuum heat pump drying. Inert gas can be used instead of air, which gives us modified atmosphere heat pump drying. Multistage heat pump drying can be used where the operating condition is set at different value in different stage. We can use solar heat pump instead of normal heat pump system. Solar collector can be incorporated into the heat pump drying system, which gives us solar-assisted heat pump drying. It can be combined with fluidized bed dryer, where heat pump system generates low temperature dry air that is charged into a fluidized bed drying chamber.

Readers (who have gone through Chapters 1 and 3) may also note that apart from freon, which is used as working fluid in normal heat pump system, carbon dioxide, ammonia, and hydrocarbon can be used as the working fluid. Of course, the safety aspect of choosing the working fluid must be taken into account in order to ensure a safe operation of heat pump-assisted drying.

5.2.1 INTERMITTENT HEAT PUMP DRYING

Generally, heat pump drying is applied intermittently where the process consists of series of active drying and tempering. During tempering, energy is saved. Compared with continuous drying, including continuous heat pump drying, intermittent drying has the advantage of reducing active drying time. Its total processing time may be longer, but its effective drying time is shorter, thus saving energy consumption (Ong et al. 2012; Yang et al. 2013). Product quality is improved as tempering allows the product to relax rather than expose to thermal stresses and mechanical stresses. In addition, nonenzymatic browning is reduced (Putranto et al. 2011). Fatouh et al. (1998) in their study of heat pump drying of onion, carrot, potato, and sweet potato reported that in continuous drying, the specific energy consumption was lower than using conventional dryers, where the intermittent heat pump drying saved about 30% of the energy consumed compared to continuous drying when the appropriate heating and resting periods were used. Yang et al. (2013) in their study of heat pump drying of cabbage reported that the total drying time of intermittent heat pump drying (intermittent ratio of 1:3) is 50% higher than continuous heat pump drying, but its energy consumption is just half of the energy consumed when compared to continuous heat pump drying.

5.2.2 Vacuum Heat Pump Drying

The purpose of conducting drying in vacuum condition is to lower the boiling point of water, which in turn allows effective drying to take place at a lower temperature. Similarly, this is also applicable to heat pump drying. Artnaseaw et al. (2010) in their study of vacuum heat pump drying of chili reported that the drying time was reduced by reducing drying pressure. It was also found that percentage shrinkage decreased while the rehydration ratio increased with a decrease in drying pressure. Drying at lower drying pressure (10 kPa) gave a smoother dried chili surface (which indicates less shrinkage on surface) than that of chili dried at higher drying pressure (40 kPa).

5.2.3 Modified Atmosphere Heat Pump Drying

Drying medium in heat pump drying is typically air which contains 21% oxygen. Therefore drying that uses air as the drying medium tends to cause oxidation during the process. One way to minimize oxidation is to replace air with inert gas such as nitrogen or carbon dioxide.

Hawlader et al. (2006) in their study of modified atmosphere heat pump drying found that heat pump drying of ginger using nitrogen or carbon dioxide gave comparable retention of 6-gingerol. The retention was slightly lower than vacuum drying but higher than freeze drying and significantly better than hot air drying. It was reported that nitrogen gave higher effective diffusivity because nitrogen has a higher specific heat (1.04 kJ/kg °C) than that of normal air (1.00 kJ/kg °C), although their densities are nearly the same (0.9737 versus 1). Whereas for carbon dioxide, its specific heat (0.85 kJ/kg °C) is lower than that of air, but its density is much higher (1.539 vs. 1). Therefore, CO_2 has a larger mass at a fixed volume compared to air, which in turn gives larger heat transfer capacity. As a result, modified atmosphere heat pump drying using either nitrogen or carbon dioxide would give higher drying rate than normal heat pump drying. Alves-Filho (2002) used a CO_2 heat pump drying to convert raw and cold-extruded cranberry and potato–turnip mixtures into high-quality instant products.

5.2.4 Two-Stage Hot Air Heat Pump Drying

Heat pump drying can be combined with other types of drying such as convective hot air drying or microwave drying. The drying process can be carried out in sequence such as heat pump drying as first stage followed by convective hot air drying or vice versa, microwave drying followed by heat pump or vice versa.

Phoungchandang and Saentaweesuk (2011) in their study of multistage heat pump drying (first stage of cabinet drying at 70°C followed by second stage of heat pump drying at 40°C–60°C) reported that first stage convective drying at 70°C followed by heat pump drying at 40°C could reduce the drying time at 40°C by 59.32% and increase 6-gingerol content by 6%, if compared with two-stage convective drying followed by tray drying. In addition, the rehydration and color change ΔE were also better.

Borompichaichartkul et al. (2013) investigated multistage heat pump drying (using heat pump drying with air or nitrogen at 40°C followed by heat pump with air or nitrogen at 50°C–60°C) of macadamia nut under modified atmosphere, reported that using 40°C in the first stage heat pump drying with nitrogen as the drying medium followed by heat pump drying at 60°C with air as the drying medium could maintain the nut's quality, and it gave the lowest energy consumption. According to their observation, the energy consumption of first stage of drying was similar regardless of using air or nitrogen as the drying medium at 40°C. On the other hand, the second stage of heat pump drying under nitrogen atmosphere, higher temperature gave faster drying kinetics, hence resulted in lower energy consumption. Furthermore, they reported that drying under nitrogen in the second stage drying seemed to consume more energy than using air as the drying medium. This may be because of the specific heat of nitrogen that is slightly higher than air.

5.2.5 SOLAR-ASSISTED HEAT PUMP DRYER

Solar collector can be installed to harness solar energy for heating up water, which is circulated to the heat pump dryer to heat up the dehumidified air in the heat pump dryer. The drying process is carried out with solar energy harnessed in daytime and a heat pump system alone at night.

There is another type of solar-assisted heat pump dryer where a solar heat pump is used instead of a normal heat pump system. A normal heat pump system has a compressor, a condenser, an expansion valve, and an evaporator. A solar heat pump has a compressor, a condenser, an expansion valve, and an evaporator collector. The evaporator collector is similar to a solar collector that harnesses solar energy to evaporate the refrigerant.

Sevik et al. (2013) in their study of solar-assisted heat pump drying of mushroom found that mushrooms that were dried at 45°C–55°C and 310 kg/h mass flow rate gave better thermal efficiency than continuous heat pump drying and solar drying. Same degree of moisture content reduction could be achieved in solar-assisted heat pump drying with less energy input as solar energy was utilized to assist the drying process.

5.3 SELECTION OF HEAT PUMP DRYER

We now know the advantages of a heat pump dryer and understand the variants of heat pump dryer. Figure 5.2 further guides us to the selection of the heat pump dryer variants/types. It should be noted that Figure 5.2 is a general guide that is concluded from the findings reported by various researchers.

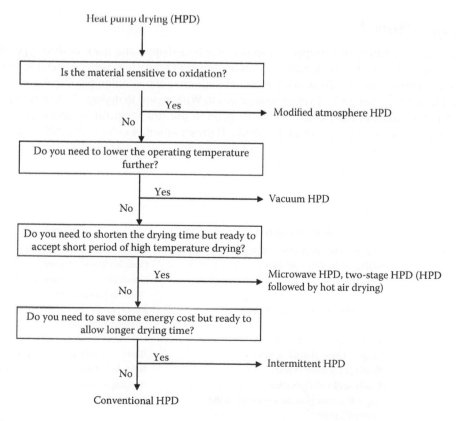

FIGURE 5.2 Selection of types of heat pump dryer.

5.4 HEAT PUMP DRYING OF VEGETABLES

Vegetables can be classified into many types. Table 5.1 lists the different classes of vegetables. Generally, vegetables are consumed in fresh; however, for preservation purpose, vegetables require dehydration in order to lower its water activity, which in turn gives longer storage period. It is well known that product size does significantly affect the drying kinetics; therefore, mechanical pretreatment such as cutting and punching hole on surface may be applied to facilitate the moisture removal during the drying process. Generally, vegetables may be cut into slices, pieces, or peeled before subjected to drying.

Vegetables are suitable to be dried by heat pump drying. This is because vegetables contain heat-sensitive bioactive ingredients that benefit health. Therefore, heat pump drying as a low temperature drying technique is suitable for vegetables. The following paragraphs briefly discuss the various vegetables that have been tested with heat pump drying. Readers should note that the heat pump drying method applied to the tested vegetables given below may not be the best or the most suitable drying method for the material or regarded as the only method that is suitable for the product. They are the drying method tested by researchers who used heat pump drying to perform the dehydration process.

5.4.1 ONION

Onion is dried for onion flakes and onion powder. Onion flakes are minced onions that are dehydrated, whereas onion powder is dehydrated onion in powder form. Dehydrated onion is a convenient food ingredient, and they are not as pungent as fresh onions.

Rossi et al. (1992) tested the drying of onion slices using a heat pump dryer, and found that heat pump drying consumed less energy in comparison to conventional hot air dryer. Furthermore, the product was better.

5.4.2 SWEET PEPPER

Sweet pepper, also called bell pepper or paprika, is a bell-shaped and thick-fleshed type of chili, which is commonly used for its characteristic sweet aroma in preparing curried vegetable and salad. It is rich in ascorbic acid as well as many essential nutrients such as chlorophyll that gives it green color, and β-carotene that gives it yellow-orange color. With regard to drying, the key is to preserve its ascorbic acid content; therefore, it would be good to use low temperature drying such as heat pump drying to dry sweet pepper in a rapid mode. If preservation of color is desirable, then it must be dried using the same drying technique as both chlorophyll and β-carotene are sensitive to heat.

TABLE 5.1
Classification of Vegetables

Vegetable Class	Brief Description	Examples
Bulbs	Consist of layers or clustered segments	Garlic, onion, and spring onion
Flowers	Edible flowers of vegetables	Cauliflower and broccoli
Fruits	Vegetable fruits are fleshy and contain seeds	Capsicum, chili, cucumber, eggplant, pumpkin, and tomato
Fungi	Macrofungi known as mushroom	Button mushroom, oyster mushroom, and shiitake mushroom
Leaves	Edible leaves of vegetables	Cabbage, lettuce, and spinach
Roots	Long or round-shaped taproot	Beetroot, carrot, daikon, radish, and turnip
Seeds	Edible seeds grow in pods	Bean, broad bean, pea, and sweet corn
Stems	Edible stalks of vegetables	Asparagus and celery
Tubers	Vegetables that grow underground on the root of a plant	Potato and yam

---

Pal et al. (2008, 2010) in their study of heat pump drying of green sweet pepper reported that the retention of total chlorophyll and ascorbic acid content was higher in heat pump-dried samples, which also gave higher rehydration ratio and better sensory score. The authors suggested that drying of green sweet pepper can be carried out using heat pump drying at 35°C judging from the energy and quality aspects.

5.4.3 CHILI

Chili or chili pepper belongs to the genus *Capsicum*, contains capsaicin (8-methyl-*N*-vanillyl-6-nonenamide) and other capsaicinoids, as well as high level of vitamins, especially vitamins C and A, minerals and flavonoids such as β-carotene. Dried chili is used as food ingredient in Indian and Chinese cuisines. In drying chili, it is important to preserve the red color and nutrients of the chili. Sun drying or hot air drying can cause damage to its flavor, color, and nutrients as the drying temperature is relatively higher compared to heat pump drying.

Artnaseaw et al. (2010a,b) applied vacuum heat pump drying for chili drying and reported that the heat pump drying gave better product quality than sun drying. In addition, reducing drying pressure could reduce the shrinkage, produce better rehydration attribute and minimize color change. Furthermore, higher drying temperature resulted in greater color degradation due to enzymatic browning reaction, which is accelerated at higher temperature.

5.4.4 TOMATO

Tomato is usually consumed in fresh form. It is also processed into different forms of food stuff such as tomato paste, sauce, ketchup, tomato juice, tomato puree, and canned. In addition, dried tomato is also produced, and it is traditionally carried out using sun drying.

Sevik (2014) tested drying of tomato using heat pump-assisted solar dryer and reported that the degradation of bioactive compounds was minimum if heat pump dryer is used, namely 3% for lycopene, 13% for ascorbic acid, and 2% for polyphenols.

5.4.5 STRAWBERRIES

Strawberries may be consumed in fresh form or used as juice, in concentrated jam, in jelly, or dry and rehydrated with yogurt and bakery products. Sevik (2014) reported the bioactive ingredients retained in dried strawberry by heat pump-assisted solar dryer were higher compared to those retained with the traditional method. The values were 150 mg ascorbic acid in dried strawberries compared to 230 mg in fresh strawberries, whereas polyphenols was 5 1g/100 mg versus 14 1g/100.

5.4.6 EDIBLE MUSHROOM

Edible mushrooms are macrofungi that bear fruiting structure. They are consumed for their nutritional values and some are consumed for their medicinal values. There are many types of edible mushroom. Shiitake mushroom is the most reported species in the literature, especially pertaining to drying. Shiitake mushrooms are highly valued as food and tonic in the East. Dried mushroom and mushroom extracts are used in herbal remedies. It contains active ingredients such as lentinan and 1,3-beta glucan, which are known as good agents for fighting tumor. It also contained eritadenine, ergosterol, and fungisterol, which provide benefits to human health. As such, proper drying methods are often required in order to preserve active ingredients in the dried mushroom.

Artnaseaw et al. (2010b) carried out vacuum heat pump drying of shiitake mushroom and found that vacuum heat pump drying gave better product quality in terms of color change and rehydration properties. They reported that the product quality would be even better if the vacuum drying of shitake mushroom is carried out at higher temperature and lower vacuum.

Sevik et al. (2013) in their study of solar-assisted heat pump drying of mushroom (the type of mushroom was not revealed) found that mushrooms that were dried at 45°C and 55°C and 310 kg/h mass flow rate gave better thermal efficiency than those dried using continuous heat pump drying and solar drying. Same degree of moisture content reduction could be achieved in solar-assisted heat pump drying with less energy input, as solar energy was utilized to assist the drying process.

5.4.7 OLIVE LEAVES

Olive leaves are rich in phenolic compounds. They include oleuropein and hydroxytyrosol, which are beneficial to human health such as preventing cardiac disease and diabetic neuropathies and protection against atherosclerosis. Olive leaves are dried for producing food ingredient in dry mixes, for extracting antioxidant phenolic compounds, and for making olive leaf tea.

Using a pilot-scale heat pump dryer, Erbay and Icier (2009) reported that the optimum operating conditions for heat pump drying of olive leaves were 53.43°C, air velocity of 0.64 m/s, process time of 288.32 min. At this condition, it would give lesser loss in total phenolic content and total antioxidant activity. The final moisture content was 6% and the exergy efficiency was 69.55%.

5.4.8 BASIL LEAVES

Basil is most commonly used fresh in cooked recipes for adding flavor to the food, and it is also a main ingredient in pesto—a green Italian oil-and-herb sauce. Fresh basil leaves are kept in refrigerator, freezer for short period storage. It can be dehydrated for longer storage. Its flavor can be easily lost during drying. Phoungchandang et al. (2003) reported that heat pump drying could retain higher amount of eugenol and methyl eugenol in dried basil leaves.

5.4.9 PARSLEY LEAVES

Parsley is an aromatic plant that is widely used in culinary and medicinal purposes. The leaves, either fresh or dried, its roots, and seeds are used to produce spice, essential oils, and drugs in food, cosmetic, and pharmaceutical industries. Dried parsley may be used for food seasoning or for making herbal tea. Parsley is a source of flavonoid, antioxidants (especially luteolin), and vitamins (such as vitamins K, C, and A).

5.4.10 MINT LEAVES

Mint is also widely used in the food, flavoring, pharmaceutical, and cosmetic industries. It is a component in many drugs. It is rich in menthol, flavonoids (timonina), and caffeic acid derivatives. Its leaves are used in a variety of cuisines as flavoring and applied medicinally for stomach soothing and anti-infective effects.

Sevik (2014) carried out the heat pump-assisted solar drying of tomato, strawberry, mint, and parsley and concluded that

- Retention of the total polyphenolic content during storage is good. Six months after drying, the retention remained unchanged. The retention of lycopene, β-carotene, and ascorbic acid in the dried products was good.
- Herbs that belong to the same family gave similar retention of polyphenols.
- In the case of tomato, strawberry, mint, and parsley, it was found that heat pump-assisted solar drying produced similar quality attributes without significant difference. The attributes are thermal damage, browning, shrinkage, and taste on dried products.

5.4.11 Asian White Radish

White radish, also known as daikon in Japan, is often used in salad, dishes, and soups. It is a good source of antioxidants (sulforaphane), electrolytes (such as potassium and sodium), minerals (copper and iron), vitamins (especially vitamin C), and dietary fiber.

Coogan and Wills (2008) studied the effect of freeze drying, heat pump drying, hot air drying, and osmotic dehydration using salt solution, and reported that retention of flavor compound (4-methylthio-3-trans-butenyl isothiocyanate—MTBITC) in dried white radish decreased during drying. Anyway, freeze drying gave the best retention. Heat pump drying at a lower temperature could give good retention, although not as high as freeze drying. The loss of MTBITC increased with increasing drying temperature for both hot air and heat pump drying. Furthermore, heat pump drying gave better kinetics when compared to hot air drying.

5.4.12 Macadamia Nuts

Macadamia nuts are rich in monounsaturated fatty acids, and it is good in reducing cholesterol and triglyceride levels, thus lowering the risk of heart disease. Macadamia nuts contain more than 70% fat, and the major fatty acids are oleic acid, palmitoleic acid, and palmitic acid. Macadamia nut quality depends significantly on moisture content and fatty acids content. High unsaturated fatty acids give rise to hydrolytic and oxidative rancidity if the free moisture is high. In addition, lipid oxidation may occur if the macadamia nuts contain high oil content (Kaijser et al., 2000).

In terms of color change, browning is likely to occur when high moisture nut is dried at higher temperature due to nonenzymatic reaction (also known as *Maillard reaction*). When water activity of macadamia nut is high, the reaction is promoted (Dominguez et al. 2007). It is desired to dry macadamia nut to have low rancidity and light color. Borompichaichartkul et al. (2009) investigated the application of two-stage drying for macadamia nuts and found that two stage strategy, which consists of heat pump drying (carried out at 40°C) in the first stage until an intermediate moisture content of 11.1% db, followed by hot air drying (carried out at 50°C) until a final moisture content of 1%–2% db gave the best quality of dried macadamia nuts.

5.4.13 Potato and Sweet Potato

Potato and sweet potato are rich in starch. In addition, they are also rich in other ingredients. For instance, potato contains vitamins (such as vitamins C and B6), minerals (such as potassium, magnesium, phosphorus, iron, and zinc), and phytochemicals (such as carotenoids and natural phenols). Sweet potato contains dietary fiber, β-carotene, and micronutrients such as vitamins B5, B6, and manganese.

Chua et al. (2000) observed that drying potato pieces in a cyclic heat pump drying under cosine drying temperature profile (started with higher temperature followed by lower temperature and the cycle continues) gave better drying kinetics without causing significant color change. They attributed it to the fact that potato contains higher initial moisture content and lower sugar content and therefore not so vulnerable to Maillard reaction.

5.4.14 Ginger

Ginger is a pale yellow pungent aromatic rhizome, containing oleoresin and essential oils. Ginger is used worldwide as an ingredient in food and medicine. It has long been used to treat many gastrointestinal disorders and is often promoted as an effective antiemetic analgesic, antipyretic, anti-inflammatory, chemopreventive, and antioxidant properties (Huang et al. 2011). Ginger rhizome contains active ingredients such as n-gingerol, zingerone, and n-shogaol (Balladin et al. 1998), and these active ingredients are antioxidant and anticancer agent. The maturity of the ginger rhizomes also affects the 6-gingerol content. Phoungchandang et al. (2009) reported the 6-gingerol content in a 12-month-old ginger rhizome is more than double the content in a 6-month-old ginger rhizome.

Phoungchandang et al. (2009) in their study of heat pump drying of ginger rhizome found that heat pump dehumidifier drying could retain higher content of 6-gingerol than hot air tray drying and sun drying. However, at higher temperature for heat pump drying such as 60°C, the retention is as low as the retention obtained in hot air tray drying and sun drying. However, it gave shorter drying time than tray drying because of lower relative humidity.

Furthermore, Phoungchandang and Saentaweesuk (2011) reported that heat pump drying carried at two stages, that is, hot air drying at 70°C followed by heat pump drying at 40°C, which could further reduce the drying time significantly and increase the retained 6-gingerol content. In addition, other quality attributes such as rehydration and ΔE^* were also better.

5.4.15 YACON

Yacon is a tuberous root that is rich in fructooligosaccharides (FOS) and inulin. FOS exhibits good health effect on humans, especially in enhancing colon health and aiding digestion, reducing blood lipid and glucose levels, and preventing and controlling constipation. However, FOS hydrolyzes rapidly after harvest, which in turn affects its functionality. Furthermore, browning and decay occur during post-harvest and transportation, which causes huge postharvest losses. Therefore, preservation is required and drying is normally applied.

Shi et al. (2013, 2014) applied heat pump drying for yacon and found that higher temperature and higher velocity gave faster drying kinetics. However, both drying temperature and air velocity have little effects ($p > .05$) on total color change, shrinkage rate, and rehydration rate of dried yacon slices.

5.5   CONCLUDING REMARKS

Although vegetables are generally consumed in fresh form, they require preservation if there is excessive supply. Vegetables contain active ingredients such as phytochemicals, antioxidants, flavonoids, and vitamins; some of these compounds are sensitive to heat, and therefore, low temperature drying may be applied when it comes to preservation and dehydration of vegetables. Heat pump drying is one of the low temperature drying candidates, whose operating cost is relatively lower than freeze drying and drying time is relatively shorter than freeze drying and vacuum drying.

Heat pump drying is generally used to dry vegetables because of the following advantages, when compared to the conventional drying methods such as hot air drying and sun drying:

- Minimize color change
- Maximize retention of heat-sensitive bioactive ingredients

Many variants of heat pump drying have been proposed in order to enhance the performance of conventional heat pump dryer. These include modified atmosphere that can induce oxidation, vacuum heat pump drying, which can further reduce the operating temperature, microwave-heat pump drying and convective heat pump drying, which can shorten the drying time (product quality is slightly compromised depending on the duration of exposure to hot air), and intermittent heat pump drying, which can save some energy cost (but total processing time is slightly increased).

Clearly, it is impossible to overview all vegetables that are suitable for heat pump drying. We can generally conclude that heat pump drying is suitable for drying of vegetables. This chapter covers various variants of heat pump drying that have been reported by researchers in the literature and briefly summarizes the findings reported on heat pump drying of vegetables. Industrial scale heat pump drying of vegetables may have already been implemented in the industry, but there is no report in the literature and public domain. It should be noted that the heat pump drying methods applied by the researchers, which are discussed in this chapter, does not imply that they are the best drying technique for the drying of the vegetables. Other drying methods may be suitable as well, and careful selection of dryer should be exercised. Readers may refer to the handbook edited by Mujumdar (2014) for detailed information on a wide range of industrial dryers, including heat pump drying.

REFERENCES

Adapa, P.K., Sokhansanj, S., Schoenau, G.J. 2002. Performance study of a re-circulating cabinet dryer using a household dehumidifier. *Drying Technology* 20(8):1673–1689.

Alves-Filho, O. 2002. Combined innovative heat pump drying technologies and new cold extrusion techniques for production of instant foods. *Drying Technology* 20(8):1541–1557.

Artnaseaw, A., Theerakulpisut, S., Benjapiyaporn, C. 2010a. Development of a vacuum heat pump dryer for drying chilli. *Biosystems Engineering* 105:130–138.

Artnaseaw, A., Theerakulpisut, S., Benjapiyaporn, C. 2010b. Drying characteristics of Shiitake mushroom and Jinda chilli during vacuum heat pump drying. *Food and Bioproducts Processing* 88:105–114.

Balladin, D.A., Headley, O., Chang-Yen, I., McGaw, D.R. 1998. High pressure liquid chromatographic analysis of the main pungent principles of solar dried West Indian ginger (*Zingiber officinale* Roscoe). *Renewable Energy* 13(4):531–536.

Borompichaichartkul, C., Chinprahast, N., Devahastin, S., Wiset, L., Poomsa-ad, N., Ratchapo, T. 2013. Multistage heat pump drying of macadamia nut under modified atmosphere. *International Food Research Journal* 20(5):2199–2203.

Borompichaichartkul, C., Luengsode, K., Chinprahast, N., Devahastin, S. 2009. Improving quality of macadamia nut (*Macadamia integrifolia*) through the use of hybrid drying process. *Journal of Food Engineering* 93:348–353.

Chua, K.J., Chou, S.K., Ho, J.C., Hawlader, M.N.A. 2002. Heat pump drying: Recent developments and future trends. *Drying Technology* 20(8):1579–1610.

Chua, K.J., Chou, S.K., Yang, W.M. 2010. Advances in heat pump systems: A review. *Applied Energy* 87(12):3611–3624.

Chua, K.J., Mujumdar, A.S., Chou, S.K., Hawlader, M.N.A., Ho, J.C. 2000. Convective drying of banana, guava and potato pieces: Effects of cyclical variations of air temperature on drying kinetics and color change. *Drying Technology* 18(4–5):907–936.

Colak, N., Hepbasli, A. 2009. A review of heat-pump drying (HPD): Part 2—Applications and performance assessments. *Energy Conversion and Management* 50(9):2187–2199.

Coogan, R.C., Wills, R.B.H. 2008 Flavor changes in Asian white radish (*Raphanus sativus*) produced by different methods of drying and salting. *International Journal of Food Properties* 11(2):253–257.

Dominguez, I.L., Azuara, E., Vernon-Carter, E.J., Beristain, C.I. 2007. Thermodynamic analysis of effect of water activity on the stability of macadamia nut. *Journal of Food Engineering* 81:566–571.

Erbay, Z., Icier, F. 2009. Optimization of drying of olive leaves in a pilot-scale heat pump dryer. *Drying Technology* 27:416–427.

Fatouh, M., Abou-Ziyan, H.Z., Metwally, M.N., Abdel-Hameed, H.M. 1998. Performance of a series air-to-air heat pump for continuous and intermittent drying. *Proceedings of the ASME Advanced Energy Systems and Division* 38:435–442.

Hawlader, M., Perera, C.O., Tian, M. 2006. Properties of modified atmosphere heat pump dried foods. *Journal of Food Engineering* 74:392–401.

Hii, C.L., Law, C.L., Suzannah, S. 2012. Drying kinetics of the individual layer of cocoa beans during heat pump drying. *Journal of Food Engineering* 108(2):276–282.

Huang, T.C., Chung, C.C., Wang, H.Y., Law, C.L., Chen, H.H. 2011. Formation of 6-shogaol of ginger oil under different drying conditions. *Drying Technology* 29:1884–1889.

Kaijser, A., Dutta, P., Savage, G. 2000. Oxidative stability and lipid composition of macadamia nuts grown in New Zealand. *Food Chemistry* 71:67–70.

Kohayakawa, M.N., Silveria-Junior, V., Telis-Romero, J. 2004. Drying of mango slices using heat pump dryer. In *Proceedings of the 14th International Drying Symposium*, August 22–25, 2004, Sao Paulo, Brazil, pp. 884–891.

Lee, K.H., Kim, O.J. 2009. Investigation on drying performance and energy savings of the batch-type heat pump dryer. *Drying Technology* 27:565–573.

Mujumdar, A.S. (ed.). 2014. *Handbook of Industrial Drying*, 4th edn., CRC Press, Boca Raton, FL.

Ong, S.P., Law, C.L., Hii, C.L. 2012. Optimization of heat pump-assisted intermittent drying. *Drying Technology* 30:1676–1687.

Pal, U.S., Khan, M.K., Mohanty, S.N. 2008. Heat pump drying of green sweet pepper. *Drying Technology* 26:1584–1590.

Pal, U.S., Khan, M.K. 2010. Performance evaluation of heat pump dryer. *Journal of Food Science and Technology* 47(2):230–234.

Patel, K.K., Kar, A. 2012. Heat pump assisted drying of agricultural produce—An overview. *Journal of Food Science and Technology* 49(2):142–160.

Perera, C.O., Rahman, M.S. 1997. Heat pump dehumidifier drying of food. *Trends in Food Science & Technology* 8(3):75–79.

Phoungchandang, S., Nongsang, S., Sanchai, P. 2009. The development of ginger drying using tray drying, heat pump dehumidified drying and mixed mode solar drying. *Drying Technology* 27(10):1123–1131.

Phoungchandang, S., Saentaweesuk, S. 2011. Effect of two stage, tray and heat pump assisted-dehumidified drying on drying characteristics and qualities of dried ginger. *Food and Bioproducts Processing* 89:429–437.

Phoungchandang, S., Sanchai, P., Chanchotikul, K. 2003. The development of dehumidifying dryer for a Thai herb drying (Kaprao leaves). *Food Journal* 33(2):146–155.

Prasertsan, S., Saen-saby, P., Nyamsritrakul, P., Prateepchaikul, G. 1996. Heat pump dryer. Part 1: Simulation of the models. *International Journal of Energy Research* 20:1067–1079.

Prasertsan, S., Saen-saby, P. 1998. Heat pump drying of agricultural materials. *Drying Technology* 16 (1 & 2):235–250.

Putranto, A., Chen, X.D., Devahastin, S., Xiao, Z., Webley, P.A. 2011. Application of the reaction engineering approach (REA) for modeling intermittent drying under time-varying humidity and temperature. *Chemical Engineering Science* 66:2149–2156.

Queiroz, R., Gabas, A.L., Telis, V.R.N. 2004. Drying kinetics of tomato by using electric resistance and heat pump dryers. *Drying Technology* 22(7):1603–1620.

Rocha, T., Lebert, A., Marty-Audouin, C. 1993. Effect of pre-treatments and drying conditions on drying rate and colour retention of basil. *Lebensmittel-Wissenschaft und-Technologie* 26:456–463.

Rossi, S.J., Neues, C., Kicokbusch, T.G. 1992. Thermodynamics and energetic evaluation of a heat pump applied to drying of vegetables. In: Mujumdar, A.S. (ed.), *Drying'92*. Elsevier Science Publishers, Amsterdam, the Netherlands, pp. 1475–1478.

Sevik, S., Aktas, M., Dogan, H., Koçak, S. 2013. Mushroom drying with solar assisted heat pump system. *Energy Conversion and Management* 72:171–178.

Sevik, S. 2014. Experimental investigation of a new design solar-heat pump dryer under the different climatic conditions and drying behavior of selected products. *Solar Energy* 105:190–205.

Shi, Q., Zheng, Y., Zhao, Y. 2013. Mathematical modeling on thin-layer heat pump drying of yacon (*Smallanthus sonchifolius*) slices. *Energy Conversion and Management* 71:208–216.

Shi, Q., Zheng, Y., Zhao, Y. 2014. Drying characteristics and quality of Yacon (*Smallanthus sonchifolius*) during heat pump drying. *Food Science* 35(3):16–22.

Sosle, V., Raghavan, G.S.V., Kittler, R. 2003. Low-temperature drying using a versatile heat pump dehumidifier. *Drying Technology* 21(3):539–554.

Yang, Z., Zhu, E., Zhu, Z., Wang, J., Li, S. 2013. A comparative study on intermittent heat pump drying process of Chinese cabbage (*Brassica campestris* L. ssp) seeds. *Food and Bioproducts Processing* 91:381–388.

6 Drying of Fruits and Vegetables
The Impact of Different Drying Methods on Product Quality

Y. Baradey, M.N.A. Hawlader, A.F. Ismail, and M. Hrairi

CONTENTS

SUMMARY

Drying is a water removal process from foodstuffs commonly used for preservation and storage purposes. Different drying methods are nowadays available, such as freeze drying, vacuum microwave drying, hot air drying, heat pump drying, and solar-assisted heat pump drying (SAHPD). Fruits and vegetables are very important products for human beings because of their high nutritional values. They are perishable products and contain high percentage of moisture. The moisture content, in some cases, may be more than 90%. Water content is considered the main reason for microorganisms' growth, which leads to putrefaction. Preservation of fruits and vegetables is considered to be a very important step to match availability in agriculture sector and consumers' utilization to keep it fresh and nutritional as long as possible, and to meet the demand for fast and instant products with high quality. Drying methods are likely to affect the physical, chemical, nutritional, and biological properties of fruits and vegetables. Despite the negative influences that could occur because of moisture removal by drying on the quality of the food, it is still an indispensable process in different sectors of industry, because it provides low packaging costs, increased shelf-life of the foodstuff, and lower shipping weights. In this chapter, the impact of different drying methods on the quality attributes of fruits and vegetables are summarized. From a comparison of all the drying methods available, it is seen that heat pump or SAHPD shows a better retention of nutrients compared to other methods at a reasonably low cost.

6.1   INTRODUCTION

Generally, sustainability, safety, and innovation have recently become the most attractive and significant goals, especially in the modern societies. According to Rahman (1999), food, including fruits and vegetables, safety is the most significant industrial topic across the world because of its importance and effects on human life. Food has been classified into harvested, fresh, raw, medical, synthetic, minimally processed, perishable (those kinds of food that do not need much time to spoil), and nonperishable. Fruits and vegetables belong to medical and perishable types. Drying usually refers to a natural process contributed by natural resources such as sun for food with more than 2.5% of water content, whereas rehydration phase refers to the artificial process that happens under specific conditions for food with less than 2.5% of water content (Vega-Mercodo et al. 2001). Food drying process is a reliable and natural way known since ancient time (Tortoe 2010). It is considered to be the oldest process and was widely used in the past to provide the troops with food in both World Wars (Ramos et al. 2003). In some countries, dried fruits and vegetables contribute to a large portion of the export goods of the country, which contributes to an annual income of the country. It is reported that the yearly income of Ghana during 2008 from dried fruits and vegetables was US$3,600,600 (Tortoe 2010).

The moisture content in fruits and vegetables reaches, in some types, more than 90%. For instance, the moisture content in the water melon is about 93% (Jangam et al. 2010). Water content is considered to be the main reason for microorganisms' growth, and this must be reduced to a desired value in order to keep it in the best quality, and prevent the microbial and enzymatic growth (Qing-guo et al. 2006). Preservation of fruits and vegetables is very important in agricultural sector, and storage is essential over a period of time to reduce the spoilage problem, keep it fresh and nutritional as long as possible (Jangam et al. 2010), and to meet the demand for fast and instant products with high quality (Qing-guo et al. 2006). Water evaporation from the product influences the solid matrix by increasing the concentration of the soluble in it. This will motivate the chemical and enzymatic reactions, which affect the structure and functionality of biopolymers (Lewicki 2006).

The aforementioned causes of deterioration and keeping the fruits and vegetables nutritional, and increasing the shelf life of the food products are considered the main reasons for the need of preservation technology (Pavan 2010). However, spoilage or putrefaction occurs due to several reasons, but the most commonly mentioned causes are chemical and microbial damages (Rahman 1999; Jangam et al. 2010). The most commonly used ways and techniques for food nutrients retention are preserving in syrup (Jangam et al. 2010), canning, food irradiation, vacuum packing, drying or dehydration, refrigeration, and freezing (Hawlader et al. 2005). Some researchers reported that drying will never be a good alternative process for freezing and canning. They claimed that these two methods give better results in terms of preserving the taste and appearances, but in reality, the drying method is more efficient in nutrition retention, more cost effective, as well as it provides much less storage space compared to freezing and canning (Hawlader et al. 2006; Jangam et al. 2010).

In developing countries, where the standard of living and energy consumption is increasing with time, a need for energy-efficient and cost-effective dryer has become necessary and inevitable. Therefore, designing and developing a totally sustainable drying technology has also become important endeavor because of increasing oil price in the past few decades till now because of wars and other political conflicts as well as global warming and ozone depletion issues. In order to achieve an efficient drying process, the product quality, the economic viability, and the environmental impacts must be taken into account (Pavan 2010). According to Hawlader et al. (2005) heat pump drying technologies are the most attractive and promising systems, because these systems lead to faster rehydration, better nutritional value of the products, better quality, better porous products, and better physical appearances of the products. Despite tremendous publications and literature about drying of fruits and vegetables, it is still a daunting task to assimilate all of it in a single chapter. In order to alleviate this problem, the most common and significant issues related to this topic are explained and mentioned in this chapter.

6.2 DRYING METHODS FOR FRUITS AND VEGETABLES

Drying of fruits and vegetables is an important industrial sector and the products have high nutritional values, which are considered very useful for human beings (Hawlader et al. 2005). It is a process of removing water content from fruits and vegetables by evaporating it (Hawlader et al. 2005). It includes a complex heat and mass transfer process leading to modified properties of the products (Hawlader et al. 2006). As stated earlier, there are many different drying methods used in fruits and vegetables preservation, such as solar drying method (direct solar dryers [DSDs], indirect solar dryers, and hybrid solar dryers), atmospheric drying method (spray dryers, kiln dryers, tower dryers, and drum dryers), novel drying method (microwave dryers, magnetic field dryers, explosion puffing dryers, osmotic dehydration, and infrared radiation dryers), freeze drying method, canning drying method, chemical drying method, and hot air drying method (Hawlader et al. 2005; Tortoe 2010). In addition, sugar and salt were used in the past to preserve the quality of the fruits and vegetable products and to preserve the water or moisture content. The most common drying methods for fruits and vegetable are summarized and explained in Sections 6.2.1 through 6.2.5.

6.2.1 Freeze Drying Method

Freeze drying (lyophilization) is a low temperature dehydration process where removal of the water content from the foodstuff is usually done by sublimation. It is considered to be one of the most common drying processes because of its wide applications. It is suitable for preservation of heat-sensitive biological materials, complex technological products, food products, fruits and vegetables, flowers, cosmetics, enzymes and ceramic powders, and microorganisms (Pavan 2010; Ciurzyńska and Lenart 2011). Freeze drying includes three basic steps: lowering the temperature of the product to very low value (below the freezing temperature) in order to increase the ice content in the product, sublimation of ice, and removing the rest of water content (the unfrozen content) by evaporation.

Freeze drying methods suffer from some disadvantages despite a good consensus on the high efficiency of this process. These demerits include economical aspect due to high fixed operating costs, complexity of the process, and large required storage space (Qing-guo et al. 2006; Rahman 2013; Xu et al. 2015). Different attempts to reduce the operating cost and energy consumption of the freeze drying method have been conducted, such as improving the mass and heat transfer in the freezing process, averting use of condenser for the sublimated vapor, making a combination between freeze drying method with other dehydration methods, and making it work at near to atmospheric pressure known as *atmospheric freeze dryer* (AFD) *system* (Rahman 2013). Some researchers prefer to create a hybrid of freeze drying and other drying methods in order to reduce the energy consumption of the process and, therefore, make it more cost effective. Microwave freeze dryer (MFD) system is an example of this combination. It is the process in which microwave is used as heat source for the freeze drying system (Xu et al. 2015). This could lead to a significant improvement in freeze drying rate and lower the required energy input. MFD has been largely tested in recent years across the world to evaluate the system performance, and its effects on the quality of dried foods.

6.2.2 VACUUM MICROWAVE DRYING METHOD

It is a combination of low temperature drying by vacuum and fast drying by microwave radiation. Vacuum microwave drying system has recently become an excellent promising way for food preservation worldwide, because it shortens the dehydration process time and enhances the heat and mass transfer gradient between inside and outside of the products (Qing-guo et al. 2006). The moisture removal process, in the traditional drying methods, happens by evaporation of water from the tissues of the product. That means it happens when the water reaches the boiling point (100°C at 1 bar). Using vacuum dryer, the evaporation will occur at lower pressure, which reduces the evaporation temperature of the water (28.96°C at 40 mbar). This is how the vacuum microwave dyers shorten the time of drying process than the conventional drying methods (Karimi 2010). Generally, a system that includes vacuum application comprises four major components: vacuum pump, heat supply, vacuum chamber, and condenser. In addition, these type of dryers are preferred to be used for the high value materials because of the high operating costs.

Hertz was the first scientist who discovered and experimentally proved the existence of microwave radiation in 1888 (Karimi 2010). The first application of microwave radiation was for communication, but its applications nowadays are widely spread in different fields of industries, especially in food preservation. Microwave dryer usually used for heat-sensitive materials. The biggest advantage of using microwave for fruits and vegetables retention compared to other drying methods is that heat production happens due to direct conversion of electromagnetic energy into kinetic molecular energy. This enables the generated heat distribution deeply within fruits and vegetables. Water has an asymmetric molecular structure, which makes it one of the microwavable materials. When the electromagnetic field from the microwave reaches the water content in the product, the molecules form electric dipoles in the same direction as the microwave field, which is the main reason for heat generation (Püschner and SiokHoon 2005). In addition, when a product is heated by microwave, the heating happens suddenly and uniformly, unlike conventional dryers or ovens, where the product is surrounded completely by hot air to heat up the product which gives better color. Attempts in designing high-powered microwave oven to reduce both manufacturing costs and electrical energy costs required for microwave ovens make microwave technology a good alternative and offer significant future improvements.

6.2.3 HOT AIR DRYING METHOD

Hot air drying method is a traditional drying method widely used in industry, because it is simple and economically feasible compared to other methods (Chin and Law 2012). Chemical and thermal treatments are usually involved before this type of drying, and the operational temperature ranges in most of the time between 50°C and 80°C. In order to avoid browning in drying, some researchers use two stage drying process. The operating temperature of the first stage is 30°C or 40°C, while 60°C–80°C for the second stage (Argyropoulos et al. 2008). According to Qing-guo et al. (2006), the most widely used system for drying fruits and vegetables is the hot-air drying method, but it is still suffering from many drawbacks, because it needs high temperature, which usually leads to many serious problems, such as reduction in the rehydration capacity and bulk density of the fruits and vegetables, reduction in the nutritional values, damage of the flavor and color of the products and biological properties of the fruits and vegetables, and movement of the solutes from inside to the surface of the drying material.

6.2.4 Osmotic Dehydration

The main purpose of drying fruits and vegetables is to decrease the water content in order to minimize or even eliminate the level of chemical reactions and microbial growth that cause deterioration and to facilitate distribution and storage (Tortoe 2010). Osmotic dehydration process is a process in which water content is removed by immersing the foodstuff in an aqueous solution of high osmotic pressure. The moisture removal happens by diffusion and capillary flow, but solute leaching happens only by diffusion. It is, however, still being considered as a new and efficient method with a potential of energy saving and enhancement of dehydrated product quality. In osmotic dehydration, when fruits and vegetables are immersed in a saline or sugar solution, three kinds of mass transfer will take place. In the first process, the water will flow from the product tissues to the solution, while in the second one, it will flow from the solution to the tissues, and finally the solutes will leach out from the tissues to the osmotic solution, but this process is still not important and is neglected by some researchers. It is reported that osmotic dehydration process is a preprocessing step to further drying method such as freeze, hot air, and vacuum dryers, because it improves the nutritional value of the product without causing any change in the physical properties. Therefore, using it as a preprocessing step will increase the percentage removal of moisture content from inside tissues and shorten the time of the whole drying process. It is also used as postprocessing step after or even during a conventional drying method (Jalaee et al. 2010).

6.2.5 Solar Drying Methods

Drying is an important process that is nowadays widely used in many industries such as chemical, timber, pharmaceuticals, food, and paper. This process usually requires great amount of energy that sometime accounts for up to 15% of all industrial energy usage, often with relatively low thermal efficiency in the range of 20%–25% (Chua et al. 2001). Solar drying process has been used since ancient times to preserve food and agricultural crops. From energy conservation point of view, solar drying is the most attractive and useful systems for food dehydration and preservation. There are plenty of advantages for solar drying systems such as saving time and energy (Karabacak and Atalay 2010), improving the quality of the product, zero impact on environment, and no global warming potential because of solar energy use instead of fossil fuel or oil (Daghigh et al. 2008). However, some researchers reported that open air sun drying (OASD) and DSD consumed frequently more energy than other drying methods such as microwave (Mechlouch et al. 2012). Despite the advantages of solar drying method stated earlier, it still suffers from some demerits, such as high labor costs for operators; risk of spoilage of some products because of wind, moisture in the air, rain, and dust; high possibility for biochemical and microbial reactions; and loss of some products because of birds and animals. Another disadvantage of solar drying is that the solar radiation intensity is a function of time, so that the implementation of this process is not possible all the time, because it depends on the availability of sun during the day or the year, and in this case, there will be a lack of process control. These processes also require high efficient collecting devices (solar collectors) to provide the system with the required energy. Flat plate and concentrating collectors are usually used in solar dryers.

There are different types of solar drying methods and designs available nowadays, such as, OASD, direct sun drying, heat pump drying, and SAHPD. Heat pump dryer consists of a condenser, evaporator, compressor, and an expansion valve connected together by using copper tubes. Solar-assisted heat pump technology is a combination of heat pump system and solar system to exploit more than one energy source simultaneously. According to Daghigh et al. (2008), hybridization or combination of solar dryer with heat pump system could bring a lot of benefits, such as making solar drying more efficient, helping overcome the difficulties of using such systems, improving the physical properties of dried foodstuff, reducing the energy consumption, improving the thermal efficiency of the drying system, and meeting the demand of the industry.

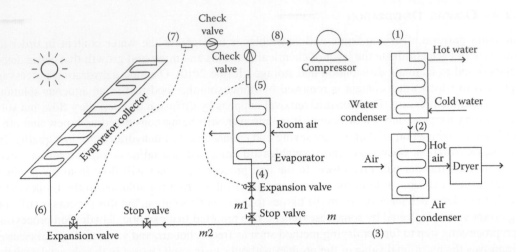

FIGURE 6.1   Schematic diagram of a solar-assisted heat pump dryer. (From Shaochun, Y., An integrated solar heat pump system for cooling, water heating and drying, A Thesis Submitted for the Degree of Doctor of Philosophy, Department of Mechanical Engineering, National University of Singapore, Singapore, 2009. With permission.)

Solar-assisted heat pump systems for drying have been widely used in industry over a period and a lot of research has been conducted on it. These systems are used and being improved in order to enhance the thermal performance of the systems and the quality of the products where low thermal and low energy systems are needed. It is considered to be one of most efficient and controllable (temperature and humidity) methods for food dehydration and the best alternative among traditional drying methods. It is reported that this technology can be combined with other drying methods, such as AFD, spray drying, and microwave drying (Patel and Kar 2012). This kind of system is categorized into active (free convection), hybrid, and passive (forced convection). The hybrid solar-assisted heat pump systems are classified into system with thermal storage (water, pool, pebble bed, phase change material [PCM], solar pond, etc.) and system without thermal storage but with backup heating (Jangam et al. 2011). An example of an SAHPD system is shown in Figure 6.1.

6.3   QUALITY OF DRIED FRUITS AND VEGETABLES

The term *quality* is defined as "a combination of characteristics, attributes, or properties that give the commodity value as a human food (Rahman 1999, p. 19)." It can be defined from either the customers' points of views or the product orientations (Abbott 1999). Customers are not only the persons who eat the final product and judge its quality. Customers, in marketing series, include farmers, packers, dispensers, sealers, retailers, managers, and shoppers. Each one of them has his or her own judgment and quality or acceptability criteria. Even among the final customers, the quality criteria could be different from one another, because customers usually use all human senses such as sight, smell, taste, touch, and sometimes hearing to evaluate the product and judge how good or bad it is. In such a case, the main questions here are what product's attributes must be measured, how to measure them, and how to determine what attributes are acceptable by customers.

Actually, most of researchers prefer and use the instrumental measurements more than the customers' senses in quality evaluation and determination, because they help in eliminating the wide variations among customers and could provide more precise healthy language among researchers accurately. Electromagnetic or optical instruments are being used to measure the appearance, texture measured by mechanical instruments, flavor which is taste and aroma is measured using chemical properties, ripeness by empirical methods, pigment content measured by light absorbance, color measured by colorimeters. However, in most cases, a relationship between the customer's wishes and acceptability criteria and the instrumental measurements must be established and taken into account by designing the measurements in ways that imitate the customer senses or by some statistical studies. The instrumental methods to measure and evaluate the quality attributes of fruits and vegetables have been largely improved worldwide in the past century.

Despite the negative influences that could occur because of moisture removal by drying on the quality of the food, it is still an indispensable process in different sectors of industry, because it provides low packaging costs, increased shelf-life of the foodstuff, and lower shipping weights. It is reported that designing the dryer carefully and consciously is considered to be one of the common solutions used in order to enhance the quality of dried fruits and vegetables. The product usage and its desired quality face two important issues (maintaining the product at its best quality and reducing the cost of drying process), which must be taken into account in designing the dryer (Lewicki 2006). However, generally, quality can be maintained at the best values but cannot be improved by any drying processes (Rahman 1999). Different factors affect the quality issue; these factors include preharvest factors (such as cultural practices, genetic makeup, and climatic factors), harvesting factors (such as maturity at harvest and harvesting methods factors), and postharvest factors (such as humidity, temperature, atmospheric gas composition, light, mechanical injury, and postharvest diseases and infection factors).

The dried fruits and vegetables can be evaluated from physical parameters (such as color, texture, shrinkage, porosity, rehydration ratio, and drying ratio), chemical parameters (such as flavor, water activity, and shelf life), biological parameters, and nutritional properties (such as antioxidants and food nutrients). Overall shrinkage ratio, rehydration ratio, and drying ratio can be calculated by using Equations 6.1 through 6.3. Other parameters influencing the quality of dried products include starch, vitamin C, chlorophyll content, bulk density, proteins, total sugar, reducing sugar, fat, and energy value. Some researchers concluded that these parameters usually were not significantly affected by the dehydration process. For instance, Rahman et al. (2010) concluded that protein retention, starch, total sugar, reducing sugar, fat, vitamin C, and energy value for dried carrot were significant and were slightly affected by the drying process, as shown in Figure 6.2.

$$\text{Overall shrinkage ratio} = \frac{\text{Weight of unprepared raw food (g)}}{\text{Net weight of the acceptable dried product (g)}} \tag{6.1}$$

$$\text{Rehydration ratio} = \frac{\text{Weight of rehydrated product (g)}}{\text{Net weight of dried product (g)}} \tag{6.2}$$

$$\text{Drying ratio} = \frac{\text{Weight of prepared raw carrot (g)}}{\text{Net weight of the acceptable dried carrot (g)}} \tag{6.3}$$

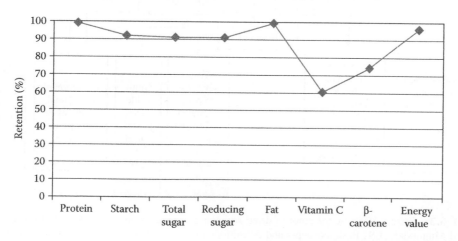

FIGURE 6.2 Retention of some parameter for dried carrot. (Data from Rahman, M.M. et al., *Bangl. J. Sci. Ind. Res.*, 45, 359–362, 2010. With permission.)

6.3.1 Physical Properties

Physical properties of food products include color, texture, shrinkage, porosity, and rehydration. Changes in one or all of these attributes affect the final quality of the dried product. In this section, each attribute is explained and the drying process effect on it is summarized.

6.3.1.1 Color

Color is considered the most significant factor that plays an important role in determining the level of acceptability of the dried fruits and vegetables (F&V) by the customers (Hawlader et al. 2005). Color changes are usually influenced by several factors, such as Maillard reactions, enzymatic browning, and color pigments, as shown in Figure 6.3 (Jangam et al. 2010). Destructive and nondestructive are the two main methods used widely by researchers for color evaluation. Color evaluation can be mathematically determined using CIE ($L^*a^*b^*$) method, where L^* is the light-dark spectrum with a range from 0 (black) to 100 (white), a^* is the green-red spectrum with a range from −60 (green) to +60 (red), and b^* is the blue-yellow spectrum with a range from −60 (blue) to +60 (yellow). High value of hue angle (which reflects the level of browning on the product), chroma (which is a measurement of the strength of color), and total color change can be calculated using Equations 6.4 through 6.6, respectively (Abbott 1999). Definitely, a big difference between the color of the dried product and the fresh product is not desirable by customer as the degree of browning would be higher, as well as they prefer to see similar color.

$$\text{Hue angle}\,(H^*) = \arctan\left(\frac{b^*}{a^*}\right) \tag{6.4}$$

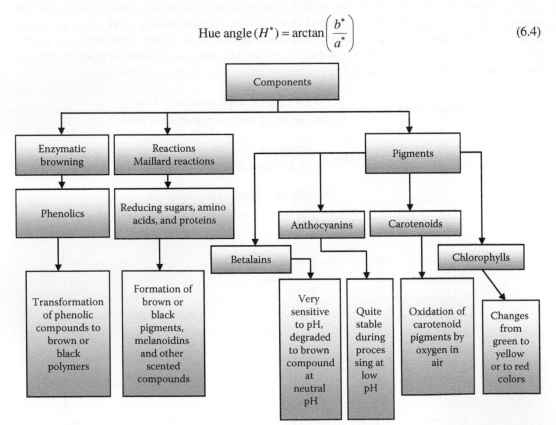

FIGURE 6.3 Factors affecting the color of dried fruits and vegetables during drying. (Data from Jangam, S.V., and Mujumdar, A.S., Basic concepts and definitions in drying of foods, vegetables and fruits, *Drying of Foods, Vegetables and Fruits*, Vol. 1, Jangam, S.V., Law, C.L., and Mujumdar, A.S (eds.), Published in Singapore, pp. 1–30, 2010. With permission.)

$$\text{Chroma}(C*) = \sqrt{a*^2 + b*^2}$$ (6.5)

$$\text{Total color change}(\Delta\Delta E) = \sqrt{(L*-L\text{ ref})2 + (a*-a*\text{ ref})2 + (b*-b*\text{ ref})2}$$ (6.6)

Browning usually impairs the sensory properties of products because of associated changes in the color, flavor, and softening besides nutritional properties. The browning in foods can occur by enzymatic and nonenzymatic (Maillard) reactions. The polyphenoloxidase (PPO) group of enzymes catalyzes the oxidation of phenolic compounds in the plants to o-quinones. Immediately, the quinines condense and react nonenzymatically with other phenolic compounds and amino acids to produce dark brown, black, or red pigments of indeterminate structure, which is better illustrated in Figure 6.4 (Hawlader et al. 2004). Consequently, exclusion of oxygen and/or application of low PH environment can ease browning. So far, sulfating agents are widely used in the drying industry, but they are considered harmful to health, which is unacceptable to some consumers. Therefore, lemon juice was used in this project as a natural inhibitor, which contains plenty of citric acid (Hawlader et al. 2004).

6.3.1.2 Texture

Massive changes could occur in the structure of fruits and vegetables during and after drying process because of moisture removal, which leads to shrinkage and other changes in porosity of the product (Jangam et al. 2010). Texture, from a conscious point of view, is defined as the overall feeling or taste that occurs in the mouth due to consumption of food and is usually composed of foods' properties that can be estimated by touch. This property includes different textural characteristics related to mechanical property, which involves viscosity, chewiness, hardness, chemical properties (moisture and lipids contents), and geometrical property, which includes the particle size and shape (Guine and Barroca 2011). These attributes are important in making a comparison between fresh and dried fruits and vegetables. Texture is influenced by some properties of fruits and vegetables such as water content, biochemical constituents, cellular organelles, and cell wall structure. During the dehydration process, all drying conditions can affect these properties, which usually lead to changes in the quality of the final products. The evaluation of the dried food by the consumer depends on the organoleptic qualities such as appearance, flavor, and texture, so that plenty of researches have been focusing on improving and controlling the textural property of the dried fruits and vegetables (Guine and Barroca 2011). Textural changes can be calculated by Equation 6.7, where F is the textural property, t is the time, n is the kinetic order of textural changes, and k is a constant.

$$\frac{dF}{dt} = kF^n$$ (6.7)

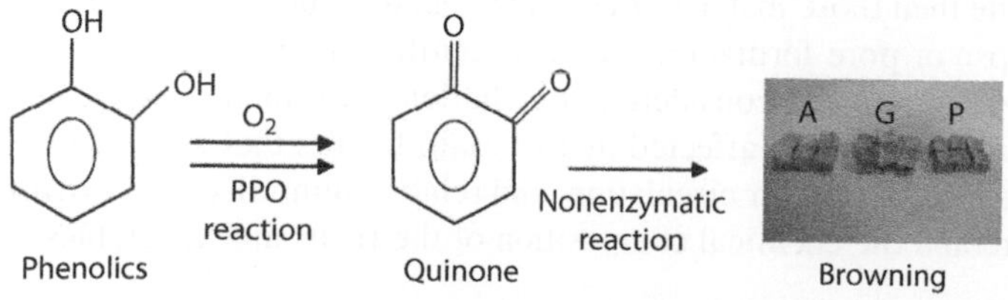

FIGURE 6.4 Schematic diagram of browning. (From Hawlader, M.N.A. et al., Heat pump drying under inert atmosphere, *Proceedings of the 14th International Drying Symposium*, vol. A, Sao Paulo, Brazil, August 22–25, 2004, pp. 309–316. With permission.)

6.3.1.3 Shrinkage

Removing water content from fruits and vegetables products during the drying process usually generates a pressure difference between inside and outside of the product, which leads to cracking in the product (also called *surface cracking*) and shrinkage phenomenon (Jangam et al. 2010). When the inside volume of the product get smaller after removing moisture content, the outer surfaces or layer of the product will have some shrinkage, which will negatively affect the quality of the product, increase the hardness of the product, and then the customer judgment. Another demerit of shrinkage reported by some researchers is that it decreases the rehydration ratio of the dried fruits and vegetables (Mayor and Sereno 2004). Mcminn and Magee (1997) concluded that the most shrunk dried potato slices had the lower dehydration ratio and capacity. Shrinkage has proportional relationship with water volume inside the product. The shrinkage level (high or low) depends on the volume of water removed from the product. Drying rate play a significant role in affecting the magnitude of shrinkage at the outer surface of the product. Fast drying rate makes the water content at the outer surface decrease quickly to a very low level because of rapid evaporation process, which will form a rigid crust (also called a shell) on the surface of the product. This happens because the shrinkage is not uniform at all parts of the material in a rapid dehydration conditions. Wang and Brennan (1995) confirmed the same results and observed a formation of that shell on the potato slices. They concluded that the level of shrinkage at high temperature drying is lower than that at a low temperature. Level of shrinkage affected both density and porosity of the potato. Surface cracking also occurs when the shrinkage of the outer surface of the dried product is not uniform along the surface. However, coupling the mass and heat transfer equations is the most common method used by different researchers for prediction of cracking phenomenon.

Calculation of shrinkage was almost negligible in modeling of food drying process. However, some researchers reported that including the shrinkage value in modeling of the drying methods leads to a result that meet the experimental results more than excluding it from consideration (Mayor and Sereno 2004). Shrinkage could be calculated and modeled using two methods, which are empirical and fundamental relations. In the empirical method, the shrinkage is a function of moisture content, whereas in the fundamental method, it is based on prediction of the structural changes (mass and volume) in the product.

6.3.1.4 Porosity

Porosity is defined as the fraction of the volume of air (pores or empty space) to the total volume of dried fruits and vegetables (Ramos et al. 2003) (see Equation 6.8, where V_a is the volume of air or pores expressed in m^3 and V_t—the total volume of dried product, also expressed in m^3) (Jangam et al. 2010). The porosity value increase when water content is removed during the drying process. Some researchers reported that this attribute could be controlled by choosing the suitable drying method (Ramos et al. 2003). Porosity phenomenon affects both mechanical and textural properties (Jangam et al. 2010). Porous fruits and vegetables shows better rehydration rate, but they have shorter shelf life (Ramos et al. 2003). The pore size and its distribution in the material structure play an important role in affecting porosity. However, it is reported that dried products by the hot air method have lower porosity value than those that are dried using vacuum microwave or freeze dryers.

The pore collapse or pore formation is a very significant and complicated process that occurs during drying process, and it is considered as a challenge for many researchers to investigate the porosity issue. Pore formation is affected by two main factors that are intrinsic (such as temperature, pressure, gas atmosphere, air circulation, and relative humidity) and extrinsic, which involves the initial structure and the chemical composition of the fruits and vegetables (Rahman 2000).

$$\epsilon = 1 - V_a / V_t \tag{6.8}$$

6.3.1.5 Rehydration

Most of the dried fruits and vegetables need to be rehydrated to make them more suitable for consumption. Rehydration process is defined as a process of moisturizing dried foodstuffs by submerging them in water or by using flowing air with high relative humidity and specific speed. It usually depends on the porosity and the level of structural changes of the dried product. It is considered to be a quite complex process that involves imbibition of water, and the swelling and the leaching of soluble. It also includes some physical mechanisms such as relaxation of the solid matrix, water imbibition, internal diffusion, and convention at the surface. In this process, hairy imbibition of water is important for rapid absorption of water by the product.

6.3.2 Chemical Properties

Chemical attributes are important properties of dried fruits and vegetables. They have significant influence on the dried product quality. Flavor, water activity, and shelf life are the main issues that play a role in determining the level of quality. Flavor is defined as "the experience of the combined perception of compounds responsible for taste and aroma" (Ibrahim et al. 2012, p. 1). Flavor of dried fruits and vegetables is considered as one of most important attributes for customer (or consumer) together with color and final appearance. Like other properties, the change in shape, size, and the structure of the product because of moisture removal during drying process influences the flavor. Some parts of flavor are removed by drying, and this is called *volatile flavor*. Chemical analysis and sensory evaluation are the two main methods used to evaluate the changes in flavor because of drying. Chemical analysis provides quantitative data, whereas sensory evaluation is usually performed by comparing the dried product with the fresh product (Jangam et al. 2010). In order to maintain the quality of flavor as better as possible, it is recommended to store the dried fruits and vegetables in a place where either the level of oxygen is lower than 1% or carbon dioxide is higher than 8% to avoid rancidity and other oxidative reactions (Perera 2005).

Water activity in most dried fruits and vegetables is less than 60% because of water removal during drying. This is the recommended percentage for it to avoid oxidation and enzymatic reactions. High percentage increases the microorganisms and bacterial growth in the dried product, which is the main cause for spoilage. It is recommended to store the dried food in a place where the relative humidity is very low in order to avoid high water activity value in the products. It is reported that the best percentage of relative humidity must be between 55% and 70% in the store in order to extend the shelf life of the dried products (Perera 2005). Type of packaging also plays a significant role in extension of the shelf life of the products, because the non suitable type may allow the oxygen to reach the product, which will reduce the shelf life.

6.3.3 Biological Properties

Biological aspects of fresh and dried foodstuffs are very important issues related to safety of the food. Moisture content allows for microbial growth in the material, which leads to many serious problems, such as mold and yeast in dried products. The mycotoxins present in food pose a very dangerous threat to human health and sometimes can cause a wide range of diseases. In order to avoid such problems, water content must not exceed the level of mould growth, which is 0.65. Some examples of mycotoxins are zearalenone, aflatoxins, ochratoxins, trichothecenes, aflatoxins, zearalenone, and fumonisins. The level of biological aspects affects both the shelf life and the quality of the products. The specification of this issue may vary from one country to another depending on the specifications and the standards provided by the government health sector.

6.3.4 Nutritional Parameters

Maintaining the nutritional value of the dried products at desired value is one of most important concern in drying fruits and vegetables. It is well known that fruits and vegetables have high nutritional value, which is very useful for human health. This value decreases during and after drying process. The level to which this value decreases depends on the type of the dryer and the drying conditions. The loss in nutrition can be minimized by using an appropriate dryer and by controlling the drying conditions such as pressure and temperature. Jangam et al. (2010) summarized the possible changes in some nutritional aspects of fruits and vegetables that could occur because of the drying process. For fiber of fruits and vegetables, no changes were recorded but the value of minerals decreased, and some changes occurred during the drying process, especially if soaking water is not used (e.g., iron is not much influenced during the drying process). Meanwhile, vitamin A was well retained, especially under controlled drying conditions (temperature). For lipids, enzymatic hydrolysis may occur at the first stage of the drying process, and rancidity in food could happen at low water activity. Enzymatic degradation and susceptibility to light oxidation could occur for proteins in fruits and vegetables during drying.

6.4 EFFECTS OF DIFFERENT DRYING METHODS ON PRODUCTS' QUALITY

6.4.1 Effect of Freeze Dehydration Process

Several studies have been conducted by researchers in order to evaluate the effect of freeze drying method on the quality of the dried fruits and vegetables. Guine and Barroca (2011) investigated the influence of freeze drying technique on the textural properties of mushrooms and onion. For mushrooms, it is concluded that as the water content decreased from 90.25% to 7.01%, hardness decreased with drying, cohesiveness did not change, but chewiness changed significantly. For onion, water content was reduced from 90.02% to 5.19%, springiness was not influenced by the drying, cohesiveness was slightly increased, and chewiness changed significantly. Different attempts to reduce the operating cost and energy consumption of the freeze drying method have also been made. These include improvement of the mass and heat transfer in the freezing process, averting to use condenser for the sublimated vapor, making a combination of freeze drying methods with other dehydration method, and making it work at near to atmospheric pressure (Rahman 2013).

Meryman (1959) was the first researcher who proposed the AFD system in 1959 when she experimentally investigated the procedures of the freeze drying and the effect of pressure on it. Rahman and Mujumdar (2008) experimentally tested the performance of an AFD system. To enhance the performance of the dryer and to produce air at low temperature, a vortex tube cooler Model 3240 with 706 kcal/h refrigeration capacities and 7.683 m³/s was used. The system was designed in a way that allow for convection, conduction, and radiation of heat transfer at the same time between the atmospheric dryer and the products without increasing the temperature of the water content above the freezing point. The physical parameters, kinetics of the dryer, and quality level were analyzed for potato slices (15 × 5 × 1 mm). Prior to placing the potato slices in the atmospheric dryer, the slices are first immersed, for about 5 min, in a solution containing 5% bicarbonate, and at a temperature of 96°C, in order to avoid the enzymatic browning and, therefore, enhance the quality of the potato. AFD showed significant results in this study, which has made it one of good alternatives of the traditional freeze drying method. It is reported that the main justification of using AFD method is to reduce the energy required for the dehydration process, as conventional freeze drying systems consume a lot of energy to produce high-quality dried fruits and vegetables (Xu et al. 2015). Practically, still some problems related to this process exist, especially in the implementation of the theory, because this process is controlled by the mass and heat transfer process at atmospheric pressure.

MFD has been largely tested in recent years across the world to evaluate the system performance, and its effects on the quality of dried foods. Xu et al. (2015) compared the effects of three different freeze drying methods on mushrooms, which are conventional freeze dryer, AFD, and MFD. The experimental results showed that conventional freeze drying system produced best quality but consumed more energy than the other two systems. Lowest energy consumption was achieved and obtained from atmospheric drying system, but with longest drying time and produced worst quality. Therefore, MFD showed acceptable products' quality and significantly decreased the energy required.

6.4.2 EFFECT OF VACUUM MICROWAVE DRYING METHOD

Vacuum microwave drying is the idea of combining low temperature drying by using vacuum, and rapid drying by using microwave radiation. The impact of these dryers has been largely investigated by a lot of researchers for many years. The most common effect of using microwave vacuum dryer (MVD) for fruits and vegetables is that the drying process is more rapid and occurs at relatively low temperature compared to other conventional drying methods. It is reported that vacuum microwave dryer needs only 33 min to remove 99% of moisture from the product compared to freeze dryers (which need 3 days) and hot air dryers, which sometimes may require more than 8 h (Karimi 2010). Another advantage of using MVD is that the drying time is controllable by determining the microwave power and the amount of moisture content need to be removed. Therefore, chamber pressure in microwave dryer influences the drying rate because of the increase in the latent heat of water vaporization, which decreases the drying rate. The faster drying process occur at lowest operating pressure.

This type of dryer affects both the flavor and color of food, so that it is difficult to achieve the desired flavor and color compared to those products from conventional dryers, because there are some reactions (for instance, Maillard reaction) in these dryers, which do not occur when product is heated or dehydrated by microwave (Ibrahim et al. 2012). Püschner and SiokHoon (2005) investigated the effect of Püschner MVD (model WaveVac 0150) on different types of fruits and vegetables. For strawberries, the results showed that dried strawberries have excellent acceptable appearances and colors as well as strong flavors. For broccoli, it was found that microwave dryer maintained its structure and color at desired levels.

In order to improve the effect of MVD on the physical and chemical properties of fruits and vegetables, some researchers used it as a second drying stage. Figiel (2007) investigated the effect of using convective method and MVD on apple slices. The convective method was used first, and then MVD was used for the final drying. It was concluded that using convective method as pre-process led to better rehydration capacity, showed better color toward red and blue, and lowered shrinkage compared to samples dried by convective method only. Another technique was used by Clary et al. (2007) to improve the quality of grape by controlling the temperature of MVD. Infrared temperature sensor (model H-L10000 infrared detector, Mikron, Oakland, NJ) was used to control the MW power and then the temperature. The experimental results showed that riboflavin, vitamins C, A, and thiamine were higher in the grape dried by MVD than in sun-dried raisins. On the other hand, some researchers performed hybridization between MVD and another drying method such as osmotic dehydration in order to improve the drying process. Changrue's (2006) PhD thesis focused on a hybrid osmotic-vacuum microwave dryer for strawberries and carrots. Osmotic dehydration process was used as pretreatment for the products to enhance the drying. For strawberries, sucrose was used as an osmotic agent, whereas sucrose and salt were used for carrots. Hybridization of osmotic and MVD led to less drying time, less energy consumption, and better physical properties except for taste, which was influenced by salt. It also led to better quality of dried strawberries, but the drying time and energy were same with and without osmotic dehydration.

6.4.3 EFFECT OF HOT AIR DRYING METHOD

Hot air drying method is the most commonly used drying method for dehydration of fruits and vegetables. In this dryer, high temperature leads to long time of drying (Sometimes around 3 h), and it plays an important role in affecting the rate of the drying. It is reported that hot air drying method provides the hardest texture appearance compared to freeze and vacuum drying (Argyropoulos et al. 2008). According to Alibaş (2012), when temperature was reduced, the time of the drying process reduced as well. He concluded that time was found to be 340, 270, and 220 min for 55°C, 65°C, and 75°C, respectively. The drying time increases when the thickness of the slices increases, whereas the drying time increases when the relative humidity is decreased. According to Argyropoulos et al. (2008), color changes during hot air drying are because of enzymatic and nonenzymatic factors. It is also affected by the degree of the operational temperature; the discoloration happens at a higher temperature in this process.

The effect of temperature (55°C, 65°C, and 75°C) on the time of the air drying process and other quality properties such as vitamins C and D, macro minerals (Mg, Na, P, Ca, and K), nonenzymatic browning, and thiosulfinate of 2 mm onion slices were studied also by Olalusi (2014). Similar results, as shown by Alibaş (2012), about temperature and time were found; the longest time was at 55°C. The loss in vitamin C at 55°C was less than at 65°C and 75°C. Figure 6.5 summarizes effect of temperature on the physical properties of onion.

6.4.4 EFFECT OF OSMOTIC DEHYDRATION METHOD

Many factors affect this process such as temperature, solute types, solution concentration, osmotic agent, time, and tissue compactness of the material. Oladejo et al. (2013) investigated the effect of some of these factors on osmotic method. It was concluded that the choice of solutes and the concentration of the solution play a significant role in influencing the natural flavor, color retention, and textures in fruit products. It is reported that osmotic dehydration process is a preprocessing step to further drying method such as freeze, hot air, and vacuum dryers, because it improves the nutritional value of the product without causing any change in the physical properties. Therefore, using it as a preprocessing step will increase the percentage of removed moisture content from inside tissues and

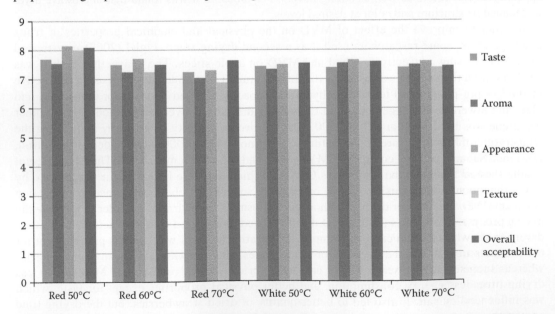

FIGURE 6.5 Effect of temperature on the physical properties of red and white onions. (Data from Olalusi, A., *J. Agric. Chem. Environ.*, 3, 13–19, 2014. With permission.)

shorten the time of the whole drying process. It could also be used as a post-processing step after or even during a conventional drying method. Dried fruits and vegetables from osmotic method are better and have lower shrinkage and browning than other products from other dryers. Osmotic dehydration process requires energy less than hot air or vacuum microwave dryers, because it operates at low temperature. It is reported that the reduction in the fruits and vegetables weight could reach 50% because of water removal by osmotic dehydration process (Jalaee et al. 2010). Minimizing the leaching of osmotic solids was the main purpose for many researches, because it alters organoleptic and nutritional values such as mineral salt and vitamins of the products.

6.4.5 Effects of Solar Drying Methods on Fruits and Vegetables

Solar drying process has been used since ancient time to preserve food and agricultural crops. Different types of solar dryers are nowadays available such as open sun drying, direct sun drying, heat pump drying, SAPHD, but the most efficient and significant one is SAHPD. These systems exploit both wasted heat from the heat pump system and solar energy from the sun. The material to be dried usually absorbs about 5%–10% of the supplied energy, so that using solar-assisted heat pump technology leads to a saving of about 90%–95% of conventional energy (Patel and Kar 2012). These systems can be used for dehydration and rehydration process together by using the hot and cold heat that produced by condenser and evaporator.

Mechlouch et al. (2012) compared the effect of three drying methods, namely OASD, direct solar drying, and microwave drying on the brix, acidity, polyphenols, drying rate, pH, moisture content, total carotenoids content, flavonoids, and antioxidant capacity of tomatoes. It was concluded that OASD and DSD need about 5 days to decrease the moisture content to 50%. The highest value of acidity was achieved by DSD, whereas the lowest value was by microwave drying method. The lowest value of brix was also obtained by the microwave method. DSD led to highest phenolics content. Therefore, Karabacak and Atalay (2010) studied the effect of drying on the drying time, appearance, energy consumption, and physical properties of tomato slices (*Lycopersicon esculentum*), with around 95% moisture content, using heat pump drying, SAHPD, and solar drying methods. Heat pump dryer showed 50% better performance than solar dryer and 30% better than SAHP in terms of time of drying. Highest drying rate was achieved by heat pump dryer, whereas the lowest one was obtained from solar drying method.

6.5 IMPACT OF HEAT PUMP DRYING ON PRODUCT QUALITY

Heat pump drying technologies have been known as energy-efficient systems when they are used for the purpose of food preservation process (Fadhel 2014). This technology has recently been developed at Norwegian University of Science and Technology (NTNU) (Jovanovic 2013). Heat pump dryer consists of condenser, evaporator, compressor, and expansion valve connected together by using copper tubes. It is an environmental-friendly technology and can provide high-quality products. Heat pump drying systems reduce the color degradation of the dried fruits and save the amount of vitamins embedded in the product by reducing their losses during and after drying. It provides longer retention of the product flavors, hygienic drying process, offer highest specific moisture extraction ratio (usually in range of 1.0–4.0), because the heat is recovered from moisture-laden air. Hence, it improves the overall energy efficiency according to Jovanovic (2013). Tremendous researches have been conducted through the years in order to investigate the effect of this technology on the products' quality. The effect of heat pump drying systems on the quality of the dried fruits and vegetables is presented in this section.

Barshlyanova et al. (2014) studied the effect of heat pump drying process on the biochemical properties (total sugars, sucrose, invert sugar, organic acids, ascorbic acid, and anthocyanin) of plums as well as the predicatory indications that help in selecting the suitable varieties of plums for drying. The experiments, in this study, were carried out on four types of plums, such as Stanley,

Gabrovo, Strinava, and Mirabelle de Nancy, in order to determine the most suitable kind for drying. A heat pump dryer with an initial temperature of 45°C, airflow velocities of 4.6 m/s, and initial air humidity of 8–10 g/kg was used. The experimental results showed that Stanley and Gabrovo had different biochemical properties except in total sugar, where the two kinds were statistically equal. Therefore, Gabrovska, Strinova, and Mirabelle de Nancy had equal invert sugar values. Heat pump drying was useful in eliminating differences between the total sugar in fresh plums (Stanley and Strinava). Nowadays, prunes like other kinds of fruits are available throughout the year because of drying technology. According to Barshlyanova et al. (2014), heat pump-dried prunes showed better quality at low energy consumption, because this technology of drying enables to provide controllable drying environment (temperature and humidity).

In the study conducted by Jovanovic (2013), a heat pump dryer was used to dry green peas. The temperature of the drying air was set at 45°C, 35°C, and 15°C with different relative humidity of 60%, 40%, and 20%. Fluidized bed mode was set and used in this study. The results showed that the effect of differences in relative humidity was much less compared to the temperature, but it, however, played an important role in the drying process. The experimental results for moisture content removal on dry basis revealed that the best drying condition is at 45°C and 20% relative humidity as well as at 35°C and 20%, as shown in Figure 6.6, where: Test 1: 45°C and 40% relative humidity, Test 2: 45°C and 20%, Test 3: 35°C and 60%, Test 4: 35°C and 40%, Test 5: 35°C and 20%, Test 6: 15°C and 60%, Test 7: 15°C and 40%, and Test 8: 15°C and 20%. Therefore, regarding the color data, experiments showed that the highest value of the brightness of dried peas was achieved from Test 2, whereas Test 3 provided the highest red-green and yellow-blue content of dried green peas, as shown in Figure 6.7. Water activity and bulk density for peas were also measured in this study and the results showed that the lowest value were obtained from Test 2, as shown in Figures 6.8 and 6.9. It means that the lowest water activity value exists at highest value of temperature, which is 45°C.

In order to optimize the performance of the heat pump system, Min (2005) proposed a heat pump system using inert gases such as carbon dioxide and nitrogen. The effect of such inert gases on energy efficiency, drying kinetics, and dried product quality of heat pump dried foods were studied. In this study, apple, guava, potato, papaya, and ginger were used as drying materials. The drying temperature was set at 45°C with 20% relative humidity and 0.7 m/s air velocity. It was found that the use of inert gases in the heat pump system accelerated the evaporation process of the product's moisture content more than in normal air. This phenomenon was observed for papaya,

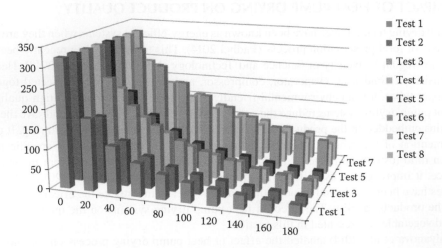

FIGURE 6.6 Moisture content on dry basis for all tests. (Data from Jovanovic, S., Quality characterization and modeling experimental kinetics in pilot scale heat pump drying of green peas, Master's thesis, Norwegian University of Science and Technology, Trondheim, Norway, 2013. With permission.)

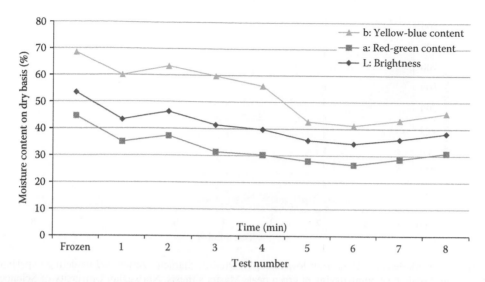

FIGURE 6.7 Color data for series of drying green peas tests. (Data from Jovanovic, S., Quality characterization and modeling experimental kinetics in pilot scale heat pump drying of green peas, Master's thesis, Norwegian University of Science and Technology, Trondheim, Norway, 2013. With permission.)

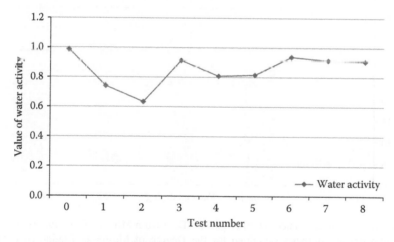

FIGURE 6.8 Water activity. (Data from Jovanovic, S., Quality characterization and modeling experimental kinetics in pilot scale heat pump drying of green peas, Master's thesis, Norwegian University of Science and Technology, Trondheim, Norway, 2013. With permission.)

guava, and ginger. Nitrogen displayed the highest drying rate, especially at the beginning of the drying process, where the moisture content at its highest value. Nitrogen also had the highest effective diffusivity followed by carbon dioxide, whereas normal air had the lowest value. The possible reason is probably due to the different physical properties of different gas. Maintaining the color of the product as near as to the original color is an important factor in drying process. In this study, it is concluded that inert gases had the lowest change on the product color, whereas the normal gas had the highest. The effect of five drying methods on colors of guava and papaya are compared in this study, as shown in Figures 6.10 and 6.11.

Numerous researchers reported that porosity is a result of moisture removal from the product. Porosity for guava, in this study, increased in the sequence of nitrogen drying, carbon dioxide drying, normal air drying, vacuum drying, and freeze drying. Therefore, the porosity of freeze-dried

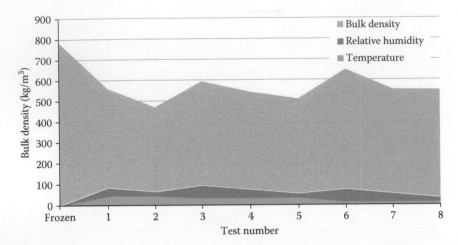

FIGURE 6.9 Bulk density. (Data from Jovanovic, S., Quality characterization and modeling experimental kinetics in pilot scale heat pump drying of green peas, Master's thesis, Norwegian University of Science and Technology, Trondheim, Norway, 2013. With permission.)

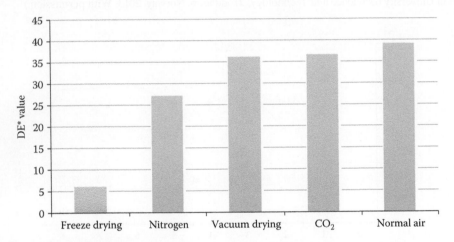

FIGURE 6.10 Total color difference (DE*) of guava. (Data from Min, T., Food quality in modified atmosphere heat pump drying, A thesis submitted for the Degree of Master of Engineering, Department of Mechanical Engineering, National University of Singapore, Singapore, 2005. With permission.)

papaya was smaller than samples dried by normal air method, while vacuum-dried papaya had the largest porosity among the five drying methods. Rehydration property of the product is related to porosity. The rate of rehydration of apple, guava, and potato dried using inert gases at 60°C and 80°C were studied and compared, as shown in Figures 6.12 and 6.13.

Texture is a very significant feature that reflects and affects the quality of dried foodstuff, because it represents the sensory acceptability factors. It is because people like to get the highest enjoyment by consuming dried foods. However, firmness, which is the main aspect of measuring the texture quality, was measured for different samples at same fixed time using different drying methods. The results showed that normal air-dried samples had higher firmness than samples dried by heat pump with nitrogen or carbon dioxide. Freeze drying provided the lowest firmness as the samples were the most porous. Furthermore, it is well known that fruits are rich of different vitamins such as A and C that are useful for human body. The experimental results proved that heat pump with nitrogen or carbon dioxide is able to preserve vitamins A and C much greater than HP with normal air

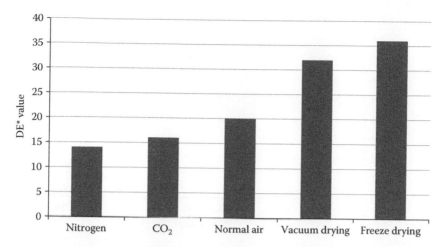

FIGURE 6.11 Total color difference (DE*) of papaya. (Data from Min, T., Food quality in modified atmosphere heat pump drying, A thesis submitted for the Degree of Master of Engineering, Department of Mechanical Engineering, National University of Singapore, Singapore, 2005. With permission.)

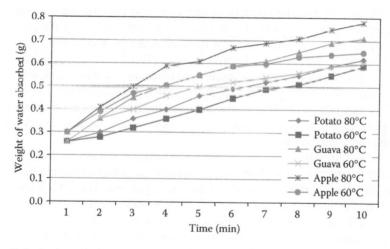

FIGURE 6.12 Rehydration rates of samples dried by heat pump with nitrogen. (Data from Min, T., Food quality in modified atmosphere heat pump drying, A thesis submitted for the Degree of Master of Engineering, Department of Mechanical Engineering, National University of Singapore, Singapore, 2005. With permission.)

but, on the other hand, less than freeze drying or vacuum drying methods. Same experiments were conducted and same results were obtained by Hawlader et al. (2004).

Other researchers such as Wiset et al. (2013) used same technique (heat pump drying under modified atmosphere) to improve the quality of heat pump-dried products but in multistage heat pump drying process. In this study, drying kinetics, moisture content, drying rate, color, and energy consumption of multistage heat pump drying of macadamia nut under modified atmosphere was examined with 40°C–60°C drying temperature, and using nitrogen and normal air as drying medium. The problem in drying macadamia nut is that drying time is usually too long (>month), so that the drying time must be reduced, and the drying process must be speeded up to meet the industry demand. It was concluded that the multistage heat pump dryer did not alter the color of the dried samples, and the drying time was significantly decreased. However, the experimental results showed that drying condition at 40°C under nitrogen followed by 60°C under air provided the lowest energy consumption.

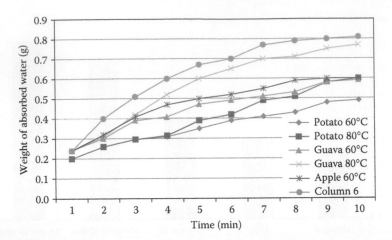

FIGURE 6.13 Rehydration rates of samples dried by heat pump with carbon dioxide. (Data from Min, T., Food quality in modified atmosphere heat pump drying, A thesis submitted for the Degree of Master of Engineering, Department of Mechanical Engineering, National University of Singapore, Singapore, 2005. With permission.)

Chemical pretreatment, freezing, osmotic dehydration, and thawing are very important process used by different researchers as pretreatment process for fruits and vegetables, as they affect the drying rate and preserve the overall quality of the final dried product. Some fruits such as grape, blueberries, and cranberries have waxy layer in the skin. This layer plays a significant role in protecting the fruits against weather, or from insects and parasites that may attack and spoil the fruits. It also plays a major role in making the drying process more difficult and increasing the drying time according to Zsivanovits et al. (2013). Therefore, pretreatment processes are necessary processes for such kind of fruits, especially in heat pump drying systems where the time of drying is significant in lowering the energy consumption.

Zsivanovits et al. (2013) investigated the efficiency to dry grape grains using heat pump dryer. Because the outer layer of the grape grain is waxy, it was dried with and without hot sodium carbonate solution in 3% concentration as osmotic pretreatment process. The drying time, in this study, was set to be 8 h. The untreated samples by the sodium solution were dried for 10 h in order to investigate the effects of osmotic dehydration as well. The results proved that using pretreatment process such as osmotic dehydration before heat pump drying method shortened the time of the drying, reduced the amount of removed moisture content during and after the drying, and produced better dried products, as shown in Figure 6.14.

Heat pump-assisted dryers are used in many food processing industries such as dehydration of fruits and milk powder. Recently, many problems appeared in heat pump-assisted drying methods in industries. One of the most significant noticeable problems is that these systems contributed to ozone depletion and global warming issues due to use of chlorofluorocarbon (CFC) and hydrochlorofluorocarbon (HCFC) refrigerants such as HFC-134a. According to Maina and Huan (2013), CO_2 refrigerant is an excellent replacement for CFCs and HCFCs in heat pump-assisted drying systems, because it is natural, inexpensive, nontoxic, nonflammable, readily available, odorless at low concentrations, have higher volumetric capacity, good heat transfer properties, and not a corrosive refrigerant. Sarkar et al. (2004) investigated and compared the performance of CO_2-, HCFC-22-, and HFC-134a-based heat pump dryers for food. The results showed that CO_2-based heat pump dryer can give better performance and improve the product quality compared to others. The results also showed that CO_2 can be considered the best alternative refrigerant for drying application, because of its high critical pressure (Sarkar et al. 2004).

The effect of a heat pump-assisted hybrid photovoltaic thermal solar dryer at different drying air temperature (40°C, 50°C, and 60°C) and two different modes of drying (with and without

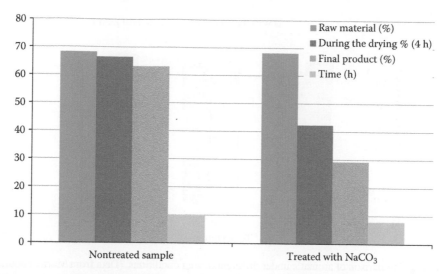

FIGURE 6.14 A comparison between nontreated and treated samples, where the *y*-axis is the moisture content %. (Data from Zsivanovits, G. et al., *J. Food Phys.*, 26, 19–28, 2013. With permission.)

heat pump system) on saffron physicochemical properties were investigated by Mortezapour et al. (2014). The use of heat pump dryer reduced the relative humidity of the drying air, improved the drying rate, decreased the drying time by 62% with increasing air temperature from 40°C to 60°C, and lowered the drying period by 40%. Color characteristics evaluation of saffron showed that color was improved with drying temperature and heat pump system, as shown in Figure 6.15. In addition, the aromatic strength increased with increasing air temperature, as shown in Figure 6.16, while no change was recorded in bitterness with temperature and heat pump system. The crocin content was significantly influenced by drying air temperature.

Xiaoyong and Luming (2015) studied the effect of using heat pump with and without far infrared (FIR) radiation at 500, 1500, and 3000 W on shrinkage, color, texture, percentage of rehydration, and moisture content of iron yam chips having initial moisture content of 76% at 50°C and 1.0 m/s. A comparison between the effects of different amounts of infrared radiations was prepared as well. It is concluded that heat pump drying with FIR shortened the drying time and, therefore, increased the drying rate. Samples dried with heap pump with FIR had a higher value of lightness and better values of redness and yellowness compared to products of heat pump system only. Iron yam chips dried by HP+1500FIR provided samples with lower shrinkage, lower hardness, higher brittleness, and improved rehydration ability than other samples dried by HP, HP+500FIR, and HP+3000FIR (Xiaoyong and Luming 2015).

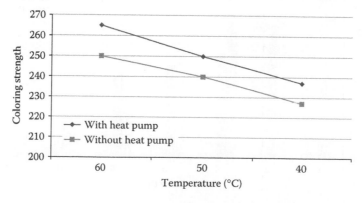

FIGURE 6.15 Comparison of color under different drying conditions. (Data from Mortezapour, H. et al., *J. Agric. Sci. Technol.*, 16, 33–45, 2014. With permission.)

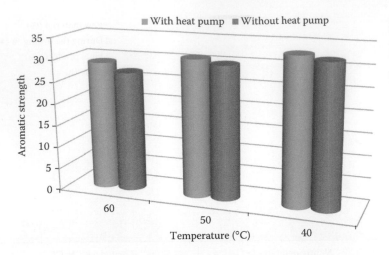

FIGURE 6.16 Comparison of aromatic under different drying conditions. (Data from Mortezapour, H. et al., *J. Agric. Sci. Technol.*, 16, 33–45, 2014. With permission.)

Therefore, Paakkonen (2002) designed and developed a small-scale heat pump rotary dryer with infrared radiation to study its effect on color, water content, and rehydration ratios of slices of red beet, carrot, rosebay willow herb, leaves of birch, and dandelion. Relative humidity and temperature of air drying were recorded as well. The experiments were conducted at constant airflow rate and the drying temperature inside the drying chamber was maintained at 40°C and 50°C. This type of dryer required about 12 h to reduce the moisture content to 12% at drying temperature of 40°C, as shown in Figure 6.17, which reflects the moisture content of birch leaves versus time, where CI and CII are results of two different experiments at same temperatures (40°C and 50°C). The results of color analysis (redness, lightness, and yellowness) for all samples at 40°C and 50°C are shown in Figure 6.18. It is concluded that the temperature difference did not affect the color of birch leaves. The color differences for rehydrated samples of red beet and carrot slices were lower than for dried samples. The differences in color between fresh and rehydrated carrot samples were much lower than the differences between fresh and dried samples.

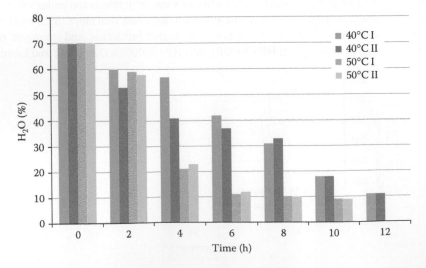

FIGURE 6.17 Moisture content versus drying time of birch leaves at 40°C and 50°C; CI and CII are results of two different experiments. (Data from Paakkonen, K., *Agric. Food Sci. Finland*, 11, 209–218, 2002. With permission.)

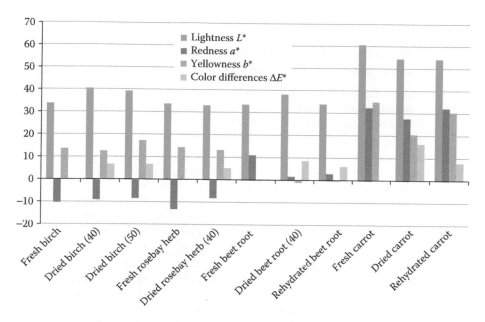

FIGURE 6.18   Color analysis of different samples. (Data from Paakkonen, K., *Agric. Food Sci. Finland*, 11, 209–218, 2002. With permission; Note: Some values of color differences are not available from the source.)

6.6   EFFECT OF STORAGE CONDITIONS ON THE MOISTURE CONTENT OF DRIED FRUITS AND VEGETABLES

Fruits and vegetables are highly perishable commodities. This kind of food cannot be stored for long time at the normal ambient temperature, especially in the tropical countries, where the humidity percentage is very high. Fresh carrot, for instance, can only be stored for maximum 3 days at ordinary temperature, whereas it can be stored for about 2 weeks at zero temperature (Rahman et al. 2010). Storage conditions such as temperature, pressure, and storage period of time play a crucial role in preserving dried fruits and vegetables and keeping the water content as low as possible. During the storage period, some increment in moisture contents of dried product can be attributed to unsuitable storage conditions. The effect of storage period and temperature on the moisture content of the osmo-dehydrated-microwave carrot slices was investigated by Gani and Avanish Kumar (Karimi 2010). It was concluded that the moisture content of the carrot increased with increasing the storage period and temperatures, as shown in Table 6.1. However, the same conclusion was reported by Rahman et al. (2010), who concluded that dried carrot by solar dryer can be stored for maximum 6 months (Kumar and Sagar 2008; Rahman et al. 2012).

6.7   SHORTCOMINGS OF CONVENTIONAL FOOD DRYERS

Most drying methods used for fruits and vegetables suffer from quality loss related to color, flavor, hardening or shrinkage, nutrient content, and rehydration (Qing-guo et al. 2006). Drying required a large amount of energy to remove moisture content from materials that are derived from conventional energy sources such as fossil fuel and oil, which make the drying process more expensive. Use of conventional energy resources in drying has contributed to problems such as global warming and ozone depletion; as such, different researchers have proposed renewable energy sources as a green alternative solution and cost-effective energy resource for drying process. Solar-assisted heat pump systems for drying has been recently proposed by different researches and considered one of most promising alternative to overcome such problems, because these systems offer energy efficiency, produce better quality, and have zero impact on environments (Rahman et al. 2013).

TABLE 6.1

Effect of Storage Period Time and Temperature on Moisture Content of Osmo-Dehydrated Carrot Slices

Moisture Content (% w.b.) at Different Temperatures and Concentrations/Storage Period	Day 0	Day 15	Day 30	Day 45
Moisture content at $T1$ and 30°B	12.3	12.45	12.53	12.6
Moisture content at $T2$ and 30°B	10.8	10.9	11	11.1
Moisture content at $T3$ and 40°B	11	11.1	11.2	11.3
Moisture content at $T4$ and 40°B	10	10	10.1	10.1
Moisture content at $T5$ and 50°B	9.6	9.8	9.9	10
Moisture content at $T6$ and 50°B	9.4	9.5	9.5	9.6
Moisture content at $T7$ and 30°B	13.8	13.9	13.96	14
Moisture content at $T8$ and 30°B	11.8	11.85	11.93	12
Moisture content at $T9$ and 40°B	12	12.1	12.32	12.27
Moisture content at $T10$ and 40°B	11	11.1	11.2	11.22
Moisture content at $T11$ and 50°B	10.8	10.9	11	11.1
Moisture content at $T12$ and 50°B	10	10	10.1	10.3

Source: Gani, G. and Avanish Kumar, A., *Int. J. Sci. Res. Publ.*, 3, 1–11, 2013. With permission.

Temperature of T1, T3, and T5 was 30°C for each but treated with 30°B, 40°B, and 50°B, respectively, and temperature of T2, T4, and T6 was 40°C and treated with same concentration as T1, T3, and T5 (30°B, 40°B, and 50°B, respectively). Syrup temperature of T7, T9, and T11 was 30°C for each but treated with 30°B, 40°B, and 50°B, respectively, and temperature of T8, T10, and T12 was 40°C and treated with same concentration as T1, T3, and T5 (30°B, 40°B, and 50°B, respectively). B represents the concentration.

6.8   CONCLUSIONS

The safety of fruits and vegetables has become the most significant industrial topic across the world because of its importance and effects on human life. Food drying process is a reliable and natural way to preserve and store fruits and vegetables. Drying is a moisture removal process from food products. It affects the physical, chemical, biological, and nutritional parameters of fruits and vegetables. This has great influence on the quality aspects of dried products. Therefore, most conventional drying methods contribute to global warming and ozone depletion by using conventional energy resources such as fossil fuel and oil. Increase of oil price throughout the past few decades has made the drying process more expensive than ever, because it requires huge amount of energy. The impact of freeze drying, vacuum microwave drying, hot air drying, and osmotic dehydration method on the quality of fruits and vegetables has been summarized. It is essential to design the dryer carefully and consciously and select the appropriate drying method and control to enhance the quality of dried fruits and vegetables. It is found that the solar-assisted heat pump system for drying fruits and vegetables is considered to be one of best solutions to overcome difficulties encountered in conventional drying.

REFERENCES

Abbott, J.A. 1999. Quality measurement of fruits and vegetables. *Postharvest Biology and Technology*, 15:207–225.

Alibaş, I. 2012. Determination of vacuum and air drying characteristics of celeriac slices. *Biological and Environmental Sciences*, 6(16):1–13.

Argyropoulos, D., A. Heindl, J. Müller. 2008. Evaluation of processing parameters for hot-air drying to obtain high quality dried mushrooms in the Mediterranean region. In *Conference on International Research on Food Security, Natural Resource Management and Rural Development*, Tropentag 2008, Stuttgart, Germany, University of Hohenheim, October 7–9.

Brashlyanova, B.P., P.H. Ivanova, D.P. Georgiev, M.T. Georgieva. 2014. Changes in the biochemical indices of plums as a result of drying at low positive temperature. *Journal of BioScience and Biotechnology* SE/ONLINE:47–50.

Changrue, V. 2006. Hybrid (osmotic, microwave-vacuum) drying of strawberries and carrots. A thesis submitted to McGill University in partial fulfillment of the requirements for the degree of Doctor of Philosophy, Montreal, Quebec, Canada.

Chin, S.K., C.L. Law. 2012. Optimization of convective hot air drying of Ganodermalucidum slices using response surface methodology. *International Journal of Scientific and Research Publications*, 2(5): 2250–3153.

Chua, K.J., A.S. Mujumdar, M.N.A. Hawlader, S.K. Chou, J.C. Ho. 2001. Batch drying of banana pieces—Effect of stepwise change in drying air temperature on drying kinetics and product colour. *Food Research International*, 34:721–731.

Ciurzyńska, A., A. Lenart. 2011. Freeze-drying—Application in food processing and biotechnology—A review. *Polish Journal of Food and Nutrition Science*, 61(3):165–171.

Clary, C.D., E. Mejia-Meza, S. Wang, A.E. Petrucci. 2007. Improving grape quality using microwave vacuum drying associated with temperature control. *Journal of Food Science*, 72(1):E023–E028.

Daghigh, R., K. Spoian, M.H. Ruslan, A.M.A. Alghoul, C.H. Lim, S. Mat, B. Ali, M. Yahya, A. Zahraim, M.Y. Sulaiman. 2008. Survey of hybrid solar heat pump drying systems. In *Proceedings of the 4th IASME/WSEAS International Conference on Energy & Environment*, Campus of the University of Cambridge, Cambridge, UK, February 23–25, pp. 411–418.

Fadhel, M.I. 2014. Drying characteristics of lemongrass in solar assisted chemical heat pump dryer. In *2nd International Conference on Food and Agricultural Sciences IPCBEE*, vol. 77, Singapore.

Figiel, A. 2007. Dehydration of apples by a combination of convective and vacuum microwave drying. *Polish Journal of Food and Nutrition Sciences*, 57(4(A)):131–135.

Gani, G., A. Avanish Kumar. 2013. Effect of drying temperature and microwave power on the physic-chemical characteristics of OSMO-dehydrated carrot slices. *International Journal of Scientific and Research Publications*, 3(11):1–11.

Guine, R.P.F., M.J. Barroca. 2011. Influence of freeze-drying treatment on the texture of mushrooms and onions. *Croatian Journal of Food Science and Technology*, 3(2):26–31.

Hawlader, M.N.A., C.O. Perera, M. Tian. 2004. Heat pump drying under inert atmosphere. In *Proceedings of the 14th International Drying Symposium*, vol. A, Sao Paulo, Brazil, August 22–25, pp. 309–316.

Hawlader, M.N.A., C.O Perera, M. Tian. 2005. Influence of different drying methods on fruits' quality. *8th Annual IEA Heat Pump Conference*, Las Vegas, NV, May 30–June 2.

Hawlader, M.N.A., C.O. Perera, M. Tian. 2006a. Properties of modified atmosphere heat pump dried foods. *Journal of Food Engineering*, 74:392–401.

Hawlader, M.N.A., C.O. Perera, M. Tian. 2006b. Comparison of the retention of 6-gingerol in drying of ginger under modified atmosphere heat pump drying and other drying methods. *Drying Technology*, 24:51–56.

Ibrahim, G.E., A.H. El-Ghorab, K.F. El-Massry, F. Osman. 2012. Effect of microwave heating on flavor generation and food processing, Chapter 2. Retrieved in June 2015 from http://cdn.intechopen.com/pdfs-wm/40691.pdf.

Jalaee, F., A. Fazeil, H. Fatemain, H.Tavakolipour. 2010. Mass transfer coefficient and the characteristics of coated apples in osmotic dehydrating. *Food and Bioproducts Processing* 89:367–374.

Jangam, S.V., Mujumdar, A.S. 2010. Basic concepts and definitions in drying of foods, vegetables and fruits. In:Jangam, S.V., Law, C.L., and Mujumdar, A.S (eds.), *Drying of Foods, Vegetables and Fruits*, Vol. 1, Published in Singapore, pp. 1–30.

Jangam, S.V., G.L. Visavale, A.S. Mujumdar. 2011. Use of renewable source of energy for drying of FVF. *Drying of Foods, Vegetables and Fruits*, 3:103–126.

Jovanovic, S. 2013. Quality characterization and modeling experimental kinetics in pilot scale heat pump drying of green peas. Master's thesis, Norwegian University of Science and Technology, Trondheim, Norway.

Karabacak, R., Ö. Atalay. 2010. Comparison of drying characteristics of tomatoes with heat pump dehumidifier system, solar-assisted system and natural drying. *Journal of Food, Agriculture & Environment*, 8(2):190–194.

Karimi, F. 2010. Properties of the drying of agricultural products in microwave vacuum: A review article. *Journal of Agricultural Technology*, 6(2):269–287.

Kumar, P.S., V.R. Sagar. 2008. Quality of osmo-vac dehydrated ripe mango slices influenced by packaging material and storage temperature. *Journal of Scientific and Industrial Research*, 67:1108–1114.

Lewicki, P.P. 2006. Design of hot air drying for better foods. *Trends in Food Science & Technology*, 17:153–163.

Maina, P., Huan, Z. 2013. Effects of Refrigerant Charge in the Output of a CO_2 Heat Pump. *African Journal of Science, Technology, Innovation and development*, 5(4):303–311.

Mayor, L., A.M. Sereno. 2004. Modelling shrinkage during convective drying of food materials: A review. *Journal of Food Engineering*, 61:373–386.

Mcminn, W.A.M., T.R.A. Magee. 1997. Physical characteristics of dehydrated potatoes—Part I. *Journal of Food Engineering*, 33:37–48.

Mechlouch, R.F., W. Elfalleh, M. Ziadi, H. Hannachi, M. Chwikhi, A. Ben Aoun, I. Elakesh, F. Cheour. 2012. Effect of different drying methods on the physico-chemical properties of tomato variety "Rio Grande." *International Journal of Food Engineering*, 8(2):Article 4.

Meryman, H.T. 1959. Sublimation freeze drying without vacuum. *Science*, 130:628–629.

Min, T. 2005. Food quality in modified atmosphere heat pump drying. A thesis submitted for the Degree of Master of Engineering, Department of Mechanical Engineering, National University of Singapore, Singapore.

Mortezapour, H., B. Ghobadian, M.H. Khoshtaghaza, S. Minaei. 2014. Drying kinetics and quality characteristics of saffron dried with a heat pump assisted hybrid photovoltaic-thermal solar dryer. *Journal of Agricultural Science and Technology*, 16:33–45.

Oladejo, D., B.I.O. Ade-Omowaye, O. AbioyeAdekanmi. 2013. Experimental study on kinetics, modeling and optimisation of osmotic dehydration of mango (*Mangifera indica* L). *The International Journal of Engineering and Science*, 2(4):01–08.

Olalusi, A. 2014. Hot air drying and quality of red and white varieties of onion (*Allium cepa*). *Journal of Agricultural Chemistry and Environment* 3:13–19.

Paakkonen, K. 2002. A combined infrared/heat pump drying technology applied to a rotary dryer. *Agricultural and Food Science in Finland*, 11:209–218.

Patel, K.K., A. Kar. 2012. Heat pump assisted drying of agricultural produce: An overview. *Journal of Food Science and Technology*, 49(2):142–160.

Pavan, M.A. 2010. Effect of freeze drying, refractance window drying, and hot-air drying on the quality parameters of acai. Thesis submitted in partial fulfillment of the requirements for the degree of Master of Science in Food Science and Human Nutrition in the Graduate College of the University of Illinois at Urbana-Champaign, Champaign, IL.

Perera, C.O. 2005. Selected quality attributes of dried foods. *Drying Technology: An International Journal*, 23(4):717–730.

Püschner, P., L.L. SiokHoon. 2005. Microwave vacuum drying of fruits & vegetables. Retrieved on May 25, from https://www.pueschner.com/downloads/publications/article_mw-vacuum-drying_hk2007-short_pap.pdf.

Qing-guo, H., Z. Min, A.S. Mujumdar, D. Wei-hua, S. Jin-cai. 2006. Effects of different drying methods on the quality changes of granular edamame. *Drying Technology*, 24:1025–1032.

Rahman, M.D.M., M.D. Miaruddin, M.G.F. Chowdhury, H.F. Khan, M.D.M. Rahman. 2012. Preservation of jackfruit (*Artocarpus heterophyllus*) by osmotic dehydration. *Bangladesh Journal of Agricultural Research*, 37(1):67–75.

Rahman, M.M., G. Kibria, Q.R. Karim, S.A. Khanom, L. Islam, M.F. Islam, M. Begum. 2010. Retention of nutritional quality of solar dried carrot (*Daucus carota* L.) during storage. *Bangladesh Journal of Scientific and Industrial Research*, 45(4):359–362.

Rahman, M.S. 1999. *Handbook of Food Preservation*, Marcel Dekker, Inc. New York.

Rahman, M.S. 2000. Mechanism of pore formation in foods during drying: Present status. In *Proceedings of the 8th International Congress on Engineering and Foods*, Puebla, Mexico.

Rahman, S.M.A. 2013. Study of an integrated atmospheric freeze drying and hot air drying system using a vortex chiller. In: Hii, C.L., Jangam, S.V., Chiang, C.L., Mujumdar, A.S. (eds.) *Processing and Drying of Foods, Vegetables and Fruits*, Published in Singapore, pp. 13–38.

Rahman, S.M.A., A.S. Mujumdar. 2008. A novel atmospheric freeze drying system using a vortex tube and multimode heat supply. *International Journal of Postharvest Technology and Innovation*, 1(3):249–266.

Rahman, S.M.A., R. Saidur, M.N.A. Hawlader. 2013. An economic optimization of evaporator and air collector area in a solar assisted heat pump drying system. *Energy Conversion and Management* 76:377–384.

Ramos, I.N., T.R.S. Brandao, C.L.M. Silva. 2003. Structural changes during air drying of fruits and vegetables. *Food Science and Technology International*, 9(3):201–206.

Sarkar, J., S. Bhattacharyya, R. Gopal. 2004. Carbon dioxide based heat pump dryer in food industry. *International Journal of Refrigeration*, 27(8):830–838.

Shaochun, Y. 2009. An integrated solar heat pump system for cooling, water heating and drying. A Thesis Submitted for the Degree of Doctor of Philosophy, Department of Mechanical Engineering, National University of Singapore, Singapore.

Tortoe, C. 2010. A review of osmodehydration for food industry. *African Journal of Food Science*, 4(6):303–324.

Vega-Mercodo, H., M. Gongora-Nerto, G.N. Barbora-Canovas. 2001. Advances in dehydration of food. *Journal of Food Engineering*, 49:271–289.

Wang, N., J.G. Brennan. 1995. Changes in structure, density and porosity of potato during dehydration. *Journal of Food Engineering*, 24(1):61–76.

Wiset, L., N. Poomsa-Ad, T. Ratchapo. 2013. Multistage heat pump drying of macadamia nut under modified atmosphere. *International Food Research Journal*, 20(5):2199–2203.

Xiaoyong, S., C. Luming. 2015. Study of iron yam-chip (*Dioscorea opposita Thunb.* cv. *Tiegun*) dehydration using far-infrared radiation assisted heat pump drying. *Journal of Food and Nutrition Research*, 3(1):20–25.

Xu, D., L. Wei, R. Guangyue, L. Wenchao, L. Yunhong. 2015. Comparative study on the effects and efficiencies of three sublimation drying methods for mushrooms. *International Journal of Agricultural and Biological Engineering*, 8(1):91.

Zsivanovits, G., B. Brashlyanova, O. Karabadzhov, M. Marudova-Zsivanovits. 2013. Heat pump drying of red grape. *Journal of Food Physics*, 26:19–28.

7 Numerical Modeling of Heat Pump-Assisted Contact Drying

Zhifa Sun and Will Catton

CONTENTS

7.1 INTRODUCTION

By recycling the heat of evaporation, heat pump drying (HPD) systems can offer significant energy savings. Although conventional heat-and-vent drying systems usually operate at a specific moisture extraction rate (SMER, kg moisture extracted per kWh total energy consumed) within about 0.2–0.6 kg kWh^{-1}, HPDs have exhibited SMERs as high as 7.94 kg kWh^{-1} [1]. High energy efficiencies of HPDs can yield substantial reductions both in primary energy consumption and in greenhouse gas emissions, even after allowing for the thermal generation of electricity [1]. This fuel-conservation and emission-mitigation potential has been appreciated for decades [2,3]. For small drying operations, HPDs may also offer capital cost savings, for example, by eliminating the need for boiler systems, which may constitute a high fraction of the capital cost of the drying system. HPD systems are capable of continuous operation, allowing maximal product throughput and short payback times, even in cases where the capital cost is higher than for other systems [4].

This chapter centers on an idea for increasing the energy efficiency of heat pump driers, based on the results and discussions in a number of papers published by the authors and colleagues (in particular, [5–7]*) and some unpublished results from Catton [8]. The idea arises naturally from consideration of the Gouy–Stodola law [9], which is expressed in local form (for a multicomponent fluid mixture) by the exergy balance equation [10]. This is the result that the exergy destruction associated with a process is given by the environmental temperature times the rate of entropy generation in the process [10].

---

* Materials from Catton et al. (2011) have been used by permission of the Elsevier Limited.

In an HPD system, recycled heat is typically returned to the drying process convectively, by heating the dehumidified air stream as it reapproaches the product. But from a second law viewpoint, this method—using air for heat transfer—appears wasteful. Heat transfer both into the air and from the air to the drying process is responsible for a substantial part of the entropy creation in such a system [11,12]. Losses are also owing to the airflow resistance of the system and fan friction [13]. Taken together, these losses result not only from the low thermal conductivity of air and thus large temperature differences for heat transfer but also from the large mass circulation rate necessitated by the relatively low specific heat capacity of air. The straightforward idea underlying this chapter is that these entropy creation mechanisms can be avoided, or at least mitigated, by instead providing heat directly to the drying process, through a conductive heating plate separating the refrigerant and the product.

Figure 7.1 schematically illustrates a contact heat pump dryer (CHPD), considered in this chapter, which combines the heat pump (the compressor CMP with a shaft power input $\dot{W}_{cmp}$, condenser CD1 and condenser-type heating plates CD2 with embedded refrigerant condenser tubes, evaporator EV, and throttling valve TV), the airflow circulation and the circulation fan with a shaft power input $\dot{W}_{fan}$, and the product layers on the heating plates. The circulated air passes across a stack of product layers and heating plates. The total face area of all air ducts in the stack is taken to be identical to the condenser face area. The mass flow rate of dry air is $\dot{m}_a$, with a ratio of b bypassing the evaporator. In an ideal case, the surface temperature of the heating plates shown in Figure 7.1 is isothermal. The novel feature of the *isothermal contact HPD* (ICHPD) is the conductive plate thermally linking hot refrigerant with the drying process. In this chapter, the term ICHPD is used for contact HPDs, although the surfaces of the conductive heating plates may depart from the ideal isothermal case.

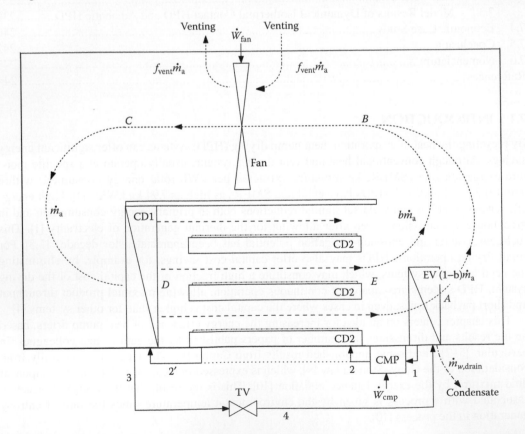

FIGURE 7.1 Schematic diagram of the CHPD system. Locations are denoted by letters and numbers on the air cycle and refrigerant cycle, respectively. (Modified from Catton, W. et al., *Energy*, 36, 4616–4624, Figure 1, 2011. With permission.)

The system depicted in Figure 7.1 becomes equivalent to an adiabatic HPD (AHPD) if the refrigerant bypasses the condenser plates (dotted line 2–2′). This AHPD case is similar to the dehumidifier wood drying systems theoretically investigated by the Otago research group [14–18].

An ICHPD system appears most likely to find applications among the medium-temperature drying processes in which conductive heating is already currently employed, particularly those that demand or would benefit from high energy efficiency or closed air-cycle operation. For instance, contact heat transfer is widely used in the food industry to treat heat-sensitive products. In food drying applications, closed air-cycle HPD can prevent the emission of volatile organic compounds such as solvents or eliminate combustion products that would be released by fuel-fired dryers, and may eliminate the need for product additives such as sulfites by allowing the use of a modified drying atmosphere [19]. Probably the most likely near-term candidate for contact HPDs is the drying of industrial sludges such as filter-cake sludges or wastewater sludges. Contact heat transfer is already routinely employed in industrial sludge drying, where closed air-cycle operation is also commonly required to meet local environmental standards, and where energy efficiency is a significant determinant of costs, and thus a key factor in the ongoing evolution of the drying technology [20]. The most important constraints on ICHPDs that have arisen are that the product must be spread into thin layers to enable good heat transfer, and should be able to be dried under fairly high-humidity conditions. An interesting potential application of ICHPDs is in the drying of microalgae pastes (after mechanical dewatering) prior to biodiesel production, because energy considerations play a central role in biofuel viability, and drying may turn out to be a crucial, energy-intensive process step [21,22]. To a plant operator, the applicability of HPD technology to any particular use would be an economic matter, dependent, for instance, upon a favorable balance between the savings or benefits and the cost of operations and capital outlay.

The idea of ICHPD has itself been advanced in the literature previously [19,23,24], but the modeling of CHPDs is not straightforward, because of the direct linkage between the refrigerant circuit, the drying process, and the air circuit, as shown in Figure 7.1. A preliminary assessment of an ICHPD system was previously conducted by this research group [5,6]. The results indicated that such a system configuration offers the potential for energy performance benefits of two to three times compared with an AHPD [6].

The aim of this chapter is to present a numerical model of an ICHPD whole system, to demonstrate the potential of such a system to improve on the energy efficiencies currently attained in HDP, and to develop an ability to optimize such a system, for example, so that the demands of the heat pump are matched to the properties and demands of the product being dried. In order to obtain a model of the widest applicability, the goal was to develop a dynamic dryer model, incorporating a drying process model based on the general drying equations developed by Whitaker [25,26]. The whole-system model that has been developed can be used to investigate the effects of a variety of parameters, for instance, the effects of product thickness, air velocity, and temperature set point, on the system performance. The model can be applied to a wide variety of porous materials, each of which is characterized by a different set of constitutive relations. In addition, although the model has only so far been used to examine the dynamics of batch-mode HPD operation, it could readily be extended to make predictions of system performance for continuous-operation driers.

7.2 DYNAMICAL HPD MODEL

As discussed above, the key difficulty that is associated with developing a whole-system model of an ICHPD is the tight linkage between the drying process and working-fluid condensation and evaporation on the refrigerant-cycle side. The novel feature of the model presented here is essentially that it deals with this complex linkage between the condenser-type heating plates and drying process, which has prevented previously developed models from being applied to contact HPD, and which has added considerable complexity to the simulation. The detailed timber stack model described by Sun et al. [14–17] has provided a basis for developing the air side dryer model, while for simulating the remaining system components, the empirical HPD model developed by Carrington and Bannister [27] has provided a large number of functions that were directly applicable in the present work. The internal drying process model is similar to a model described by Stanish et al. [28], although based on the constitutive relations employed by Wang and Chen [29].

The detailed interactions between the airflow, product, and refrigerant in the stack of drier ducts for the ICHPD are shown in Figure 7.2. In addition to the information of the geometry of the drying section, the temperature profiles of the airflow, product, heating plate and refrigerant, and the profiles of the water saturation, s, of the product and the vapor mass fraction, ω_v, of the moist air are schematically shown in this figure. Note that the orientation of refrigerant flow is at right angles to the flow direction assumed in our previous papers [5,6]. It is assumed here that the bottom surface of the heating plate is perfectly insulated. According to the curves of the temperature profiles across the refrigerant, heating plate, and product, heat energy is transferred from the hot refrigerant in the tubes to the product over the heating plate to evaporate the moisture contained in the product. However, heat energy can also be supplied from the airflow to the product when the airflow temperature is higher than the product temperature. Moisture is transferred from the product to the airflow.

7.2.1 Model of the Moist Air Stream

A detailed airflow model has been developed, which solves the mass, momentum, and energy balances within the air ducts. Employing Fick's and Fourier's laws, and the Newtonian model of viscous stress, and adopting a one-dimensional flow model for the control volume of the airflow surrounded by the boundary surfaces S_w, S_e, S_n, and S_m, the mass, momentum, and energy balance equations can be expressed for the air-side control volume depicted in Figure 7.2 as follows [14,15,17]:

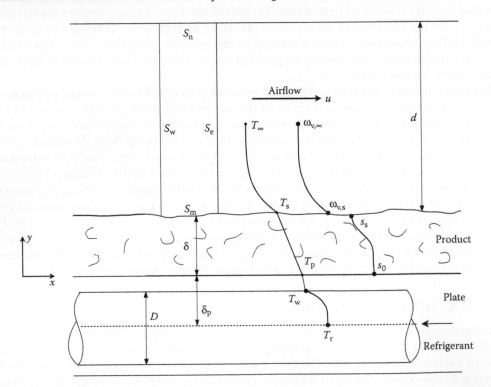

FIGURE 7.2 Airflow, product, and refrigerant in the stack of drier ducts in the CHPD system. Surfaces (S_w, S_e, S_n, and S_m) of an air-side control volume are indicated. (Modified from Catton, W. et al., *Energy*, 36, 4616–4624, Figure 2, 2011. With permission.)

Control volume mass balance for species k ($k = a,v$):

$$\frac{\mathrm{d}}{\mathrm{d}t} \int_{V_{(g)}} \rho\omega_k \mathrm{d}V = \int_{S_w} \left(\rho_k u - \rho D_{va} \frac{\partial \omega_k}{\partial x} \right) \mathrm{d}S$$

$$- \int_{S_e} \left(\rho_k u - \rho D_{va} \frac{\partial \omega_k}{\partial x} \right) \mathrm{d}S + \dot{m}_k^{(m)}$$

(7.1)

Control volume momentum balance:

$$\frac{\mathrm{d}}{\mathrm{d}t} \int_{V_{(g)}} \rho u \, \mathrm{d}V = \int_{S_w} \left(\rho u^2 + p - 2\mu \frac{\partial u}{\partial x} \right) \mathrm{d}S$$

$$- \int_{S_e} \left(\rho u^2 + p - 2\mu \frac{\partial u}{\partial x} \right) \mathrm{d}S - F_f$$

(7.2)

Control volume energy balance:

$$\frac{\mathrm{d}}{\mathrm{d}t} \int_{V_{(g)}} \rho \left(\hat{u} + \frac{1}{2} u^2 \right) \mathrm{d}V = \int_{S_w} \left[\rho \left(\hat{h} + \frac{1}{2} u^2 \right) u - k \frac{\partial T}{\partial x} - 2\mu \frac{\partial}{\partial x} \left(\frac{1}{2} u^2 \right) \right] \mathrm{d}S$$

$$- \int_{S_e} \left[\rho \left(\hat{h} + \frac{1}{2} u^2 \right) u - k \frac{\partial T}{\partial x} - 2\mu \frac{\partial}{\partial x} \left(\frac{1}{2} u^2 \right) \right] \mathrm{d}S$$

(7.3)

$$+ \dot{Q}^{(m)} + \dot{Q}$$

In Equations 7.1 through 7.3, $\dot{m}_k^{(m)}$ is the surface mass transfer of species k (where $k = a,v$) from the mass transfer surface S_m to the airflow, F_f is the friction force, $\dot{Q}$ is the heat transfer rate from the mass transfer surface to the airflow, and $\dot{Q}^{(m)}$ is the energy transfer rate accompanying the mass transfer from the mass transfer surface. These are evaluated by using the following equations, respectively [6,17]:

$$\dot{m}_k^{(m)} = \int_{S_m} n_k \mathrm{d}S = \int_{S_m} \left[(\omega_k \rho v)_s + k_\omega^\bullet (\omega_{k,s} - \omega_{k,\infty}) \right] \mathrm{d}S$$

(7.4)

$$F_f = \int_{S_m} f^\bullet \frac{1}{2} \rho u^2 \mathrm{d}S$$

(7.5)

$$\dot{Q} = \int_{S_m} h^\bullet (T_s - T_\infty) \mathrm{d}S$$

(7.6)

$$\dot{Q}^{(m)} = \sum_{k=a,v} \dot{m}_k^{(m)} \hat{h}_{k,s}$$

(7.7)

In Equations 7.4 through 7.6, $k_\omega^\bullet$, $f^\bullet$, and $h^\bullet$ are the local mass transfer coefficient, friction factor, and heat transfer coefficient, respectively, all adjusted for high mass transfer rates by the method of Bird et al. [30]. These have been calculated using the Dittus–Boelter equation [31] and the Chilton–Colburn analogy [30]. In this chapter, we have ignored the effects of the boundary-layer separation and reattachment at the entry region of the drier ducts on the heat and mass transfer in the drier ducts [32,33].

The dry-air mass transfer rate into the duct will be near zero at the surface of a porous product undergoing drying, because dry airflows into the product to fill the volume previously occupied by liquid water, whose density is roughly 10^3 times greater than that of dry air. Thus, the mass transfer flux of water vapor at the product surface can be expected to be $O(10^3)$ times larger than that of dry air. Neglecting the dry-air mass flow rate across the top surface of the product, Equations 7.4 and 7.7 are simplified to [17]:

$$\dot{m}_v^{(m)} = \int_{S_m} n_v dS = \int_{S_m} \frac{k_\omega^\bullet(\omega_{v,s} - \omega_{v,\infty})}{1 - \omega_{v,s}} dS \tag{7.8}$$

and

$$\dot{Q}^{(m)} = \dot{m}_v^{(m)} \hat{h}_{v,s} \tag{7.9}$$

Equations 7.1 through 7.3 have been discretized on a one-dimensional finite element grid along the x-direction, shown in Figure 7.2, by using the methods of Patankar [34]. The discretized equations have been solved using Patankar's staggered-grid SIMPLER algorithm, subject to the initial and boundary conditions given by the inlet airflow velocity and psychrometric state at location D (Figure 7.1), the outlet air pressure p_E at location E (Figure 7.1), similar to those discussed in Sun et al. [17]. Based on Equations 7.1 through 7.9, a dynamic wood drying kiln model has been developed and the model has been validated using experimental data [15–17].

A maximum value of 10^{-4} in any control volume for the relative source magnitude (relative to the quantity in the given control volume, as defined by Patankar [34]) of the three conserved quantities has been adopted as an acceptable balance of convenience and accuracy. Thus, convergence of the model implies that the discretized mass, momentum, and energy conservation equations have been satisfied to within this convergence criterion. A number of $N_x = 50$ equal size control volumes have been used in the airflow direction (x-direction in Figure 7.2) in the analysis below. It was found that no significant change occurs when N_x is increased above this number.

7.2.2 MODEL OF THE PRODUCT

The majority of dynamical drying simulations reported during the past two decades have been based either on the volume averaging method introduced by Whitaker [25,26], on a simpler diffusion model of moisture transport [35,36], or on the simplified assumption of a discrete drying front [37]. The trade-offs between comprehensiveness, computational tractability, and ease of implementation influence the best choice of model for a particular context. Detailed models are significantly more demanding to produce and run than diffusion models, which can be used when low drying intensity allows the equations to be recast in the form of a diffusion equation. However, previous researchers have found that temperature gradients may significantly influence internal moisture transport in convective drying [29]. The dynamical internal process drying model, if it is to successfully represent the drying process, will need to be consistent with the influence of temperature gradients on moisture transport as predicted by the detailed theory of drying. For a slab without conductive contact heat transfer, Wang and Chen [29] have incorporated temperature effects into the diffusion equation. The resulting effective moisture diffusivity, which varies to incorporate the linkage between heat and mass transport, can be predicted from local conditions using the volume averaging theory of Whitaker [25,26]. Such a diffusion model thus has the potential to combine comprehensiveness under appropriate conditions with simplicity and computational speed. A straightforward generalization of Wang and Chen's equations (described below) allows a sufficiently detailed representation of the porous product layer to capture plate heat transfer and temperature-gradient effects, while at the same time being relatively simple and easy to implement within the ICHPD model.

For low-intensity drying, by neglecting variation in the total pressure within the pores, neglecting heat transfer associated with convection mass transfer, and neglecting accumulation of water vapor within the pores of the product medium, the following simplified set of equations can be obtained from the classic volume-averaged equations describing the transfer processes within a porous medium undergoing drying [29]:

$$u_{sat} \frac{\partial s}{\partial t} = \frac{\partial}{\partial y}\left(D_T \frac{\partial T}{\partial y} + D_s \frac{\partial s}{\partial y} \right) \tag{7.10}$$

$$\rho \hat{c}_p \frac{\partial T}{\partial t} = -\frac{\partial n_v}{\partial y} \Delta \hat{h}_{vap} + \frac{\partial}{\partial y}\left(k_{eff} \frac{\partial T}{\partial y} \right) \tag{7.11}$$

where u_{sat} is the moisture density of the product when fully saturated, and where

$$D_T = -\frac{k_\beta^r k_\beta^0}{\mu_\beta} \rho_\beta \frac{\partial p_c}{\partial T} + \varepsilon(1-s)D_{va} \frac{\partial \rho_v}{\partial T} \tag{7.12}$$

$$D_s = -\frac{k_\beta^r k_\beta^0}{\mu_\beta} \rho_\beta \frac{\partial p_c}{\partial s} + \varepsilon(1-s)D_{va} \frac{\partial \rho_v}{\partial s} \tag{7.13}$$

The boundary conditions used for the internal process drying model, in the context of the ICHPD system, are as follows:

$$\left(D_T \frac{\partial T}{\partial y} + D_s \frac{\partial s}{\partial y} \right)_{y=0} = 0 \tag{7.14}$$

$$\left(D_T \frac{\partial T}{\partial y} + D_s \frac{\partial s}{\partial y} \right)_{y=\delta} = k_\omega^\bullet(\omega_{v,s} - \omega_{v,\infty}) \tag{7.15}$$

$$\left(n_v \Delta h_{vap} - k_{eff} \frac{\partial T}{\partial y} \right)_{y=\delta} = h^\bullet(T_s - T_\infty) \tag{7.16}$$

$$\left(k_{eff} \frac{\partial T}{\partial y} \right)_{y=0} = \dot{q}_p \tag{7.17}$$

In Equation 7.17, $\dot{q}_p$ denotes the heat transfer flux from the refrigerant to the base of the product across the plate, which is discussed in details in the subsection follows. The adiabatic mode, for which plate heat transfer is absent, can be simulated by simply setting $\dot{q}_{plate}$ to be equal to zero. The above-mentioned equations are slightly more general than those of Wang and Chen [29], because they include the effect of plate heat transfer, and the mass and energy balance equations, Equations 7.10 and 7.11, retain the mass and energy accumulation terms [7,8].

The balance equations for the product have been discretized on a one-dimensional finite element grid along the y-direction, as shown in Figure 7.2, using Patankar's methods [34]. The discretized equations have been solved using Patankar's SIMPLER algorithm, with modifications. A 2D model, comprising a series of side-by-side (in the x-direction shown in Figure 7.2), noninteracting

1D models, each with its own independent gas-side and heating plate-side boundary conditions, is a straightforward extension of the 1D model. That is, we use the idealization of a thin product layer within which lateral moisture transport can be neglected. This 2D model is then combined with the duct airflow model discussed in Section 7.2.1, and the heating plate model discussed below through the links of the corresponding boundary conditions, to form a dynamical duct model. Finally, this duct model is incorporated into the whole-system HPD model to produce a dynamical whole-system model. The convergence criterion that has been adopted is a maximum absolute relative change of less than 10^{-4} in the updated estimate of the temperature or saturation in any control volume. A number of $N_y = 16$ equal size control volumes have been used in the y-direction shown in Figure 7.2, to obtain the model results presented in this chapter. Using the product model developed, we have obtained the results obtained for the adiabatic mode [7], which are in good agreement (in terms of both the saturation profiles and overall drying time) with those presented by Wang and Chen [29].

7.2.3 Model of the Heating Plates

In each drier duct, a product layer is sandwiched between an air stream (passing over the product) and a heating plate. As depicted in Figure 7.3, several parallel circuits of refrigerant flow pass through each plate, each running back and forth through the length of the plate multiple times. The orientation of refrigerant flow through the heating plate is parallel to the airflow direction. This configuration has the effect of averaging the wall temperature experienced near a given location z along each refrigerant circuit, and also of averaging the heat transfer coefficient associated with a given location x from the leading edge of the plate. This allows us to assume an averaged plate temperature in the energy balance at all locations along the refrigerant flow, and then to calculate an effective (average) heat transfer coefficient, which applies at all locations on the plate. Our purpose is not to establish a detailed model of the refrigerant flow within the plate, but to obtain an estimate for the mean plate refrigerant heat transfer coefficient. A simple refrigerant flow model has been developed to estimate the effective refrigerant heat transfer coefficient $U_{p,eff}$, for use in determining the temperature at the top surface of the heating plate.

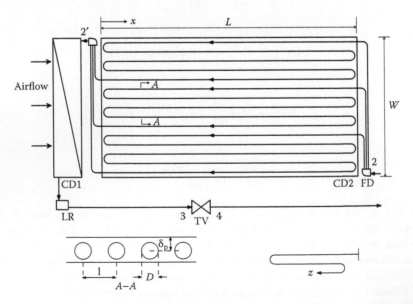

FIGURE 7.3 Conceptual schematic diagram of multi-pass condensing plate with three loops. FD, refrigerant flow distributor and LR, liquid receiver. Directions along the dimensions x (air stream) and z (refrigerant stream) are shown. (Modified from Catton, W., Numerical study of heat pump contact drying. A thesis submitted for the degree of Doctor of Philosophy at the University of Otago, Dunedin, New Zealand, Figure 3.3, 2011. With permission.)

Since the heat transfer processes in the refrigerant and in the metal components of the heating plates are much faster than the heat and mass transfer processes in the air stream and the product, the mass and energy accumulations of the refrigerant and the energy accumulation of the metal parts in the heating plate can be neglected. Therefore, the dynamic mass and energy balances of the refrigerant in the heating plate have been simplified to

$$\int_{z_1}^{z_2} \mathrm{d}\dot{m}_r = 0 \tag{7.18}$$

and

$$\dot{m}_r \int_{z_1}^{z_2} \mathrm{d}\hat{h}_r = \dot{Q}_{r,z_2-z_1} \tag{7.19}$$

where:

$\dot{m}_r$ is the mass flow rate of the refrigerant in a single tube

$\hat{h}_r$ is the specific enthalpy of the refrigerant, which is the function of the vapor quality and the temperature of the refrigerant

$\dot{Q}_{r,z_2-z_1}$ is the heat transfer rate from the refrigerant in a single tube to the top surface of the heating plate along the tube length, $z_2 - z_1$, which is given by

$$\dot{Q}_{r,z_2-z_1} = \int_{z_1}^{z_2} U_{p,\mathrm{eff}}(T_r - T_p)\mathrm{d}z \tag{7.20}$$

where the overall effective heat transfer coefficient per unit length of the tube, $U_{p,\mathrm{eff}}$, can be evaluated using

$$\frac{1}{U_{p,\mathrm{eff}}} = \frac{1}{h_{r,\mathrm{eff}}} + \frac{1}{h_{p,\mathrm{eff}}} \tag{7.21}$$

where:

$h_{r,\mathrm{eff}}$ is the effective refrigerant convective heat transfer coefficient per unit length of the tube

$h_{p,\mathrm{eff}}$ is the effective heat transfer coefficient of the heating plate associated with the condenser tube considered [31, pp. 3–121], as follows:

$$h_{r,\mathrm{eff}} = \pi D h_r \tag{7.22}$$

$$h_{p,\mathrm{eff}} = \frac{2\pi k_p}{\ln\left[(2l/\pi D)\sinh\left(2\pi\delta_p/l\right)\right]} \tag{7.23}$$

In Equations 7.18 and 7.19, the variable z is used to represent the distance along a given refrigerant flow path through the plate (Figure 7.3). The dimensions D, l, and δ_p that appear in Equations 7.22 and 7.23 are illustrated in Figure 7.3 (A–A view). The thermal conductivity k_p of the heating plate is assumed to be the thermal conductivity of copper.

In Equation 7.19, the mass specific enthalpy of the refrigerant at the inlet of the heating plate is determined by the performance of the compressor of the heat pump, and the mass flow rate $\dot{m}_r$ of the each refrigerant flow circuits passing through a single plate can be evaluated by

$$\dot{m}_r = \frac{\dot{m}_{r,t}}{n_b N_d} \tag{7.24}$$

where:

$\dot{m}_{r,t}$ is the whole-system refrigerant mass flow rate

N_d is the number of heating plates

n_b is the number of refrigerant flow circuits in each plate (Figure 7.3)

From Equation 7.20, an average heat flux from the refrigerant to the base of the product can be obtained as follows:

$$\dot{q}_p = \frac{\dot{Q}_p}{WL} = \frac{n_b}{WL} \int_0^{n_p L} U_{p,\text{eff}} (T_r - T_p)dz \tag{7.25}$$

where:

n_p is the number of passes through the plate for each refrigerant flow circuit

W and L are the width and the length of the heating plate (Figure 7.3)

The local heat transfer coefficient of the refrigerant h_r in Equation 7.22 can be calculated as a function of the specific enthalpy as follows. The enthalpy and pressure of the refrigerant are used to establish the vapor quality X and refrigerant physical properties. In the single-phase regions, the heat transfer coefficient is given by the following equations [38]:

$$h_r = 0.023 \left(\frac{k_r}{D} \right) \text{Pr}^{0.4} \text{Re}^{0.8} \tag{7.26}$$

In the two-phase region, the heat transfer coefficient is evaluated using Cavallini et al.'s method [39]. The dimensionless gas velocity (or the modified Froude number) J_v can be calculated by

$$J_v = \frac{XG}{[gD\rho_v(\rho_l - \rho_v)]^{0.5}} \tag{7.27}$$

where G is the mass velocity, in kg m^{-2} s^{-1}

$$G = \frac{4\dot{m}_r}{\pi D^2} \tag{7.28}$$

The transition dimensionless vapor velocity J_v^T is estimated using the following empirical correlation, where for HFCs such as R-134a, $C_T = 2.6$:

$$J_v^T = \left(\left[\frac{75}{(4.3\chi_{tt}^{1.111} + 1)} \right]^{-3} + C_T^{-3} \right)^{-1/3} \tag{7.29}$$

In this equation, the Martinelli parameter, χ_{tt}, is given by

$$\chi_{tt} = \left(\frac{\mu_l}{\mu_v} \right)^{0.1} \left(\frac{1-X}{X} \right)^{0.9} \left(\frac{\rho_v}{\rho_l} \right)^{0.5} \tag{7.30}$$

When the dimensionless vapor velocity $J_v > J_v^T$ (ΔT-independent flow):

$$h_{r, J_v > J_v^T} = h_{lo} \left[1 + 1.128 X^{0.8170} \left(\frac{\rho_l}{\rho_v} \right)^{0.3685} \left(\frac{\mu_l}{\mu_v} \right)^{0.2363} \left(\frac{\mu_l - \mu_v}{\mu_l} \right)^{2.144} \left(\frac{\mu_l \hat{c}_{p,l}}{k_l} \right)^{-0.100} \right] \tag{7.31}$$

When the dimensionless vapor velocity $J_v \leq J_v^T$ (ΔT-dependent flow):

$$h_{r,J_v \leq J_v^T} = \left[h_{r,J_v > J_v^T} \left(\frac{J_v^T}{J_v} \right)^{0.8} - h_{strat} \right] \left(\frac{J_v}{J_v^T} \right) + h_{strat} \qquad (7.32)$$

In Equations 7.31 and 7.32

$$h_{lo} = 0.023 \frac{k_l}{D} Pr_l^{0.4} Re_{lo}^{0.8} \qquad (7.33)$$

$$h_{strat} = 0.725 \left[1 + 0.741 \left(\frac{1-X}{X} \right)^{0.3321} \right]^{-1} \left[\frac{k_l^3 \rho_l (\rho_l - \rho_v) g h_{lv}}{\mu_l D \Delta T} \right]^{0.25} + (1 - X^{0.087}) h_{lo} \qquad (7.34)$$

and the liquid-only Reynolds number in Equation 7.33 is evaluated in terms of the total mass velocity and the liquid-phase viscosity:

$$Re_{lo} = \frac{GD}{\mu_l} \qquad (7.35)$$

The above functions have been tested by Catton [8]. The results agree with those of Cavallini et al. [39].

The pressure drop in the refrigerant tubes embedded in the heating plates is evaluated by numerically integrating the pressure derivative from z_1 to z_2. Cavallini et al. [40] suggest using the annular-flow pressure drop correlation to evaluate pressure drops in all the flow regimes. We follow this approach, employing the pressure drop relation of Traviss et al. [41]. Neglecting the external force of gravity, since the tubes are horizontal, the pressure gradient along the tubes is given by [41]:

$$\frac{dp_{r,p}}{dz} = \left(\frac{dp_{r,p}}{dz} \right)_f + \left(\frac{dp_{r,p}}{dz} \right)_m \qquad (7.36)$$

In Equation 7.36, the pressure gradient is decomposed into a friction part, subscript f, and a part due to momentum change, subscript m, which is negligible in the one-phase region.

In the one-phase region, the friction part is given by the correlations [38]:

$$\left(\frac{dp_{r,p}}{dz} \right)_f = 2 \frac{fG^2}{D\rho} \qquad (7.37)$$

where:

$$f = \begin{cases} 0.3164 Re^{-0.25} & (Re \leq 10^5) \\ 0.0032 + 0.221 Re^{0.237} & (10^5 < Re < 3 \times 10^6) \end{cases} \qquad (7.38)$$

The Reynolds number in Equation 7.38 is given by its usual definition $Re = GD/\mu$.

In the two-phase region, the friction part of the pressure gradient is evaluated using [41]:

$$\left(\frac{dp_{r,p}}{dz} \right)_f = \frac{G^2}{D\rho_v} \left[-0.09 \left(\frac{GD}{\mu_v} \right)^{-0.2} \phi_v^2 \right] \qquad (7.39)$$

$$\phi_v = 1 + 2.85 \chi_{tt} \qquad (7.40)$$

The pressure gradient because of momentum changes is as follows [41]:

$$\left(\frac{dp_{r,p}}{dz}\right)_m = \frac{G^2}{\rho_v}\left(\frac{dX}{dz}\right)f(X) \tag{7.41}$$

$$f(X) = 2X + (1-2X)\left(\frac{\rho_v}{\rho_l}\right)^{1/3} + (1-2X)\left(\frac{\rho_v}{\rho_l}\right)^{2/3} - 2(1-X)\left(\frac{\rho_v}{\rho_l}\right) \tag{7.42}$$

Equation 7.19 has been numerically solved using the Runge–Kutta integration method by Catton [8] to obtain the values of the vapor quality, pressure, and effective overall heat transfer coefficient to estimate the heat transfer rate from the hot condensing refrigerant to the product and the state of the refrigerant at the exit of the heating plates.

7.2.4 Model of the Whole-System ICHPD

The whole-system model used in the present work, which is the model of Carrington and Bannister [27] modified to include the detailed drier-duct model described in Sections 7.2.1 through 7.2.3, incorporates the isentropic and volumetric efficiencies of a ZR61K2-TFD scroll compressor with R-134a as the refrigerant [42]. The detailed discussions on the mass and energy balance equations and heat transfer rates and pressure drops in the various components in the heat pump of the whole-system ICHPD can be found in [27]. Locations E, A, C, and D of the moist air stream in Figure 7.1 in this work correspond to locations (1), (7), (8), and (9) of the airflow in Figure 1 of Carrington and Bannister [27], respectively. The heating plates (CD2) are connected to both the compressor (CMP) and the condenser (CD1) in this work, whereas the compressor was connected to the condenser directly in [27].

Air-side pressure drops in the evaporator (EV) and condenser (CD1) in Figure 7.1 are evaluated using the heat exchanger pressure-drop correlations developed by Turaga et al. [43]. The pressure drop across each heat exchanger is given by the following equation:

$$\Delta p_a = f_a \cdot \frac{2 L_d G_{m,f}^2}{D_h \rho_a} \tag{7.43}$$

where:

$G_{m,f} = \dot{m}_a/A_{m,f}$ is the mass velocity evaluated using the minimum flow cross section within the heat exchanger, $A_{m,f}$

L_d is the coil depth [43]

The air-side hydraulic diameter of the heat exchanger is defined as follows:

$$D_h = \frac{4 A_{m,f} L_d}{A_0} \tag{7.44}$$

where A_0 is the total air-side heat exchange area. We use the following correlations for the condenser and evaporator friction factors [43]:

$$f_{a,CO} = 0.589\left(\frac{A_0}{A_p}\right)^{-0.28} Re_a^{-0.27} \tag{7.45}$$

$$f_{a,EV} = 0.325\left(\frac{A_0}{A_p}\right)^{0.01}\left(\frac{S_f}{Y_f}\right)^{0.4} Re_a^{-0.41} \tag{7.46}$$

In Equations 7.45 and 7.46, Re_a is the air-side Reynolds number calculated using the hydraulic diameter of the heat exchanger:

$$Re_a = \frac{G_{m,f} D_h}{\mu_a}$$

(7.47)

In Equations 7.45 and 7.46, the ratio (A_0/A_p) represents the ratio of the total air-side heat exchange area to the primary (tube) area, and the ratio (S_f/Y_f) represents the ratio of the fin spacing to the fin thickness. Air-side pressure drops for other air ducts are estimated using the dynamic loss coefficients (k-factors) obtained by Carrington et al. [44]:

$$\Delta p_a = k \cdot \frac{1}{2} \rho_a v_a^2$$

(7.48)

To solve the model equations of the whole-system CHPD, the Newton–Raphson method is applied to the state vector $x = (x_1, x_2, x_3)$, where, in terms of the locations shown in Figure 7.1,

$$x_1 = T_{rsat,1}$$

(7.49)

$$x_2 = T_{rsat,3}$$

(7.50)

$$x_3 = \omega_D$$

(7.51)

Together with the compressor model, the fixed air temperature at location D, and the correlations for the heat pump components, a unique value of this state vector can consistently describe the system state, and the three parameters are sufficient to deduce the condition at all system locations [6].

Pressure drops of the refrigerant stream are evaluated using the functions for the condenser heating plate (CD2) described in Section 7.2.3 and the empirical correlations for other components in the heat pump [27]. Using the compressor model and assuming isenthalpic throttling, the two saturated refrigerant states $T_{rsat,1}$ and $T_{rsat,3}$ are then used to update the estimated total refrigerant mass flow rate $\dot{m}_{r,t}$ and the operation of the thermodynamic cycle. The refrigerant state at location 2′ is estimated using the heating plate model in Section 7.2.3.

On the air-side, the air pressure at location E is set equal to the ambient pressure: $p_E = 1$ atm. In order to evaluate conditions within the dryer duct, the air duct, product, and heating plate models are iterated toward solutions as discussed above. As venting is controlled to maintain T_D fixed, the air state at location D is determined by ω_D and p_D. Air states around the system are obtained using moisture and energy balances for the whole system.

In the whole-system ICHPD model, it is assumed that the total dry-air mass flow rate $\dot{m}_a$ is constant, that is, the quantity of the dry-air vented out at location C is the same as that of the dry air entering the system at location B. Thus, the amount of water vapor vented out from the system is evaluated by

$$\dot{m}_{w,vent} = f_{vent} \dot{m}_a (\omega_C - \omega_{env})$$

(7.52)

where f_{vent} represents the fraction of the dry air that is removed from the system through venting. The value of f_{vent} is specified by the required dry-bulb temperature, T_D. Under normal operating conditions, water vapor condensed as hot moist air encounters the evaporator cold surface. Condensate is drained from the system at the rate $\dot{m}_{w,drain}$, which is given by

$$\dot{m}_{w,drain} = (1-b) \dot{m}_a (\omega_E - \omega_A)$$

(7.53)

where:

$$\omega_A = \omega_{sat}(\hat{h}_A, p_A) \qquad (7.54)$$

In the condenser (CD1) of the heat pump, no condensate is formed, thus,

$$\omega_D = \omega_C \qquad (7.55)$$

Moisture balance equations for the mixing processes of the humid air streams at locations B and C in the whole-system HPD can be written as follows:

$$\omega_B = b\omega_E + (1-b)\omega_A \qquad (7.56)$$

$$\omega_C = \frac{\omega_B + f_{vent}\omega_{env}}{1 + f_{vent}} \qquad (7.57)$$

Energy balance equations, in terms of $\hat{h} = \hat{h}_a + \omega\hat{h}_v$, for the evaporator, condenser, air-circulation fan, and the air-stream mixing can be written as

$$\hat{h}_A = \hat{h}_E - \frac{\left[\dot{Q}_{ev} + \dot{m}_{w,drain}\hat{h}_w(T_{sat,A})\right]}{(1-b)\dot{m}_a} \text{ (evaporator)} \qquad (7.58)$$

$$\hat{h}_B = b\hat{h}_E + (1-b)\hat{h}_A \text{ (mixing at B)} \qquad (7.59)$$

$$\hat{h}_C = \frac{\hat{h}_B + f_{vent}\hat{h}_{env}}{1 + f_{vent}} + \frac{\dot{W}_{fan}}{(1 + f_{vent})\dot{m}_a} \text{ (fan and venting)} \qquad (7.60)$$

$$\hat{h}_D = \hat{h}_C + \frac{\dot{Q}_{co}}{\dot{m}_a} \text{ (condenser)} \qquad (7.61)$$

In Equations 7.56 and 7.59, ω_E and $\hat{h}_E$ are estimated from the state-vector x using the air duct model. The fan power $\dot{W}_{fan}$ in Equation 7.60 is estimated using the following equation [45]:

$$\dot{W}_{fan} = \frac{1}{\varepsilon_{fan}}\dot{m}_a v_{a,B}\Delta p_{fan} \qquad (7.62)$$

where a constant fan efficiency ε_{fan} of 50% has again been assumed [45].

The energy balance for the refrigerant in the evaporator (EV) and condenser (CD1) give the following two equations:

$$\dot{m}_{r,t}(\hat{h}_1 - \hat{h}_4) = \dot{Q}_{ev} \qquad (7.63)$$

$$\dot{m}_{r,t}(\hat{h}_{2'} - \hat{h}_3) = \dot{Q}_{co} \qquad (7.64)$$

The heat transfer rates between the air-side and the refrigerant side in the evaporator and condenser are estimated using Carrington and Bannister's model [27]:

$$\dot{Q}_{ev} = A_{ev}[g(T_{wb,E})f_{ev}(v_{ev})(T_{wb,E} - T_4) - 2.69] \times 10^3 \qquad (7.65)$$

$$\dot{Q}_{co} = A_{co}[f_{co}(v_{co})(T_3 - T_C) - 2.50] \times 10^3 \tag{7.66}$$

where:

A_{ev} and A_{co} are the face areas of the evaporator and condenser heat exchangers, respectively the polynomials g, f_{ev}, and f_{co} are as follows:

$$g(T_{wb,E}) = 0.212 + 0.1283T_{wb,E} - 0.001181T^2_{wb,E} \tag{7.67}$$

$$f_{ev}(v_{ev}) = -0.017 + 1.486v_{ev} - 0.5145v^2_{ev} \tag{7.68}$$

$$f_{co}(v_{co}) = -0.0065 + 1.1535v_{co} - 0.11079v^2_{co} \tag{7.69}$$

where v_{ev} and v_{co} are the face velocities at the entries of the evaporator and condenser, respectively.

Based on Equations 7.55, 7.63, and 7.64, the error vector Δ that is used in the Newton–Raphson method is specified as follows:

$$\Delta_1 = \dot{m}_{r,t}(\hat{h}_1 - \hat{h}_4) - A_{ev}[g(T_{wb,E})f_{ev}(v_{ev})(T_{wb,E} - T_4) - 2.69] \times 10^3 \tag{7.70}$$

$$\Delta_2 = \dot{m}_{r,t}(\hat{h}_{2'} - \hat{h}_3) - A_{co}[f_{co}(v_{co})(T_3 - T_C) - 2.50] \times 10^3 \tag{7.71}$$

$$\Delta_3 = \omega_D - \omega_C \tag{7.72}$$

Each iteration of the Newton–Raphson method occurs as follows. From the current estimate of the state vector, x, an updated estimate of the states throughout the system is formed, and the error vector Δ is evaluated. An estimated value $\mathbf{J}$ of the Jacobian matrix for the system equations (Equations 7.70 through 7.72) is then used to update the state-vector estimate using

$$x_1 = x_0 - \mathbf{r} \cdot \mathbf{J}^{-1}\Delta \tag{7.73}$$

where:

r represents the relaxation factor vector that is applied for the convergence of the solutions
x_1 is the new estimate of the system state, which is obtained from the previous estimate x_0

The convergence criterion for the whole-system model is that the maximum error function, across all submodels of the whole-system model, must be less than a threshold, which is typically set to 10^{-5}. Since Equations 7.63 and 7.64 represent the heat transfer at the condenser and evaporator that is a function of the conditions of the refrigerant and the airflow, the components of the error vector Δ approach zero only if the heat transfer at these heat exchangers is consistent in both the refrigerant side and the air-flow side. In addition, the global convergence criterion requires that the mass, momentum, and energy balances within the drier ducts be satisfied, and that the heating plate surface temperature estimate be satisfied, to within this convergence criterion. Thus, on convergence, the prediction is stable and is guaranteed to satisfy mass and energy balances across all system components. Figure 7.4 schematically shows the model's flowchart. The validation of this whole-system model has been discussed by Catton et al. [6].

In summary, there are four components that form the basic building blocks that are constructed, tested, and assembled together for the whole-system ICHPD model presented in this chapter. These are (1) a dynamic model of the airflow through the drier, (2) a dynamical model of the transport processes occurring within the product, (3) a model of the refrigerant flow through the heating plates, and (4) a model of the remaining heat pump components, which is based on correlations taken from the empirical heat pump model developed by Carrington and Bannister [27].

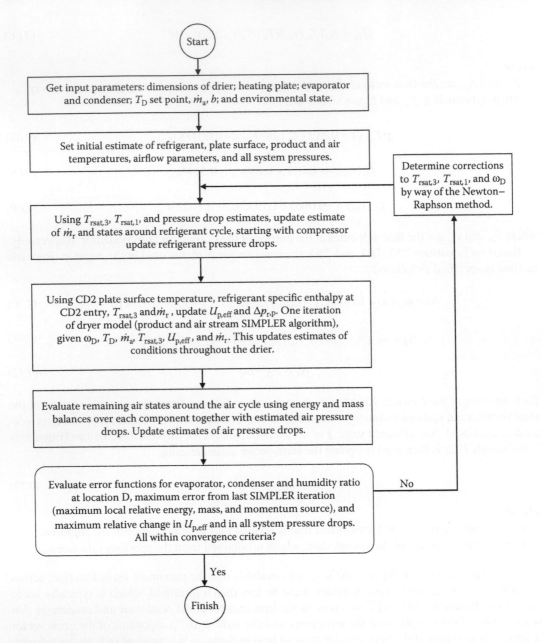

FIGURE 7.4 Flowchart for the whole-system model.

Taken together, components (1)–(3) form a detailed dynamical model of the contact dryer, subject to the inlet conditions of the refrigerant and the initial and boundary conditions of the airflow and the product. The whole-system ICHPD model is capable of describing the evolution of a non-steady batch drying operation; for instance, predicting the decline in energy efficiency that occurs during the falling-rate period of drying. It can also be used to model the operation of a continuously operated system at steady state [6]. Although no new experimental work is presented here in support of this whole-system ICHPD model, each component of the model has been tested against previously established numerical models, against empirical data, or against ideal scenarios to which theoretical solutions apply.

7.3 RESULTS AND DISCUSSION

7.3.1 MODEL RESULTS OF STEADY-STATE ICHPD AND AHPD DURING THE CONSTANT DRYING RATE PERIOD

The performances of an ICHPD system and an AHPD system have been studied under the steady-state condition. Here, we consider the constant drying rate period of the product drying, and thus we assume that the product layer is saturated with water and that evaporation takes place entirely at the product surface. We further assume that the effective heat conductivity k_{eff} of the product is equal to the thermal conductivity of water and the effective heat transfer coefficient through the product is given by $h_{eff} = k_{eff}/\delta$. Thus, for the isothermal mode, the heat transfer flux across the product is given by $\dot{q} = h_{eff}(T_p - T_s)$. For the adiabatic mode, the plate thermal conductivity, k_p, is set to zero. In this adiabatic case, product surface temperature T_s is given by the wet-bulb temperature of the moist air stream in the local control volume [46]. The two modes have identical specifications aside from plate heat transfer. In particular, $T_D = 55°C$; evaporator bypass ratio $b = 0$. The dimensions and other operating conditions of the system are listed in Table 7.1. In addition, no subcooling and 5°C superheating are assumed. The adopted number n_b of refrigerant circuits per plate is 3, and the number p of passes through the plate for each circuit is 5. These values lead to a tube separation l of 6.7 cm in the plate (Figure 7.3).

Figure 7.5 shows the refrigerant thermodynamic state-cycle under typical ICHPD ($1'-2'-3'-4'$) and AHPD ($1-2-3-4$) conditions. The condenser refrigerant pressure drop for the isothermal mode (between the locations labeled $2'$ and $3'$) is somewhat greater than that for the adiabatic mode (between the locations labeled 2 and 3), owing to the pressure drop of the refrigerant within the heating plates. Despite this relatively small additional pressure drop, the isothermal mode can be seen to enable the system to operate over a significantly smaller pressure range. The heating plate pressure

---

TABLE 7.1

Values of Key System Parameters

Parameter (Unit)	Baseline Value
Shaft power of compressor $\dot{W}_P$ (W)*	5.0×10^3
Condenser face area A_{co} (m²)	1.0
Evaporator face area A_{ev} (m²)	1.0
Number of ducts N_D	10
Product airflow $\dot{m}_{a,co}$ (kg s^{-1})	1.0
Relative humidity at drier inlet ϕ_D (%)*	30.0
Tray drier length L (m)	5.0
Tray drier width w (m)	1.0
Air duct depth d (m)	20.0×10^{-3}
Product thickness δ (m)	1.0×10^{-3}
Heating plate condenser tube diameter D (m)	10.0×10^{-3}
Heating plate condenser tube midline depth x_p (m)	6.0×10^{-3}
Refrigerant circuits per plate n_b (–)	3
Passes through plate per circuit n_p (–)	5
Total refrigerant mass flow rate $\dot{m}_r$ (kg s^{-1})*	1.0
Refrigerant saturated condensing temperature $T_{rsat,3}$ (°C)*	60.0
Drier maximum temperature T_D (°C)	55.0

Source: Catton, W. et al., *Energy*, 36, 4616–4624, Table 1, 2011. With permission.
* Subject to variation in the whole-system model.

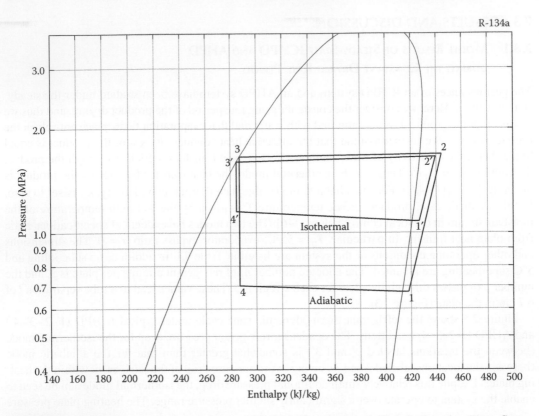

FIGURE 7.5 R-134a state-cycle in the isothermal (1′–2′–3′–4′) and adiabatic (1–2–3–4) modes. (From Catton, W., Numerical study of heat pump contact drying. A thesis submitted for the degree of Doctor of Philosophy at the University of Otago, Dunedin, New Zealand, 2011. With permission.)

drop model thus leads to the prediction that the refrigerant-side tradeoff between heat- and momentum transfer irreversibilities does not significantly impact on performance in the isothermal mode. The system MER and SMER for the isothermal mode are 55.2 kg h^{-1} and 13.6 kg kWh^{-1}, respectively, whereas the system MER and SMER for the adiabatic mode are 20.5 kg h^{-1} and 4.0 kg kWh^{-1}, respectively. The isothermal mode is found to yield a SMER gain of around three times, compared with the adiabatic mode. An exergy analysis conducted by Catton et al. [6] shows that about half of the energy efficiency gain (reduction in irreversibility) is associated with the condenser and the drying process. Since most of the exergy destruction in the condenser and product is associated with the transfer of heat [11,12], this portion of the avoided irreversibility can be attributed chiefly to the isothermal mode's avoidance of the poor heat transfer of air [6]. Most of the rest of the energy efficiency gain is at the compressor and throttle, and can be attributed to the narrower temperature and pressure range of the refrigerant cycle for the isothermal mode. Since the majority of the narrowing of the temperature range in the isothermal mode is because of its avoidance of air cooling in the drier ducts, we can associate most of this latter improvement with the fact that the isothermal mode avoids using air (with its small specific heat capacity) as a heat carrier [5].

The psychrometric state cycles taken by the air are shown in Figure 7.6. The evaporating and condensing temperatures T_{ev} and T_{co} are represented by the vertical bars to the left and right of the cycles. Locations 0, 1, 2, 3, 4, and 5 m into the air duct from location D to location E are indicated using circles. The MER is given by the vertical displacement of the curve D–E (from the inlet to the exit of the air duct) for each cycle, multiplied by the airflow rate $\dot{m}_a$. As the figure illustrates, the isothermal mode yields a significantly greater drying rate than the adiabatic mode, subject to the other

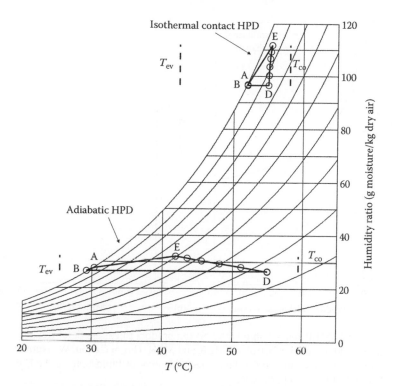

FIGURE 7.6 Psychrometric chart showing air property paths for adiabatic and isothermal heat pump dryers with zero evaporator bypass. (From Catton, W., Numerical study of heat pump contact drying. A thesis submitted for the degree of Doctor of Philosophy at the University of Otago, Dunedin, New Zealand, 2011. With permission.)

constraints, with a lower heat pump temperature lift, $T_D - T_B$. This smaller temperature lift leads to a significantly higher isothermal COP, which contributes to the improved isothermal energy performance. This lower temperature lift arises mainly because heat is provided progressively throughout the drying process, rather than to the air entirely before the process. The heat provided in the isothermal dryer thus effectively drives the moisture directly into the air stream.

Figure 7.7 shows the effect that product thickness has on the energy performance of the isothermal and adiabatic modes. Since we are considering the constant drying rate period, convective heat transfer from the hot air stream to the product for the adiabatic mode is not affected by its thickness, and thus the adiabatic performance is unaffected by product thickness. In contrast, a thick product layer presents a significant thermal resistance to the heat transfer from the refrigerant stream to the product, and rapidly nullifies the benefit of the isothermal mode. As discussed above, the requirement that the product must be able to be spread to a thin layer limits products for which isothermal drying will be appropriate. Figure 7.7a suggests that the isothermal contact HPD system being modeled would yield a significant performance advantage only in the drying of products that can be spread into layers less than about 1 cm thick. However, Figure 7.7b shows a significant potential benefit of the isothermal HPD mode, in cases where it may be applicable. The figure has been produced by varying the product thickness, and plotting the resulting MER and SMER against one another. As the figure indicates, isothermal HPD may enable maximization of energy performance and product throughput simultaneously (by ensuring a thin product layer) [6]. This absence of a trade-off between the SMER and MER contrasts with adiabatic timber HPD systems, which must be operated at relatively low drying rates to obtain good energy performance [5].

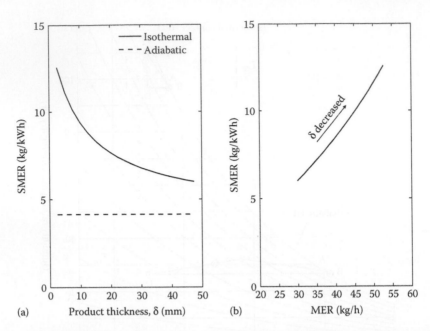

FIGURE 7.7 MER and SMER relationship with varying product thickness δ. (a) Relation between SMER and product thickness and (b) relation between SMER and MER. (From Catton, W., Numerical study of heat pump contact drying. A thesis submitted for the degree of Doctor of Philosophy at the University of Otago, Dunedin, New Zealand, 2011. With permission.)

7.3.2 Model Results of Dynamical Isothermal Contact HPD and Adiabatic HPD

The modeling results presented in Section 7.3.1 has been restricted to steady-state conditions during the constant drying rate period. In this section, results from the whole-system HPD model are presented, including the effects of the dynamics of the air stream and the product in the drier duct on overall system performance. A 1 mm product layer has been used here. For simplicity, we have ignored the heating-up period for the product. The initial temperature of the product is assumed to be equal to the dry-bulb temperature of the inlet air, and the initial product saturation is 0.99. Other operating conditions are the same as those for the steady-state ICHPD and AHPD during the constant drying rate period, in Section 7.3.1.

Figure 7.8 shows the psychrometric state profiles of the airflow within the dynamical whole HPD system when operated in the isothermal mode. The simulation has been run until the overall drying rate has fallen to 10^{-8} kg s^{-1}, although in practice, it may not be necessary for the drying operation to reach this small drying rate. The air flow cycles, in which the air stream passes the drier, evaporator and condenser have been represented by snapshots, as shown in Figure 7.8. In this figure, the snapshots are separated by 15-min intervals. Initially, the humidity in the system is high, and the humidity increase that occurs within the drier duct is appreciable (curved right-hand side of state-cycle triangles). As the drying process proceeds, the humidity in the system falls. The line connecting the diamonds in the figure represents the time-evolution of the air at the drier-duct outlet (location E in Figure 7.1). Because the temperature is constrained to equal 55°C at the air duct inlet, the system tends to maintain a constant high temperature at the condenser, but the low temperature at the evaporator falls as the system humidity drops. After about 75 min, moisture condensation ceases in the evaporator of the heat pump as humidity falls and the evaporator surface is no longer below the dew point temperature. Thus, the final state-cycle shown in Figure 7.8 does not reach the saturation line. At this point, the ICHPD system effectively becomes a conventional drying system, with all moisture removed from the product being vented to the environment as vapor.

Figure 7.9 shows the time evolution of the system refrigerant state-cycle. The snapshots are again separated by 15-min intervals. The condensing temperature can be seen to be fairly stable through the

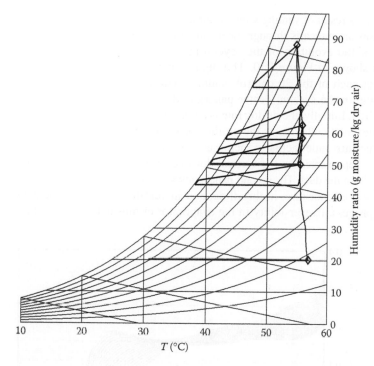

FIGURE 7.8 Psychrometric path traced by air cycle with time. Isothermal mode, δ – 1 mm. Cycles are snapshots of the system, separated by 15 minutes. (From Catton, W., Numerical study of heat pump contact drying. A thesis submitted for the degree of Doctor of Philosophy at the University of Otago, Dunedin, New Zealand, 2011. With permission.)

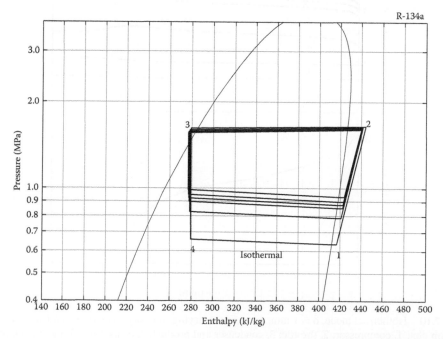

FIGURE 7.9 Evolution of R-134a state-cycle. Isothermal mode, δ = 1 mm. Cycles are snapshots of the system, separated by 30 minutes. (From Catton, W., Numerical study of heat pump contact drying. A thesis submitted for the degree of Doctor of Philosophy at the University of Otago, Dunedin, New Zealand, 2011. With permission.)

drying process, as a result of venting which maintains constant T_D. In contrast, the evaporating temperature decreases dramatically through the drying process, as the relative humidity falls (Figure 7.8).

The impact of the evolution of the psychrometric and refrigerant state-cycles on the system performance is shown in Figure 7.10. This figure shows the time evolution of the drying intensity occurring at each of the 20 control volumes within the duct (top panel), the breakdown of irreversibility within the system (center panel), and the system specific moisture extraction rate (SMER, kg kWh^{-1}). Initially, the drying rate is a maximum near the duct entrance. Later, the drying rate near the duct inlet collapses as the product near the inlet reaches completion, and the drying rate near the duct air outlet actually increases. By the time when moisture condensation ceases (about 75 min as represented by the vertical dot line in Figure 7.10), the evaporation intensity has collapsed throughout the duct. The second panel shows that exergy destroyed in the compressor, by the draining of condensed moisture, in throttling, and in venting, do not change appreciably throughout the drying process. Irreversibility in the evaporator climbs noticeably at the end of the drying

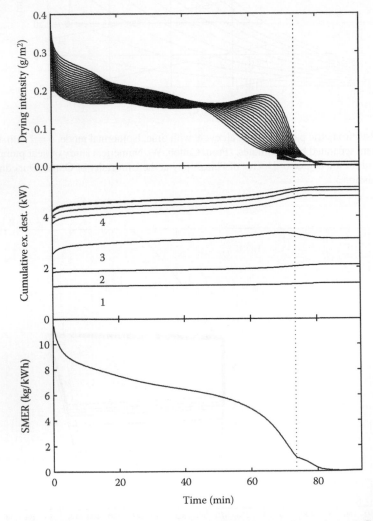

FIGURE 7.10 Isothermal mode, $\delta = 1$ mm. *First panel*: drying intensity. *Second panel*: Cumulative exergy destruction plot. 1, compressor; 2, throttle; 3, condenser and product; 4, evaporator; 5, fan; 6, venting; and 7, condensate (very thin band). *Third panel*: system SMER. (From Catton, W., Numerical study of heat pump contact drying. A thesis submitted for the degree of Doctor of Philosophy at the University of Otago, Dunedin, New Zealand, 2011. With permission.)

process. Irreversibility occurring at the condenser and product decreases at the end of the drying process as the product temperature increases. The net effect is an approximately 20% increase in total system irreversibility from the start of drying to the end of the process. Combined with the decrease in the drying intensity (top panel), this leads to the evolution of the system SMER shown in the bottom panel of Figure 7.10, which shows a significant decrease with the drying time.

Figures 7.11 and 7.12 illustrate the psychrometric paths and the evolution of the refrigerant state-cycle of the system when it is operated in the adiabatic mode. The impact of the variation on the system performance is shown in Figure 7.13, which depicts the drying intensity, exergy destruction rates, and SMER. The drying process takes about twice as long as in the isothermal mode, with an approximate factor of 2 difference in average SMER. The most striking differences, when comparing the adiabatic mode outputs with those of the isothermal mode, are the path taken by the air at the air duct outlet (solid line connecting the diamonds in Figure 7.11) and the spread of drying times in the duct. Although the drying process is seen to end fairly abrupt in the isothermal mode, with product at all locations within the duct reaching completion simultaneously, this does not occur in the adiabatic mode. Instead, a drying front is seen to advance through the duct, with a substantial delay separating completion of drying at the two ends of the duct. The behavior seen in Figure 7.13 appears qualitatively similar to that produced by the integrated timber-stack HPD model [14,17].

The influence of the product thickness on the performance of the whole dynamic HPD systems has been investigated using the whole-system model. The obtained model results show that the mean SMERs (prior to condensation halting) in the isothermal cases are 6.41, 5.48, and 4.48 kg kWh^{-1} for $\delta = 1$ mm, $\delta = 2.5$ mm, and $\delta = 5$ mm, respectively, or 36%, 27%, and 25%, respectively, less than the SMERs predicted for the constant drying rate period of the drying system (Figure 7.7). For comparison, the mean SMERS in the adiabatic cases are 2.66, 2.49, and 2.00 kg kWh^{-1} for $\delta = 1$ mm, $\delta = 2.5$ mm, and $\delta = 5$ mm, respectively. These results suggest that the SMERs obtained from the steady-state

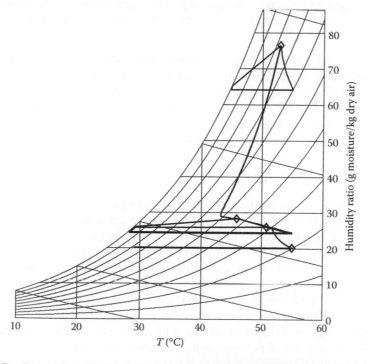

FIGURE 7.11 Psychrometric path traced by air cycle with time. Adiabatic mode, $\delta = 1$ mm. Cycles are snapshots of the system separated by 75 min. (From Catton, W., Numerical study of heat pump contact drying. A thesis submitted for the degree of Doctor of Philosophy at the University of Otago, Dunedin, New Zealand, 2011. With permission.)

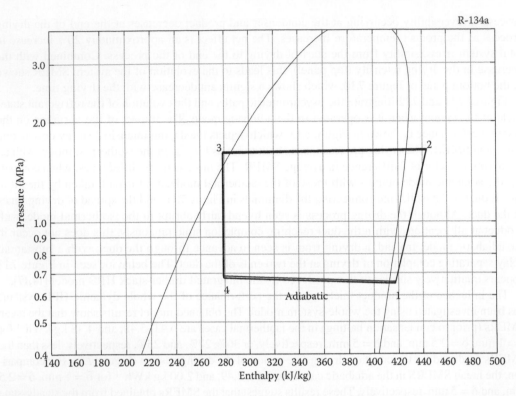

FIGURE 7.12 Evolution of R-134a state-cycle. Adiabatic mode, $\delta = 1$ mm. Cycles are snapshots of the system separated by 30 min. (From Catton, W., Numerical study of heat pump contact drying. A thesis submitted for the degree of Doctor of Philosophy at the University of Otago, Dunedin, New Zealand, 2011. With permission.)

HPD systems during the constant drying rate period overestimate the energy performance of the system by approximately 30%. One interesting feature of the model outputs is the fact that in the isothermal mode, the decline in drying rate and energy performance at the end of the process is more sudden than in the adiabatic mode. This leads to a smaller performance penalty due to batch operation than in the adiabatic mode. Another finding is that the time-averaged performance appears to be less sensitive to product thickness than that predicted by the steady-state model.

7.4 ECONOMIC CASE STUDY

The system performance values for the constant drying rate period discussed in Section 7.3.1 have been employed by Catton et al. [6] to conduct a tentative analysis of the relative economics of ICHPD. An operation that produces 1000 kg of waste sludge daily, with an initial moisture content of 0.65 kg kg^{-1} (dry-mass basis), has been analyzed. A waste sludge drying operation has been selected for the following reasons: (1) adiabatic HPD of filter-cake sludge is used today; (2) the value that is added to the product by drying is typically not large compared with the energy cost involved; and (3) waste sludges could be dried under ICHPD conditions that are set to optimize energy performance. This sludge is to be dried to a final moisture content of 0.1 kg kg^{-1}, in order to reduce transport and landfilling costs [47]. Using the rule of thumb that the capital cost of adiabatic HPD is approximately $1 per watt of heat provision at the condenser, we estimate a capital cost of $33,250 for the adiabatic case. We assume that this cost is financed at an annualized rate of 7%.

Table 7.2 shows the value added per day (after electricity costs have been met), the payback time, and the net present value (NPV) [48] of installation for adiabatic and isothermal HPD systems over a range of scenarios. We do not consider any costs associated with depreciation, servicing, or labor,

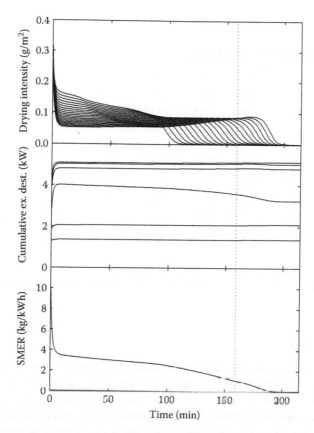

FIGURE 7.13 Adiabatic mode, $\delta = 1$ mm. First panel: drying intensity. Second panel: Cumulative exergy destruction plot. 1, compressor; 2, throttle; 3, condenser and product; 4, evaporator; 5, fan; 6, venting; and 7, condensate (very thin band). Third panel: system SMER. (From Catton, W., Numerical study of heat pump contact drying. A thesis submitted for the degree of Doctor of Philosophy at the University of Otago, Dunedin, New Zealand, 2011. With permission.)

and in evaluating the NPV, we assume a long project lifetime. The parameters that are varied are the cost per unit of electricity, the cost of sludge transport and landfilling, and the relative capital cost of an ICHPD system, compared to AHPD. In each scenario, each independent parameter is either high or low. Scenarios 1–4 reflect current electricity prices of roughly 10¢ kWh^{-1}. Scenarios 5–8 correspond to a significantly higher electricity price of 30¢ kWh^{-1}, and could represent a future scenario characterized by energy shortages and/or a strong CO_2-emission price signal [49,50]. Scenario 6 (in which electricity is expensive, sludge disposal is cheap, and ICHPD is costly to install) is the only scenario in which neither AHPD nor ICHPD is economically feasible. ICHPD appears to be a sensible investment in scenarios 1, 3, 5, 7, and 8. Both scenarios 3 and 4 have high disposal costs. The economics favor AHPD somewhat in scenario 3, and ICHPD somewhat in scenario 4. In all of the high electricity cost scenarios 5–8, ICHPD is strongly favored. In particular, in scenario 5, ICHPD is a sensible investment, whereas AHPD yields a negative net value, and in scenario 8, ICPHD is preferable to AHPD despite its much higher up-front cost. Electricity and waste disposal costs vary with region, whereas the relative capital cost of ICHPD is currently unknown. Our results indicate that at present, the economic viability of ICHPD strongly depends on its capital cost being less than three times that of AHPD, but that this dependency could be lessened or reversed by increases in the costs of sludge disposal or of electricity. Viewed alternatively, Table 7.2 shows that the isothermal mode's high energy efficiency makes its economics relatively less sensitive to the price of electricity, a result that may be significant in a time of energy price uncertainty.

TABLE 7.2

Economics of Adiabatic HPD and ICHPD

Scenario		1	2	3	4	5	6	7	8
Electricity cost, $/kWh		0.1	0.1	0.1	0.1	0.3	0.3	0.3	0.3
Disposal costs, $/kg		0.05	0.05	0.15	0.15	0.05	0.05	0.15	0.15
Relative capital cost, ISO		1	3	1	3	1	3	1	3
Value added, $/day	ADI	16.1	16.1	77.6	77.6	−13.2	−13.2	48.3	48.3
	ISO	26.3	26.3	87.8	87.8	17.3	17.3	78.8	78.8
Payback time, years	ADI	7.2	7.2	1.2	1.2	–	–	2.0	2.0
	ISO	4.0	18.1	1.1	3.5	6.6	–	1.2	4.0
Net present value, k$	ADI	53.4	53.4	384.5	384.5	−104.2	−104.2	226.9	226.9
	ISO	108.1	41.6	439.2	372.7	59.8	−6.7	390.9	324.4

Source: Catton, W. et al., *Energy*, 36, 4616, Table 5, 2011. With permission.

7.5 CONCLUSION

In this chapter, we have presented a numerical model of an ICHPD. The model has the capacity to assess the potential of ICHPDs to improve on the energy efficiencies currently attained in HPD, and also has the capacity to optimize such a system, for example, so that the operation and demands of the heat pump are matched to the properties and demands of the product being dried. The model that has been developed can be used to investigate the effects of a variety of parameters, for instance, the effects of product thickness, air and refrigerant mass flow rates, and temperatures on the system behavior. The model could be applied to a wide variety of porous materials, each characterized by a different set of constitutive relations. The model has established that for applicable products, the ICHPD configuration may substantially increase the energy efficiency of HPD, by a factor of 3 compared to conventional adiabatic HPDs. This ICHPD energy efficiency gain is, however, highly sensitive to the product thickness. The energy efficiency gain of ICHPD is also sensitive to any constraint on the temperature and the maximum allowable relative humidity above the product. Isothermal HPD is thus likely to be most applicable in the drying of those products, such as sludges and pastes, which can be spread into thin layers, in particular those that are also least vulnerable to quality deterioration at high temperature and humidity. Product throughput has been shown to be simultaneously maximized at low product thickness, implying that ICHPD provides an opportunity to avoid the adiabatic mode's trade-off between drying rate and energy efficiency, by using a thin product layer. ICHPD is found to reduce irreversibility occurring within the refrigerant cycle by roughly the same amount as that occurring in heat transfer from the condenser to the drying process and in the drying process itself, highlighting the synergistic nature of HPD systems.

Although the analysis that is presented in this chapter remains fairly general, modeling an HPD system and its components has required explicit assumptions to be made on parameters such as component sizes and system configuration. This is somewhat unfortunate, as in particular, there has not been any reported attempt to build and test a prototype ICHPD system. Since the research described here is a theoretical investigation of an untried design possibility for ICHPD, it seems appropriate for the analysis to remain at a fairly high level. However, the value of this work would undoubtedly be strengthened if it could be linked into a practical research program toward a working system design, for a specified application. In the absence of such a program, the appropriate path has seemed to lie in producing a working model of a likely system design, with parameters selected to maximize the energy performance predicted by the models. Although the model has only been used so far to examine the dynamics of batch-mode HPD operation, it could be readily extended to make predictions of system performance for continuous-operation driers.

7.6 NOMENCLATURE

A	Area [m^2]
b	Ratio of the air mass flow rate being bypassed the evaporator to the total air mass flow rate [–]
$\hat{c}_p$	Heat capacity [J kg^{-1} K^{-1}]
D	Heating plate refrigerant tube internal diameter [m]
D_{va}	Binary diffusivity for system of vapor and dry air [m^2 s^{-1}]
D_s	Mass *diffusivity* caused by the saturation distribution [kg m^{-1} s^{-1}]
D_T	Mass *diffusivity* caused by the temperature distribution [kg m^{-1} K^{-1} s^{-1}]
F_f	Friction force [N]
f	Friction factor [–]
h	Heat transfer coefficient [J m^{-2} K^{-1} s^{-1}]
h	Heat transfer coefficient per unit length of the tube [J m^{-1} K^{-1} s^{-1}]
$\hat{h}$	Specific enthalpy [J kg^{-1}]
$\hat{h}_{vap}$	Latent heat of vaporization of water [J kg^{-1}]
k	Thermal conductivity [J m^{-1} s^{-1} K^{-1}]
k_{eff}	Effective thermal conductivity [J m^{-1} K^{-1} s^{-1}]
k_β^0	Intrinsic permeability of medium to free water [m^2]
k_β^r	Relative permeability of medium to free water [–]
$k\omega$	Mass transfer coefficient [kg m^{-2} s^{-1}]
MER	Moisture extraction rate [kg s^{-1}]
$\dot{m}$	Mass flow rate [kg s^{-1}]
n	Mass transfer flux [kg m^{-2} s^{-1}]
Pr	Prandtl number [–]
p	Pressure [Pa]
p_c	Capillary pressure [Pa]
$\dot{Q}$	Rate of heat transfer [J s^{-1}]
$\dot{q}$	Heat transfer flux [J m^{-2} s^{-1}]
Re	Reynolds number [–]
S	Surface area [m^2]
SMER	Specific moisture extraction rate [kg kWh^{-1}]
s	Saturation [–]
t	Time [s]
T	Temperature [K]
U	Overall heat transfer coefficient per unit length of the tube [J m^{-1} K^{-1} s^{-1}]
u	Velocity in the x-direction [m s^{-1}]
u_{sat}	Saturated liquid density of porous medium [kg m^{-3}]
$\hat{u}$	Specific energy [J kg^{-1}]
v	Velocity [m s^{-1}]
V	Volume [m^3]
$\dot{W}$	Rate of work [J s^{-1}]
X	Refrigerant vapor quality [–]
x	x-coordinate [m]
y	y-coordinate [m]
z	z-coordinate [m]

Greek symbols

δ Product thickness (m)

ε Porosity [–]

μ Dynamic viscosity of fluid [N s m^{-2}]

ρ Density [kg m^{-3}]

ω_k Mass fraction of species k [–]

ω Humidity ratio [kg vapor/kg dry-air]

Subscripts

a   Dry air

co   Condenser

env   Environment condition

ev   Evaporator

g   Gas phase

l   Liquid phase

p   Heating plate

r   Refrigerant

s   Surface of product to be dried

sat   Saturation condition

v   Vapor

wb   Wet-bulb

β Free water phase

∞ Air bulk flow

Superscripts

•   Transfer coefficient corrected for finite mass-transfer rate

(*m*)   Quantity crossing a mass-transfer surface

REFERENCES

1. Bansal, B., Bannister, P., and Carrington, G. (1997). Performance of a geared dehumidifier. *Int. J. Energy Res., 21*, 1257–1260.
2. Hodgett, D. (1976). Efficient drying using a heat pump. *Chem. Eng., 311*, 510–512.
3. Kolbusz, P. (1976). Industrial applications of heat pumps. In Camatini, E. and Kester, T., eds., *Heat Pumps and Their Contribution to Energy Conservation*. NATO Advanced Study Institute Series, Series E, Applied Sciences, No. 15, Noordhoff, Leiden, the Netherlands.
4. Chua, K., Chou, S., Ho, J., and Hawlader, M. (2002). Heat pump drying: Recent development and future trends. *Drying Technol., 20*, 1579–1610.
5. Catton, W., Carrington, G., and Sun, Z. (2011). Performance assessment of contact heat pump drying. *Int. J. Energy Res., 35*, 489–500.
6. Catton, W., Carrington, G., and Sun, Z. (2011). Exergy analysis of an isothermal heat pump dryer. *Energy, 36*, 4616–4624.
7. Catton, W., Sun, Z., and Carrington, G. (2011). Dynamical modelling of an isothermal contact heat pump dryer. *Fifth Nordic Drying Conference*, Helsinki, Finland.
8. Catton, W. (2011). Numerical study of heat pump contact drying. A thesis submitted for the degree of Doctor of Philosophy at the University of Otago, Dunedin, New Zealand.
9. Bejan, A. (1997). *Advanced Engineering Thermodynamics*, 2nd edn. John Wiley & Sons, New York.
10. Sun, Z. and Carrington, G. (1991). Application of non-equilibrium thermodynamics in heat and mass transfer phenomena. *Trans. ASME: J. Energy Resour. Technol., 113*, 33–39.
11. Carrington, G. and Baines, P. (1988). Second law limits in convective heat pump driers. *Int. J. Energy Res., 12*, 481–494.
12. Vaughan, G., Carrington, G., and Sun, Z. (2007). Exergy analysis of a wood-stack during dehumidifier drying. *Int. J. Exergy, 4*, 151–167.
13. Carrington, G., Sun, Z., Sun, Q., Bannister, P., and Chen, G. (2000). Optimizing efficiency and productivity of a dehumidifier batch dryer. Part 1: Capacity and airflow. *Int. J. Energy Res., 24*, 187–204.

14. Sun, Z. and Carrington, G. (1995). Modelling of a wood-drying kiln. Technical report, Department of Physics, University of Otago, Dunedin, New Zealand.

15. Sun, Z. and Carrington, G. (1999). Dynamic modelling of a dehumidifier wood drying kiln. *Drying Technol., 17*, 711–729.

16. Sun, Z. F., Carrington, C. G., and Bannister, P. (1999). Validation of a dynamic model for a dehumidifier wood drying kiln. *Drying Technol., 17*, 731–743.

17. Sun, Z. F., Carrington, C. G., and Bannister, P. (2000). Dynamic modelling of the wood stack in a wood drying kiln. *Trans. IChemE, 78* (Part A), 107–117.

18. Sun, Z. F., Carrington, C. G., Anderson, J. A., and Sun, Q. (2004). Air flow patterns in dehumidifier wood drying kilns. *Chem. Eng. Res. Des., 82*, 1344–1352.

19. Carrington, G. (2007). Heat pump and dehumidification drying. In Hui, Y. H., Clary, C., Farid, M. M., Fasina, O. O., Noomhorm, A., and Welti-Chanes J., eds., *Food Drying Science and Technology: Microbiology, Chemistry, Applications.* DEStech Publications, Inc., pp. 249–273.

20. Chen, G., Yue, P. L., and Mujumdar, A. S. (2002). Sludge dewatering and drying. *Drying Technol., 20*, 883–916.

21. Ehiaze, E., Sun, Z., and Carrington, G. (2010). Variables affecting the in situ transesterification of microalgae lipids. *Fuel, 89*, 677–684.

22. Ehiaze, E., Sun, Z., and Carrington, G. (2012). Renewability analysis of microalgae biodiesel using different transesterification processes. *Int. J. Energy Environ. Econ., 20*, 135–156.

23. Jonassen, O., Kramer, K., Strømmen, I., and Vagle, E. (1994). Non-adiabatic two-stage counter-current fluidized bed drier with heat pump. *Ninth International Drying Symposium*, Gold Coast, Queensland, Australia, pp. 511–517.

24. Strømmen, I., Eikevik, T., Neksa, P., Pettersen, J., and Aarlien, R. (1999). Heat pumping systems for the next century. *Twentieth International Congress of Refrigeration, IIR/IIF*, Sydney, New South Wales, Australia.

25. Whitaker, S. (1977). Simultaneous heat, mass and momentum transfer in porous media: A theory of drying. *Adv. Heat Mass Transfer, 13*, 110–203.

26. Whitaker, S. (1988). Coupled transport in multiphase systems: A theory of drying. *Adv. Heat Transfer, 31*, 1–104.

27. Carrington, G. and Bannister, P. (1996). An empirical model for a heat pump dehumidifier drier. *Int. J. Energy Res., 20*, 853–869.

28. Stanish, M., Schajer, G., and Kayihan, F. (1986). A mathematical model of drying for hygroscopic porous media. *AIChE J., 32*, 1301–1311.

29. Wang, Z. and Chen, G. (1999). Heat and mass transfer during low intensity convection drying. *Chem. Eng. Sci., 54*, 3899–3908.

30. Bird, R. B., Stewart, W. E., and Lightfoot, E. N. (2007). *Transport Phenomena*, 2nd edn. Wiley, New York.

31. Rohsenow, W. and Hartnett, J. (1972). *Handbook of Heat Transfer.* McGraw-Hill, New York.

32. Sun, Z. (2001). Numerical simulation of flow in an array of in-line blunt boards: Mass transfer and flow patterns. *Chem. Eng. Sci., 56*, 1883–1896.

33. Sun, Z. (2002). Correlations for mass transfer coefficients over blunt boards based on modified boundary layer theories. *Chem. Eng. Sci., 57*, 2029–2033.

34. Patankar, S. (1981). *Numerical Heat Transfer and Fluid Flow.* McGraw-Hill, New York.

35. Bowser, T. and Wilhelm, L. (1995). Modeling simultaneous shrinkage and heat and mass transfer of a thin, nonporous film during drying. *J. Food Sci., 60*, 753–757.

36. Islam, R., Ho, J., and Mujumdar, A. (2003). Simulation of liquid diffusion-controlled drying of shrinking thin slabs subjected to multiple heat sources. *Dry. Technol., 21*, 413–438.

37. Keey, R., Langrish, T., and Walker, J. (2000). *Kiln-Drying of Lumber.* Springer Series in Wood Science, Springer, Berlin, Germany.

38. Jolly, P., Jia, X., and Clements, S. (1990). Heat-pump assisted continuous drying. Part 1: Simulation model. *Int. J. Energy Res., 14*, 757–770.

39. Cavallini, A., Col, D. D., Doretti, L., Matkovic, M., Rossetto, L., and Zilio, C. (2006). Condensation in horizontal smooth tubes: A new heat transfer model for heat exchanger design. *Heat Transfer Eng., 27*, 31–38.

40. Cavallini, A., Censi, G., Del Col, D., Doretti, L., Longo, G., Rossetto, L., and Zilio, C. (2003). Condensation inside and outside smooth and enhanced tubes—A review of recent research. *Int. J. Refrig., 26*, 373–392.

41. Traviss, D., Rohsenow, W., and Baron, A. (1973). Forced-convection condensation inside tubes: A heat transfer equation for condenser design. *ASHRAE Trans. Part 2A, 79*, 157–165.

42. Carrington, G., Bannister, P., and Liu, Q. (1996). Performance of a scroll compressor with R134a at medium temperature heat pump conditions. *Int. J. Energy Res., 20,* 733–743.

43. Turaga, M., Lin, S., and Fazio, P. (1988). Correlation for heat transfer and pressure drop factors for direct expansion air cooling and dehumidifying coils. *ASHRAE Trans., 94,* 616–629.

44. Carrington, G., Wells, C., Sun, Z., and Chen, G. (2002). Use of dynamic modelling for the design of batch-mode dehumidifier dryers. *Drying Technol., 20,* 1645–1657.

45. Daly, B. B. (1979). *Woods Practical Guide to Fan Engineering.* Woods of Colchester Limited, Colchester, England.

46. Menon, A. and Mujumdar, A. (1987). Drying of solids: Principles, classification, and selection of dryers. In Mujumdar, A., ed., *Handbook of Industrial Drying.* Marcel Dekker, New York.

47. Macolino, P., Bianco, B., and Veglio, F. (2009). Drying process of a biological industrial sludge: Experimental and process analysis. *Chem. Eng. Trans., 17,* 699–704.

48. Brealey, R. A., Myers, S. C., and Marcus, A. J. (2004). *Fundamentals of Corporate Finance,* 4th edn. McGraw-Hill, Boston, MA.

49. Ghoshray, A. and Johnson, B. (2010). Trends in world energy prices. *Energy Econ., 32,* 1147–1156.

50. Hauch, J. (2003). Electricity trade and CO_2 emission reductions in the Nordic countries. *Energy Econ., 25,* 509–526.

8 Advances in Dehumidifier Timber Drying in New Zealand

Gerald Carrington

CONTENTS

8.1   HEAT PUMP TIMBER DRYING

The use of heat pumps for timber drying has long been discussed in the literature (Geeraert 1976). Bannister et al. (2002) noted that the technology had not achieved its potential in this application and urged the timber industry "to adopt a more integrated approach to dehumidifier technologies, taking into account the potential of unique drying environments for new products and added value, high system efficiency and reliability, low environmental impact, and ease of use." In pursuit of these ambitions, the author contributed to a heat pump drying research program in New Zealand for a number of years.

This chapter presents results from that research program with the aim of increasing understanding of the use of heat pumps for softwood drying. It reviews timber drying in New Zealand's industrial and energetic context, and details the technological basis of a dehumidifier dryer supplied by Delta S Technologies Ltd to a timber processing firm in Whangarei (New Zealand) in 2003. Van der Pal et al. (2005) and Burnett et al. (2015) have previously described the performance of that dryer in their papers.[*] This chapter also examines the commercial environment within which Delta S operated, drawing primarily on Bell et al. (2014). It analyzes the current energy culture of the New Zealand timber industry and considers the opportunities for heat pump dryers to help stimulate business in that sector.

8.2   INDUSTRY CONTEXT

8.2.1   New Zealand Timber Industry

New Zealand produced 3.6 Mm^3 of sawn softwood from plantation sources in 2009 (MPI 2015a), some 50% of which was exported. Radiata pine dominated both domestic and export timber sales. Based on the data provided by the Energy Efficiency and Conservation Authority (EECA) (2005) and the Ministry of Economic Development (2012), the timber processing industry used approximately 2% of New Zealand's consumer energy and 4% of electricity production. The average export value of logs and sawn timber was NZ\$113/m^3 and NZ\$397/m^3, respectively in 2009 (MPI 2015b).

---

[*] This chapter includes extracts from the paper, Operational performance of a batch-mode dehumidifier timber dryer, by Burnett, M.D. et al., *Drying Technol.*, 33, 455–465, 2015. Used with permission of the publisher, Taylor & Francis Group.

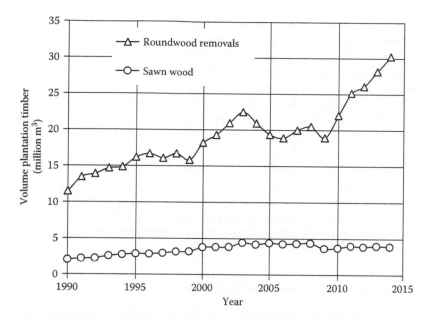

FIGURE 8.1 Annual volume (million m³) of New Zealand plantation sawn wood and roundwood removals 1990–2014. (Based on data from MPI, Production of rough sawn timber in New Zealand, 1970–2014. Estimated roundwood removals from New Zealand forests, year ended 31 March, 1951 to 2014, Ministry of Primary Industries, Wellington, New Zealand, 2015a.)

Figure 8.1 shows that the volume of sawn wood produced annually in New Zealand did not change significantly between 2000 and 2014, while the roundwood harvest increased by some 66%. Thus, the sawn wood industry did not respond to growth in the roundwood harvest over a period of 15 years. Moreover, in 2014, the volume of logs and poles exported was more than four times the total production of sawn wood; and some 40 sawmills closed between 2003 and 2014 (Morrison 2014). Evidently, the New Zealand sawn wood industry has been in difficulty for a considerable time (NBR 2008).

8.2.2 TIMBER DRYING IN NEW ZEALAND

Drying is important because it reduces biological attack, distortion, and crack development in sawn timber (Perré and Keey 2006), and decreases transport costs. The drying process has a major impact on the market value of the sawn wood (Alexiadis 2003).

Most combustion-heated timber kilns in New Zealand operate at dry-bulb temperatures of 70°C–120°C and the energy used for drying is typically 3.2 GJ/m³ (EECA 2005). About 7% of the energy input to these kilns is supplied as electricity, the balance being in the form of heat produced by combustion of coal, natural gas, or sawmill residues. Heat pump dryers use only electricity, typically 0.5 GJ/m³ (Van der Pal et al. 2005). Based on these data, the energy use of a heat pump dryer is some 16% of a combustion-heated dryer, although heat pumps typically use twice the electrical energy per cubic meter of sawn wood. Because heat pumps operate at lower temperatures than combustion-heated dryers, 50°C–60°C, drying times are longer. It is also necessary to observe relevant environmental regulations when disposing of heat pump dryer condensate.

Atmospheric emissions represent a significant point of difference between the two technologies. The emissions of combustion-heated dryers include particulates and carbon dioxide, generating 345 kg CO_{2-e} per cubic meter of sawn timber, based on the emission factors for wood combustion and electricity for 2009 (MBIE 2012). If all of New Zealand's sawn softwood were dried with combustion-heated dryers using wood-waste for heating, the atmospheric emissions would be

1.1 Mtonne CO_{2-e} per annum, 3.9% of national combustion emissions (MBIE 2012). By comparison, the emissions arising from drying with a heat pump kiln are approximately 25 kg CO_{2-e} per cubic meter of sawn timber, 7.2% of the emissions for wood-fired conventional kilns. This comparison does not take into account that wood residues, used as fuel by many timber firms, come from carbon-neutral sources and typically have other carbon-neutral uses if not used for drying. This assessment would change if the demand for wood waste to displace fossil fuels were to grow. Approximately 75% of New Zealand's electricity is generated from renewable resources, so the emission benefit of heat pump dryers is larger than it would be in countries where a greater fraction of electricity is generated by fossil fuels.

Particulate emissions are regulated in New Zealand, but volatile organic compounds (VOCs), mostly monoterpenes (McDonald et al. 2002), are not. Based on the rates reported by Pang et al. (2006), national emissions of VOCs by combustion-heated timber drying kilns are approximately 300 tonne per annum. By comparison, in an unvented heat pump kiln, the release of VOCs is essentially eliminated. Simpson (2004) reported no terpenes in the condensate of an unvented heat pump kiln drying Radiata pine.

8.2.3 Comparison with Combustion-Heated Dryers

In their review of international wood drying principles and practices, Perré and Keey (2006) noted that "the drying principle of these (heat pump) kilns, however, is the same as vented conventional kilns." Certainly the drying process is the same, but the control of the process is different. In a combustion-heated dryer, the kiln wet-bulb and dry-bulb are controlled and the drying rate responds. In a dehumidifier dryer, on the other hand, the drying rate is determined primarily by the capacity of the dehumidifier and the temperature setting, and the relative humidity responds to these drivers.

Another difference between heat pump and combustion-heated dryers is the effect of the higher temperatures achieved in the latter. Perré and Keey (2006) state: "A high value of temperature is most often a positive factor. This accelerates the internal moisture transfer and activates the viscoelastic creep. However, care should be taken with sensitive species; high temperature levels can increase the risk of collapse, problems of color, or even thermal degradation of the wood constituents."

In the particular case of Radiata pine, drying high-quality grades at higher temperatures reduces the value due to darker color development (McCurdy et al. 2004), more intense kiln brown-stain (Kreber and Haslett 1997), and greater incidence of internal checking (Haslett and Dakin 2001). It is well understood that these effects are mitigated by drying at lower temperatures, such as those used in heat pump dryers (Perré and Keey 2006).

In spite of these advantages, heat pump kilns have achieved limited penetration in the New Zealand market, producing some 2.8% of the volume of dried lumber in 2001 (Bannister et al. 2002). This is similar to other countries with significant softwood processing industries. Alexiadis (2003) found that 8.8% of timber kilns in Canada were of the heat pump type. A report by Cooper (2003) on energy efficiency in wood drying kilns in Europe indicated that heat pump timber kilns were not viewed there as a mainstream technology. Perré and Keey (2006) include heat pump dryers among the "less-common drying methods." Some of the reasons heat pump drying has not yet become a mainstream technology in New Zealand are discussed in Section 8.7.

8.3 RESEARCH BACKGROUND OF DELTA S

8.3.1 Heat Pump Dryer Needs and Concerns

In the 1980s, more than 100 batch-type heat pump timber dryers were in service in New Zealand (Laytner and Carrington 1994). There was, however, general awareness at that time that heat pump dryers did not provide the consistency of performance needed to win the confidence of the industry

(Cox-Smith et al. 1987; Cox-Smith and Carrington 1991). A survey of 21 firms using heat pump dryers in New Zealand (Carrington et al. 2002a) showed that, on a scale of 0 (unsatisfactory) to 100 (very satisfactory), those drying Radiata pine rated their heat pump dryer at 55 on average, while those drying other species had an average satisfaction rating of 73.

Table 8.1 lists the survey data, showing that the Radiata pine firms had larger operations than those drying other species; they dried faster and they used less electricity per cubic meter of timber. The desired improvements are shown in Table 8.2 and the degrade concerns are listed in Table 8.3, in order of the priorities of the Radiata pine processors. The results show that the firms drying Radiata pine were much more concerned about the limitations of heat pump dryers and timber degrade than other firms. Their most pressing needs were faster drying, lower costs, higher efficiency, less warping, and more uniform final moisture content.

8.3.2 HEAT PUMP DRYER DESIGN

A difficulty in the 1980s was the absence of a consensus on good design principles and practices for heat pump timber dryers. It was decided to investigate these issues, mainly as they affected the drying of Radiata pine. It was also decided that the most promising configuration was a dehumidifier, in which the heat pump evaporator would be located within the kiln chamber (Baines and Carrington 1988).

TABLE 8.1
Average Results from a Survey of 21 Firms Using Dehumidifier Timber Dryers

Results	Radiata Pine	Other Species
Timber volume per load (m³)	55	28
Maximum temperature (°C)	53	42
Drying time per load (day)	12	26
Average production per month (m³)	175	66
Entry/exit moisture content (%)	59/13	53/12
Cost of electricity ($/m³)	23	33

Note: Moisture content is measured as percentage, dry basis.

TABLE 8.2
Desired Improvements in Dehumidifier Dryers Expressed by 21 Timber Firms

Desired Improvements	Radiata Pine	Other Species
Faster drying	75	34
Lower cost	75	38
Higher efficiency	72	54
Better reliability	59	50
Better controls	59	50
Higher temperature	44	17

Note: 100 = most important.

TABLE 8.3

Timber Drying Degrade Concerns of 21 Firms Using Dehumidifier Dryers

Timber Drying Degrade	Radiata Pine	Other Species
Warping	72	47
Nonuniform moisture	68	41
Kiln brown stain	53	8
Internal checking	53	11
Case-hardening	44	4
Surface checking	41	28

Note: 100 = most important.

Because it has a number of closed cycle processes, the state of a dehumidifier dryer is strongly influenced by feedback effects. As a consequence, a detailed model of the batch drying process is required to determine the process trajectory during drying, including the refrigeration system of the dehumidifier, the airflow, the product being dried, and the kiln chamber. An engineering model of the dehumidifier was first established (Carrington and Bannister 1996) based on laboratory component measurements (Carrington et al. 1995). A submodel for determining the volumetric and isentropic efficiency of the compressor as a function of the port pressures was also developed (Carrington et al. 1996).

The dehumidifier model was subsequently embedded in a dynamic whole-dryer model (Sun and Carrington 1999a), including a stack model based on a numerical solution of the equations for mass, momentum, and energy balance for the airflow and the timber (Sun et al. 2000). The composite model yielded the time profiles for the air temperature, humidity, pressure and velocity, as well as the wood temperature and moisture content distributions. The model results were consistent with measured data (Sun et al. 1999, 2003).

The model was used to assess how changes in the kiln, the dehumidifier, and the control strategy affected performance indicators, such as the drying time and the energy consumption. This study included the influence of the wood-stack configuration, the airspeed, various efficiency options, the dehumidifier capacity as well as the kiln heat loss and air leakage (Sun and Carrington 1999b; Carrington et al. 2000a,b). The effects of limiting the evaporating and condensing temperatures, the impact of a low humidity restriction, and the temperature set point for electric preheating were also investigated (Carrington et al. 2002b). The influence of ill-conditioned airflows was examined (Sun et al. 2004) and the model was used to investigate the use of dehumidifier dryers for products other than Radiata pine (Carrington et al. 1999; Chen et al. 2002).

8.3.3 Product Quality Issues

There is tension between the industry's desire for faster drying using higher temperatures and its wish for less kiln brown stain, lighter timber color, and less internal checking. McCurdy et al. (2004) showed that the color of Radiata pine dried at heat pump temperatures of 50°C/40°C (dry-bulb/wet-bulb) was significantly lighter than the same timber dried at conventional dryer temperatures, 90°C/60°C. The CIELAB lightness coordinate, L*, was used to indicate the lightness of the board color. This parameter has values from 0–100%, with 0% being pure black and 100% being pure white. Averaged throughout the volume of the board, the change in L* was −1% for drying at 50°C/40°C compared with −6% to −11% for drying at 90°C/60°C. McCurdy et al. (2004) and Haslett and Dakin (2001) showed that the lower temperature also reduces kiln brown stain, a dark band which forms 1–2 mm below the surface during drying. Because the brown stain layer is normally removed mechanically for appearance applications, it creates additional processing costs as well as a loss of product volume and value.

Another quality issue for Radiata pine is the presence of internal checks in the timber that become apparent only during subsequent remanufacturing. Because of this, the price obtained for appearance grade Radiata pine internationally is normally discounted relative to competing species. Haslett and Dakin (2001) showed that internal checks were substantially eliminated in samples of Radiata pine dried under a dehumidifier schedule at temperatures less than 60°C dry-bulb. In connection with the industry concern about warping, Pearson and Haslett (2002) demonstrated that Radiata pine dried using a dehumidifier schedule could be successfully reconditioned by means of high-pressure water sprays to relieve residual strain and mitigate uneven drying within boards.

8.4 THE WHANGAREI DRYER

8.4.1 BACKGROUND

In 2001, the research team received a number of requests to help supply heat pump dryers for industrial applications. In response, Delta S Technologies Ltd was established by the research stakeholders as the vehicle to commercialize the technology available at that time. Based on its understanding of the needs of firms with heat pump timber dryers in New Zealand, Delta S decided to develop a dehumidifier dryer that would operate at dry-bulb temperatures of 65°C or less. All relevant intellectual property was assigned to Delta S and only limited details of subsequent developments were published.

An enquiry from a Whangarei firm was for drying Radiata pine boards in an existing kiln chamber. In June 2003, Delta S agreed to supply that firm with a dehumidifier dryer that matched its needs, and provided a performance guarantee. The dryer was delivered and commissioned in November 2003, and it was shown to satisfy the guarantee in March 2004.

Figure 8.2 shows a schematic diagram of the Whangarei dehumidifier. Drying was achieved by cooling the fraction of the kiln airflow that passed through the evaporator, in order to condense water

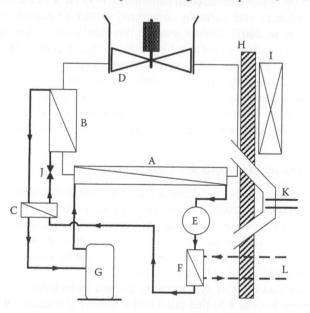

FIGURE 8.2 Schematic diagram of the Whangarei dehumidifier showing: A, condenser; B, evaporator; C, suction-line condensate heat exchanger; D, fans; E, liquid receiver; F, subcooler for heat rejection; G, compressors; H, insulating panel inserted in the kiln wall; I, electrical and controls cabinet; J, expansion valve; K, dry-bulb and wet-bulb temperature sensors; and L, cool water supply for the subcooler.

vapor to liquid which was piped to waste. The rest of the kiln airflow was heated in the condenser and the combined flow of overall warmer dryer air was returned to the timber stack. Functionally, the system was similar to the *split drying heat pump* described by Minea (2010, 2013), except that the evaporator was located inside the drying chamber. Some features of the system were the subject of patents (Scharpf and Carrington 2006a,b; Scharpf et al. 2006), which lapsed after Delta S closed down.

Because the Whangarei dryer was used for drying timber for pressure preservative treatment, the required final moisture content was relatively high, 20% dry basis. The timber quality requirements were not demanding and there was no need for reconditioning. Nevertheless, the client had many of the other needs listed in Tables 8.2 and 8.3.

The Whangarei dehumidifier was designed to meet these needs by selecting the capacity of the compressors and the heat exchangers to provide an acceptable balance between first-cost, drying speed, and operating cost. The peak drying rate of the dehumidifier, 5 tonne of moisture per day, was selected to match the capacity of the kiln, 55 m^3 of 50 mm sawn Radiata pine boards. In addition, the system enabled: (1) the airflow direction in the timber stack to be reversed periodically; (2) the dehumidifier to be prefabricated as a module; (3) the use of the wet-bulb temperature as the primary thermal control parameter; (4) capacity staging; and (5) the use of commercial hermetic scroll compressors.

8.4.2 Airflow Reversal

Compared with normal dehumidifier practice in New Zealand at the time, where the airflow was not reversed, periodic airflow reversal reduced variations in the timber moisture content across the timber stack and enabled the stack as a whole to dry faster. As a result, the drying cycle required less energy and the drying-induced stresses in the timber were also reduced (Simpson and Haque 2007). Thus, airflow reversal contributed positively to several of the needs identified in the industry survey (Section 8.3.1).

The reason the airflow was not reversed in dehumidifiers in New Zealand at that time was that the evaporator and condenser were normally configured so that air passed from the evaporator to the condenser (Ceylan et al. 2007). Delta S avoided this limitation by having the evaporator and condenser in parallel airstreams as shown in Figure 8.2 (Scharpf et al. 2006). Of course the condensing temperature was higher than in the series configuration, but this effect was small because the evaporator airflow rate was only 15% of the condenser airflow.

In practice, the airflow direction was reversed every 4 hours by shutting down the dehumidifier briefly to stop the fans and by restarting with the fan rotation reversed. Direct-drive axial multiblade fans were used, with the orientation of the blades modified to support bidirectional airflow.

8.4.3 Modular Construction and Air Circulation

To reduce the cost of on-site work, the dehumidifier was frame-mounted as a packaged module (Figures 8.3 and 8.4). This was shipped to the kiln site with the fans, refrigeration, power, control, and monitoring systems already assembled. At Whangarei, it was installed in a kiln chamber (Figure 8.5) that had been used previously as a combustion-heated dryer. The insulation of the concrete walls, floor, and ceiling of the chamber was upgraded by lining them with 50 mm thick polyisocyanurate foam panels, surfaced with metal foil.

The electrical power and controls systems were located in lockable cabinets, mounted on an insulating panel (Figures 8.4 and 8.5) that fitted into a matching aperture in one wall of the kiln. This allowed the electrical and control cabinets to be located in a cooler plant room adjacent to the drying chamber. The drying cycle was controlled by an Allen Bradley PLC and operators accessed the controls and status displays using a password protected touch screen located on an electrical cabinet.

FIGURE 8.3 The Whangarei dehumidifier package from the kiln side showing the evaporator and compressors. (Reprinted from Burnett, M.D. et al., *Drying Technol.*, 33, 455–465, 2015. With permission.)

The modular approach ensured the dehumidifier was well integrated with the kiln structure and the timber stack. It avoided the need for external airducting and reduced the risk of air leaks, while minimizing the number of fans and associated electrical equipment. Three fans, with a total airflow rate of 25 m³/s, circulated air through both the heat exchangers and the timber stack (Figure 8.5). The total fan electrical power was 9 kW.

The air speed in the timber stack was typically 1–2 m/s, but it depended on the effectiveness of baffles installed by the operator before each drying run to reduce the airflow passing around the sides of the timber stack. The baffles consisted of plywood sheets that bridged the space between the kiln walls and the edges of the stack.

8.4.4 Control

Timber was preloaded onto track-mounted trolleys outside the kiln (Figure 8.6) while the previous load was drying. A new drying run was initiated when the trolleys with fresh timber were rolled into the kiln, pushing the previous load out of the other side. This could be done quickly, the time between the end of one drying run and the start of the next being less than one hour on occasion. When a drying run started, only the fans and electric resistive heaters (3 × 25 kW) were energized before the wet-bulb temperature of the kiln reached a preset level, normally 32°C, at which time the compressors were started. This avoided the risk of evaporator freezing. Normally the resistive heaters stayed on until the wet-bulb reached the set point, about 44°C.

Heat was discharged from the kiln chamber when required for thermal control by subcooling the liquid refrigerant upstream of the expansion valves (Figure 8.2, item F). This was done using a

FIGURE 8.4 The Whangarei dehumidifier package showing the electrical and controls cabinets. The vertical black pipe contains the dry-bulb and wet-bulb temperature sensors. (Reprinted from Burnett, M.D. et al., *Drying Technol.*, 33, 455–465, 2015. With permission.)

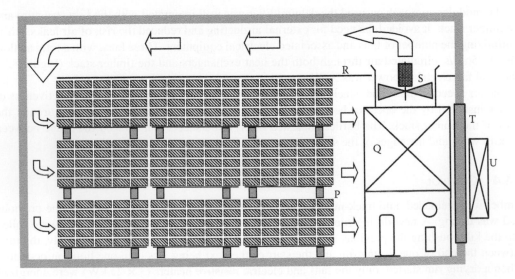

FIGURE 8.5 Schematic of the Whangarei dryer showing the kiln with anticlockwise airflow. P, stack exit; Q, dehumidifier air plenum; R, canvass floor of the return airflow duct; S, fans; T, insulated panel in the kiln wall; and U, electrical and controls cabinet.

FIGURE 8.6   A track mounted trolley at the Whangarei kiln loaded with fresh Radiata pine boards ready to enter the kiln.

refrigerant water heat exchanger with externally supplied cold water, a method also described by Minea (2010). Air venting was not used for heat rejection because the design team was concerned about the risk of a malfunctioning vent. To cover the occasions when the cooling water temperature was too high to achieve the required level of liquid subcooling, a hot gas de-superheater was included in the dehumidifier module as a backup, but this was rarely used.

For kiln temperature measurement, duplicate dry-bulb and wet-bulb sensors (Figure 8.2, item K) were located in an air duct connecting the kiln chamber with the plenum space bounded by the fans, the evaporator, and the condenser (Figure 8.5, item Q). The wet-bulb sensors consisted of Pt-100 resistant thermometers in stainless steel tubes, covered with cotton wicks that were wetted by a water bath. Referring to Figure 8.5, in clockwise air circulation the airflow direction in the sampling duct was downward; in anticlockwise circulation, it was upward. In principle, this arrangement ensured that the wet-bulb and dry-bulb temperatures were sampled in the airstream leaving the stack for both directions of air circulation. However, there were some differences, as described in Section 8.6.1.1.

The primary measurement used for kiln thermal control was the wet-bulb temperature (Scharpf and Carrington 2006a). This parameter was the main determinant of the refrigerant evaporating temperature which, in turn, had a direct influence on the capacity and efficiency of the dehumidifier. The condenser and evaporator airflow rates were not actively controlled. During commissioning, a perforated metal screen was installed on the timber side of the evaporator to reduce the evaporator airflow in order to improve the sharing of airflow between the evaporator and condenser.

8.4.5 CAPACITY STAGING

There were three refrigerant compressors connected in parallel (Copeland hermetic scroll ZR19ME-TWD). These were suitable for saturated suction temperatures up to 25°C with hydrofluorocarbon (HFC) 134a. Scroll compressors were used because there was little risk of motor overload at elevated evaporator pressures, as can happen with reciprocating compressors.

Normally all compressors ran, but the number was reduced to limit the drying rate when the wet-bulb–dry-bulb temperature difference exceeded a preset limit, usually near the end of a drying run. The purpose of this feature was to avoid low-humidity drying conditions that could result in high timber stresses at the end of the drying run, a concern of the timber industry (Table 8.3).

The area of active evaporator surface was reduced when a compressor was stopped by closing a solenoid-valve in the liquid-line feeding part of the evaporator. This helped to maintain the dryer efficiency by ensuring the evaporating temperature did not change significantly as the total compressor capacity was reduced. The refrigerant flow in each evaporator section was controlled by an associated thermostatic expansion valve, rated for saturated suction temperatures up to 30°C. The solenoid and expansion valves were located in the plant room.

Before all the compressors were stopped, the evaporator was pumped down to a preset pressure to prevent compressor lubricating oil being diluted with refrigerant as the oil cooled. In addition, a liquid-refrigerant suction-vapor heat exchanger was installed to reduce the risk of liquid refrigerant returning to the compressor (Figure 8.2, item C). This was done as the dryer was being commissioned, when it was found that the temperature of the compressor oil, which was at the suction pressure, was typically only 5°C higher than the saturated suction temperature, indicating an excess of liquid refrigerant in the oil. After the heat exchanger was installed, the difference between the oil temperature and the saturated suction temperature increased to about 30°C.

8.5 MONITORING THE WHANGAREI DRYER

8.5.1 INSTRUMENTATION

Measurements made on the Whangarei dryer included the electrical power input, the flow rate of condensed moisture, the dry-bulb and wet-bulb temperatures for air entering the dehumidifier, the compressor suction and discharge pressures, and temperatures at 22 other locations. Measurements were recorded using a Campbell Scientific CR10X data logger that sampled the inputs every 15 s; 5 min data averages were saved. Between November 2003 and March 2005 the data were regularly recovered using a dial-up modem. Subsequently the modem failed and data were then recovered only during visits.

The measured parameters were calibrated on site using reference instrumentation for electrical power, temperature, and pressure. The condensed moisture flow meter was calibrated during commissioning for a range of water flow rates, by timing multiple fillings of a reference volume. The flow meter calibration was found to be stable when checked in March and November 2004 but in May 2008 it had changed, the recorded output being down by 60%. Data recorded in 2007 also appeared to be affected, but there was no indication of any calibration error in data recorded in May 2006. For this reason, the data for May 2006 are the last used in the analysis presented here. The uncertainty in the recorded data was approximately ±1% for the electric power, ±1°C for the temperatures, ±1% for the refrigerant pressure, and ±2% for the condensate flow rate.

8.5.2 DATA ANALYSIS

The energy efficiency of the dryer was determined primarily from the volume of moisture extracted from the timber stack and the corresponding measured energy input. However, to obtain the total volume of moisture it was necessary to also include the quantity leaving the kiln through air leaks. This contribution was obtained by applying the following energy balance model to the kiln (Van der Pal et al. 2005).

The electrical energy input (E_{in}) was measured directly, but it was more complex to obtain the energy output (E_{out}). This was determined by estimating each contribution to the heat rejected from the kiln due to: (1) heat pump subcooler (E_{sub}); (2) de-superheater (E_{des}); (3) warm water condensate leaving the kiln ($E_{condens}$); (4) thermal conduction through the kiln walls (E_{cond}); (5) the hot timber and kiln structure being cooled when the kiln doors were opened at the end of drying ($E_{heat loss}$); and (6) leakage of moist air from the kiln during drying ($E_{air leak}$). These quantities are linked because energy balance over the whole drying cycle requires:

$$E_{in} = E_{out} = E_{sub} + E_{des} + E_{condens} + E_{cond} + E_{heat loss} + E_{air leak} \qquad (8.1)$$

The subcooling and de-superheating rates were calculated from the refrigerant mass flow rate, which was determined from the compressor characteristics, using the number of operating compressors and the measured compressor suction and discharge pressures. The enthalpy differences for the refrigerant in the subcooling and de-superheating heat exchangers were determined from the refrigerant properties and the measured inlet and outlet temperatures.

The heat loss due to the warm condensate leaving the kiln was calculated from the measured flow rate, the outdoor temperature, and the estimated condensate temperature. Here the outdoor dry-bulb temperature was set using a representative measured value when the kiln doors were closed and the wet-bulb was set using a fixed wet-bulb depression of 3°C. The condensate temperature was taken to be the same as the refrigerant saturated suction temperature, which was calculated from the measured compressor suction pressure.

The heat loss by thermal conduction through the kiln walls was calculated from the estimated heat conductivity of the walls, based on the construction materials, the surface area, and the difference between the outdoor air and the average air temperature in the kiln. The energy lost due to the hot timber at the end of the drying run was based on the volume of timber, the basic timber density (450 kg/m³), and the final moisture content (20% dry basis). The mass of structural material in the kiln was taken to be 12,000 kg steel based on the chamber structure and the construction of the heat pump unit. The specific heat capacity was taken to be constant and the energy calculation was based on the difference in temperature between the start and end of the drying run when all the compressors stopped.

To determine the heat loss by air leakage, the leak rate (kg-dry/s) was taken to be constant during each drying run. The enthalpy for the kiln air was calculated from the measured wet-bulb temperature and the estimated average kiln dry-bulb temperature, the latter based on the measured dry-bulb temperature for the air leaving the timber stack, with a correction determined from dehumidifier model calculations (Sun et al. 2000). The outdoor air enthalpy was determined using the outdoor air dry-bulb and wet-bulb temperatures. The leakage rate was then set to balance the measured energy input to the kiln with the sum of the estimated energy losses during the drying run.

The outputs of this procedure included the heat lost by refrigerant subcooling, de-superheating, warm condensate water, kiln wall conduction, air leaks and the thermal energy lost when the dry timber left the kiln and the kiln structure cooled down. The resulting air-leak rate was then used to calculate the rate of moisture leaving the kiln due to leaks throughout the drying run. The moisture loss due to both air leaks and condensate was then integrated over the drying cycle.

8.5.3 Data Selection

The accuracy of the method used to determine the total moisture loss was assessed using data obtained during two performance guarantee runs in March 2004, when the volume of timber was measured and it was weighed before and after drying. This enabled the moisture loss obtained using the procedure described above to be compared with the weight-loss data.

To obtain agreement between the two measurements, adjustments were made to the estimated thermal conductivity of the kiln walls. The air-leak rate, obtained by applying the energy balance criteria to these two runs, was 0.06–0.07 kg-dry/s, and contributed 6%–7% of the total moisture removed from the timber. The total mass loss determined from the condensate measurement plus the air leaks contribution was within ±0.5% of the mass obtained by weighing.

The multiple-cycle results, reported in Section 8.6.2, were obtained during two monitoring periods: December 2003 to March 2005 and February 2006 to May 2006. To analyze the data it was necessary to estimate the timber volume, since this was not normally measured, as well as the air-leakage rate. This was done by applying the energy balance criteria described to both the kiln heat-up period and the drying cycle as a whole. For some runs no reasonable solutions for the timber volume and kiln air leakage rates were obtained that led to energy imbalances of less than 2% of the input energy. These runs were not included in the data presented here. Two of the runs in the first period were unusable because the moisture flow meter provided no data and three others from this period were rejected as outliers: the cycle time was very short, the initial moisture content was very low, or the estimated air leak rate was very low. The maximum energy imbalance for the runs used here—for either the heat-up period or the whole drying run—was ±0.4%. The average value of the absolute energy imbalance was 0.1%.

When this procedure was applied to the two weighed runs, the moisture errors were ±1% and the volume errors were ±8%. Taking this result into account, as well as the uncertainty in the condensate volume measurement, the uncertainty in the total moisture mass for each of the 115 runs presented here is about ±3%. The corresponding uncertainty in the timber volume is ±15%.

In the reported performance data (Section 8.6), the dryer moisture extraction rate (MER), in kg/h, is used as a measure of the average drying rate of the system during a defined interval, including both condensate and moisture lost through kiln air leaks. The specific moisture extraction rate (SMER), in kg/kWh, is used as a measure of the energy efficiency of the dryer. This is the total moisture extracted from the stack in a defined interval divided by the total electrical energy (fans, compressor, resistive heaters, and controls) used by the dryer in that interval. The power consumption of the small water pump for the subcooler, understood to be a few hundred watts, was not measured or included.

8.6   WHANGAREI DRYER PERFORMANCE

8.6.1   Single Drying Cycle Data

8.6.1.1   Time Profiles

To illustrate the conditions in the kiln chamber, Figure 8.7 shows the dry-bulb and wet-bulb temperatures measured in the air stream at the exit of the timber stack for one of the runs where the timber was weighed before and after drying. It also shows the dry-bulb temperature at the stack entry, estimated from the measured data and the dehumidifier model. This and subsequent graphs show 30-min running-time-averages of the data recorded at 5-min intervals, apart from Figure 8.12 which has 5-min data.

For the weighed runs, special efforts were made to minimize the airflow passing around the sides of the stack by careful placement of the baffles. As discussed later, the data indicate that there was much more air bypassing during normal drying operations. For the run shown in Figure 8.7, the product consisted of 51 m³ of sawn Radiata pine boards with an average initial moisture content of 82% (dry basis). The average MER for this run was 157 kg/h and the whole cycle SMER was 3.10 kg/kWh.

Figure 8.7 shows the kiln temperature rising sharply at the start of the drying run as the timber was preheated by the fans and electric resistive heater. After 6 h, the compressors started when the wet-bulb temperature reached 32°C. The corresponding changes in electric power input are shown

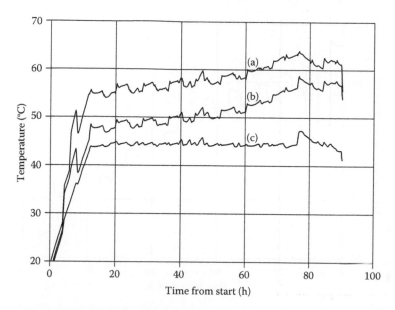

FIGURE 8.7 Air temperatures for an example drying run (March 15–19, 2004). (a) dry-bulb at the stack entry, (b) dry-bulb at the stack exit, and (c) wet-bulb at the stack exit. (Adapted from Burnett, M.D. et al., *Drying Technol.*, 33, 455–465, 2015. With permission.)

in Figure 8.8. The heater and dehumidifier both continued to run until the wet-bulb temperature reached the operating set point (43°C), 12 h after starting. The heat pump then ran for a further 79 h with the heater off. During this time the wet-bulb temperature was relatively stable, being controlled by heat rejection through the refrigerant subcooler.

As the timber dried dry-bulb temperature increased, the stack entry peaking at 63°C, and the relative humidity dropped, as shown in Figure 8.9. This is the expected response of the system to the fixed wet-bulb temperature, subject to the nominally constant kiln air space water content,

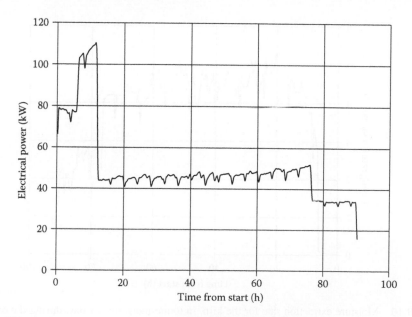

FIGURE 8.8 Electrical power input to the kiln for the illustrative drying run.

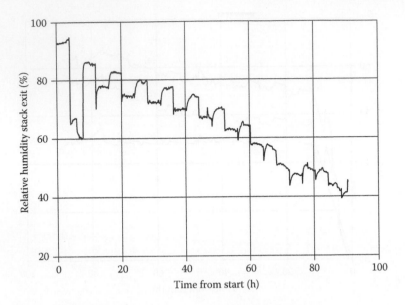

FIGURE 8.9   Relative humidity at the stack exit for the illustrative drying run.

where the rate of moisture extraction from the product is balanced by the rate of removal of moisture from the kiln by the dehumidifier. The cyclic changes shown in Figure 8.9 are due to airflow
reversal, which is discussed next.

The effect of this form of control was that the MER of the kiln (Figure 8.10) was relatively constant until 76 h when one of the three compressors stopped as the indicated wet-bulb depression at
the stack exit reached an upper set point. The reason for this limit is that the relative humidity of
air leaving the timber stack should not be less than 40% to mitigate the risk of internal checking
in the timber. The effect of this capacity change at 76 h can be seen in Figure 8.8 as the power

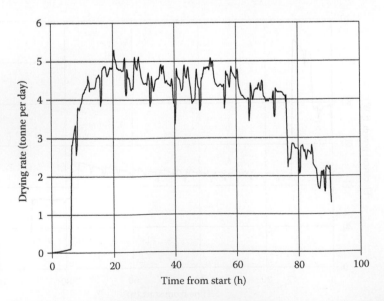

FIGURE 8.10   Moisture extraction rate for the kiln, in tonne-moisture per day, during the example drying
run. (Adapted from Burnett, M.D. et al., *Drying Technol.*, 33, 455–465, 2015. With permission.)

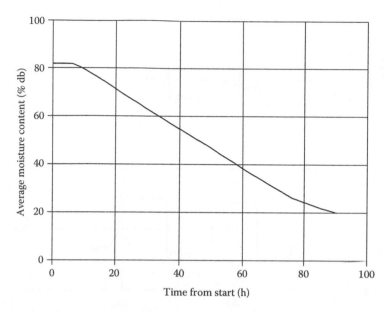

FIGURE 8.11 Average moisture content of the timber stack (% dry basis) during the example drying run. (Adapted from Burnett, M.D. et al., *Drying Technol.*, 33, 455–465, 2015. With permission.)

consumption fell, in Figure 8.10 as the drying rate dropped, and also in the average stack moisture content time profile, Figure 8.11

A feature of Figure 8.9 is the cyclic change in the indicated relative humidity each time the airflow direction reversed. With reference to Figures 8.2 and 8.5, when the airflow direction was anticlockwise, the air state was measured at an exit point near the right-hand side of the timber stack. But when the airflow was clockwise, the dehumidifier sampled a mixture of hotter air that had bypassed the timber stack and cooler air that had passed through the stack. As a consequence, during the clockwise steps, the indicated dry-bulb temperature was higher than that during the anticlockwise steps, resulting in cyclic variations in the indicated relative humidity, shown in Figure 8.9. By contrast, the wet-bulb temperature did not change significantly as the air passed through the timber stack, so remained relatively stable.

The size of this effect depended on the effectiveness of the stack bypass baffles, as shown in Figure 8.12, comparing the indicated dry-bulb temperature on March 16, 2004 (upper panel), with another run on June 17, 2004 (lower panel). Care was taken to minimize stack bypass in the first run, one of the performance guarantee runs, whereas bypass control was less effective in the second run. As a result, the variations in the dry-bulb temperature when the airflow was reversed are much larger in the second run, clearly seen on the lower panel of Figure 8.12.

The stack drying rate (Figure 8.10) shows a sharp change each time the airflow direction reversed. These changes were also reflected in the kiln efficiency, Figure 8.13, and in the temperatures of the dehumidifier refrigerant, Figure 8.14. In addition, the figures show other fluctuations, primarily due to changes in the wet-bulb temperature and the amount of subcooling used for kiln heat rejection, as discussed in Section 8.6.3.2.

8.6.1.2 Short-Term Averages

Table 8.4 summarizes the average operating parameters for the dehumidifier from 24 to 72 h during the example drying run, as the average moisture content of the timber stack fell from 68% to 29%. The uncertainty shown is the standard deviation in the data recorded at 5-min intervals. The fan power was not monitored separately and the uncertainty is for a single fan power measurement.

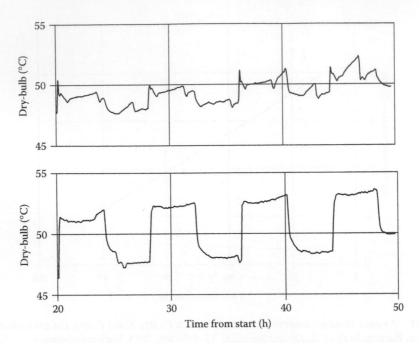

FIGURE 8.12 Comparison of the recorded dry-bulb temperature on March 16, 2004 (upper panel), when good stack bypass control was in place, with June 17, 2004 (lower panel) when bypass control was less effective.

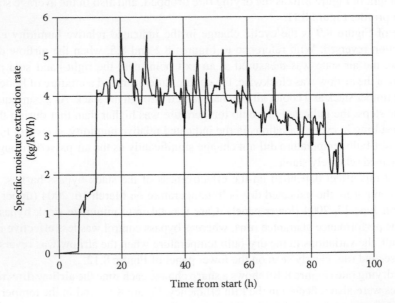

FIGURE 8.13 Specific moisture extraction rate for the kiln, in kg-moisture per kWh, for the example drying run. (Adapted from Burnett, M.D. et al., *Drying Technol.*, 33, 455–465, 2015. With permission.)

The standard deviation was determined using the STDEVA function in Excel (1997–2003), which treats its arguments as a sample of a population. The reported data precision has been set so that the standard deviation has at least one significant digit.

The wall conduction heat loss, shown in Table 8.4, represents 20% of the total heat loss of the kiln in the sample period. Of the loss mechanisms assessed this is the least productive, since it carries

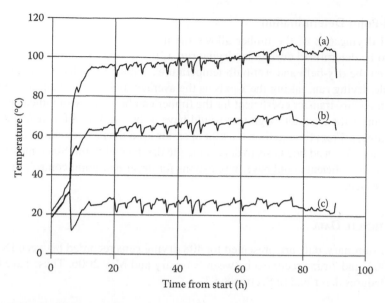

FIGURE 8.14 Refrigerant temperatures during the example drying run: (a) compressor discharge, (b) saturated discharge, and (c) saturated suction. (Adapted from Burnett, M.D. et al., *Drying Technol.*, 33, 455–465, 2015. With permission.)

no moisture from the kiln. However this was not a major loss, for if conduction losses were halved, the MER and SMER would increase by just 3%, a relatively small gain. Burnett et al. (2015) provide data for another run where the short-term average SMER was 4.7 kg/kWh. It is likely the reason this is higher than the results shown here is that the moisture content of the timber and the relative humidity were higher.

TABLE 8.4

Dehumidifier Short-Term Average Operating Parameters for the Period from 24 to 72 h during the Illustrated Single Drying Run

Single-Cycle Average Parameter	Value	Standard Deviation
Dry-bulb temperature off stack (°C)	51	2
Dry-bulb temperature on stack (°C)	58	2
Wet-bulb temperature off stack (°C)	44.4	0.3
Compressors input power (kW)	38	3
Fans input power (kW)	9	1
Heat rejected by subcooling (kW)	28	14
Heat loss due to wall conduction (kW)	9.8	0.5
Heat loss due to kiln air leaks (kW)	10.6	0.2
Moisture loss due to air leaks (kg/h)	11.1	0.4
MER (kg/h)	184	20
SMER (kg/kWh)	4.0	0.6
Compressors suction temperature (°C)	45	2
Saturated suction temperature (°C)	26	2
Compressors discharge temperature (°C)	98	3
Saturated discharge temperature (°C)	66	2
Compressors oil-sump temperature (°C)	65	2

Source: Adapted from Burnett, M.D. et al., *Drying Technol.*, 33, 455–465, 2015. With permission.

8.6.1.3 End-Point Determination

The measured drying rate of the timber allows the average moisture transfer coefficient of the timber stack to be determined during the drying run as the drying rate divided by the average difference between the dry-bulb and wet-bulb temperatures. This parameter is shown in Figure 8.15 for the example drying run, taking the x-axis as the average moisture content of the timber stack. Arguably the moisture transfer coefficient for the timber can be used as an indicator of the average moisture content of the load, although this would have to be calibrated for different timber types and board thickness. From time to time the kiln operator used the progression of the stack moisture transfer coefficient on an ad hoc basis to determine the drying end point. Since measurement of the moisture transfer coefficient could have implications for dryer end-point control, further investigation may be justified.

8.6.2 MULTICYCLE DATA

In this section summary, data are presented for 108 drying runs recorded between December 2003 and March 2005, and 7 runs recorded between February and May 2006. These have been selected using the procedures described in Section 8.5.3.

8.6.2.1 Illustrative Results

The dryer was used intensively during 2004, operating for 7759 h, almost 89% of the available time. For much of this time the system was in continuous use, with short breaks between runs. This is illustrated in Figure 8.16, which shows the dry-bulb temperature of air leaving the timber stack for 19 runs over a period of two months in 2004. The variations in the dry-bulb temperature on airflow reversal are evident.

Figure 8.17 shows the composition of the heat losses for the full set of 115 runs, each contribution represented by the cumulative percentage of the total heat loss for the run. Thus, the subcooling heat loss is typically 40% of the total, subcooling plus air leaks make up over 60%, and when wall conduction is included more than 80% of the losses are usually accounted for. The figure shows that

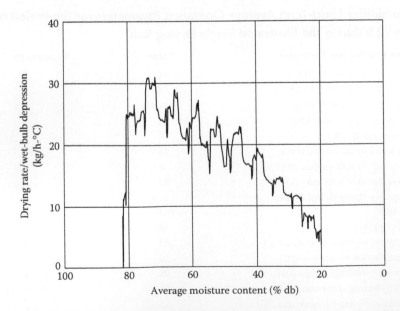

FIGURE 8.15 Average moisture transfer coefficient for the timber stack for the example drying run, as indicated by the drying rate (kg/h) divided by the average wet-bulb depression (°C) in the kiln. (Adapted from Burnett, M.D. et al., *Drying Technol.*, 33, 455–465, 2015. With permission.)

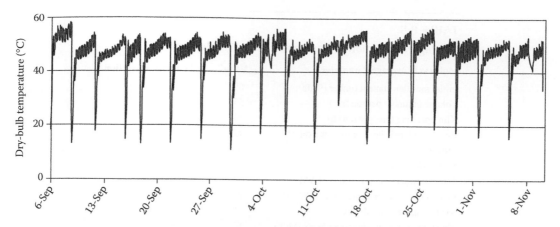

FIGURE 8.16 Dry-bulb temperatures at the stack exit of the kiln over a period of 2 months in 2004, showing a sequence of 19 drying runs.

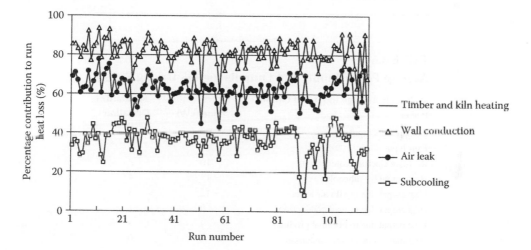

FIGURE 8.17 Cumulative percentage contributions to the average heat loss from the kiln for 115 drying runs.

the losses were reasonably consistent during the period of monitoring, with some variation in heat loss by air leaks and corresponding changes in the subcooling contribution.

8.6.2.2 Long-Term Averages

Table 8.5 summarizes average operating parameters for the 115 runs analyzed, and Table 8.6 lists the corresponding average power inputs and outputs. For each run, the parameter shown is averaged over the drying cycle and each entry in the tables is the average of the parameter over the 115 drying runs. The uncertainty shown is the standard deviation for the 115 runs, apart from the fan power, where the uncertainty is that for a single measurement. For most parameters, the precision corresponds to one significant digit in the standard deviation.

The average electricity consumption per m^3 of timber shown in Table 8.5 (79 kWh/m^3) is 30% higher than the average electricity input (61 kWh/m^3) for combustion-heated Radiata pine kilns in New Zealand, excluding combustion sourced heat (EECA 2005). It is 21% less than the electricity consumption for accelerated conventional Radiata pine schedules, examined by Ananias et al. (2012) of approximately 100 kWh/m^3. However, the Whangarei dryer was used to dry Radiata pine to a nominal 20% moisture content (db), whereas the data cited by EECA and Ananias et al. apply to a final moisture content of 10%. Had the Whangarei dehumidifier been used to achieve

TABLE 8.5

Average Performance Parameters for 115 Drying Runs

Multicycle Average Parameter	Value	Standard Deviation
Initial moisture content timber (% db)	69	24
Volume of timber per run (m³)	55	14
Kiln operating time per run (h)	83	19
Resistive electric heating time per run (h)	12	3
MER (kg/h)	131	26
SMER (kg/kWh)	2.7	0.4
Total electricity input per run (MWh)	4.0	0.9
Electricity input per cubic meter (kWh/m³)	79	32
Air-leakage rate (kg-dry/s)	0.09	0.03
Moisture output rate due to air leaks (kg/h)	12	3

Source: Adapted from Burnett, M.D. et al., *Drying Technol.*, 33, 455–465, 2015.
With permission.

TABLE 8.6

Average Power Inputs and Outputs for 115 Drying Runs

Multicycle Average Input and Output Power	Value (kW)	Standard Deviation
Compressors input	29	3
Fans input	9	1
Resistive electric heaters input	10	3
Heat output by subcooler	17	4
Heat output due to kiln air leaks	12	3
Heat output due to wall conduction	9	1
Heat output due to hot final product	9	3
Heat output due to hot condensate	0.7	0.7

Source: Adapted from Burnett, M.D. et al., *Drying Technol.*, 33, 455–465,
2015. With permission.

10% moisture content, the electricity consumption would have increased by some 20 kWh/m³ and the drying time would have been 20 h longer. The average electricity consumption per cubic meter would then be 62% higher than the EECA (2005) value, but essentially the same as the estimates by Ananias et al. (2012) for accelerated conventional schedules.

8.6.2.3 Payback Time

During 2004, the dryer completed 95 runs and the total volume of timber dried was approximately 5070 m³. Based on the average electricity consumption per cubic meter shown in Table 8.5, and the average cost of industrial electricity in 2004 of NZ8¢/kWh excluding gst (MBIE 2013), the cost of electricity per cubic meter was NZ$6.30/m³. This indicates that the cash saving in the cost of heat pump drying relative to the cost of contract drying (about NZ$70/m³) was some NZ$323k during 2004. Hence, the payback time for the Whangarei dryer was probably less than 12 months, based on the cost of the dehumidifier module (about NZ$200k) and the dryer set-up costs (on the order of NZ$100k). This conclusion should be treated cautiously, however, since the cost of contract drying comes from informal sources, and the influence of local electricity network charges is unknown.

8.6.2.4 Performance Correlations

To show the variation in performance within the multicycle data set, Figure 8.18 plots the total mass of moisture extracted from the timber stack in each drying run against the corresponding total electrical energy input, including the heat pump compressors, fans, and resistive heaters. This indicates that the average mass of moisture extracted is linearly correlated with the electrical energy used ($R^2 = 0.84$), and there is an initial input of 1.0 MWh due to the start-up heat requirement of the kiln. As the moisture load increases, the start-up requirement becomes a smaller fraction of the energy input, so the overall energy efficiency increases.

The drying cycle time, in turn, depends on the total mass of moisture but the correlation is not strong with $R^2 = 0.59$. A possible reason is that variations in the amount of moisture extracted are due to two factors, the initial moisture content and the timber volume. A better correlation was obtained by narrowing the range of timber volumes and considering just the initial moisture content, as in Figure 8.19. This shows how the drying time depends on the initial moisture content for a subset of 45 runs where the volume of timber is limited to 55 ± 6 m³.

Another correlation of interest is that between the electricity consumption per cubic meter and the average initial moisture content of the timber stack, shown in Figure 8.20 for the full set of 115 runs. Both parameters are sensitive to the estimated volume of the timber for each run and it is possible that the strength of this correlation is overstated. The dashed line in Figure 8.20 shows the estimated total energy use per cubic meter if the stack had been dried to 10% moisture content, an increase of 20 kWh/m³. The relationships shown in Figures 8.18 through 8.20 indicate that, within the range of application illustrated, the performance of the dryer is well represented by simple linear correlations. This is surprising in view of the coupled time-dependent processes within the dryer.

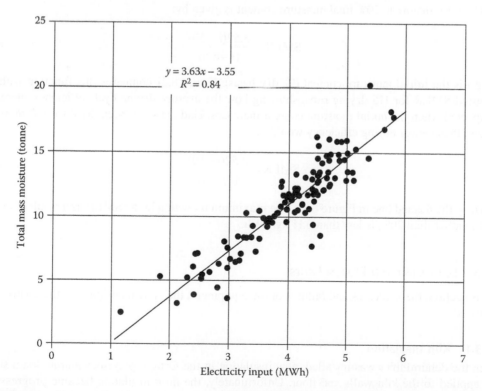

FIGURE 8.18 Scatter plot showing the correlation between the electrical energy used (MWh) for each drying run and the total mass of moisture (tonne) extracted from the timber stack.

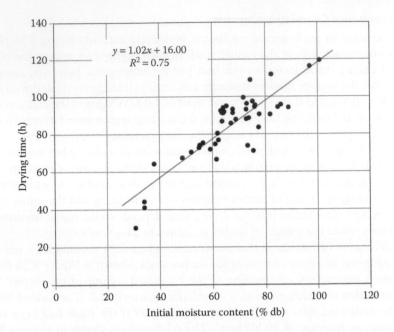

FIGURE 8.19   Correlation between the drying time and the initial moisture content for 45 drying runs where the timber volume is 55 ± 6 m³.

Using the correlation shown in Figure 8.20, and taking the basic density of Radiata pine to be 450 kg/m³, it is straightforward to establish that the average specific moisture extraction rate (kg/kWh) for drying to 20% final moisture content is given by:

$$SMER = \frac{3.52(i-20)}{(i-6.5)} \qquad (8.2)$$

where i is the initial moisture content (%, dry basis). Figure 8.21 compares this function with the measured SMER for 115 drying runs, showing how the average drying cycle efficiency increases as expected when the initial moisture content increases. Had the stack been dried to 10% moisture content, the average drying efficiency would be:

$$SMER = \frac{3.52(i-10)}{(i+9.1)} \qquad (8.3)$$

shown by the dashed line in Figure 8.21. The maximum reduction in the SMER as a result of drying to 10% rather than 20%, is less than 0.11 kg/kWh.

8.6.3   EQUIPMENT AND DESIGN ISSUES

In this section, the successes and failures of the equipment, the system design, and the controls are discussed.

8.6.3.1   Kiln Insulation

When the dehumidifier was installed, new insulation in the form of polyisocyanurate foam sheets was applied to the kiln walls and floor. Unfortunately, the floor insulation became progressively damaged during kiln loading, so it was removed after about four months. This is an example of the difficulties that arise when retrofitting a dehumidifier to a preexisting chamber.

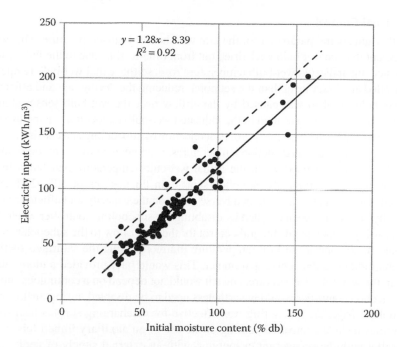

FIGURE 8.20 Correlation between the electricity consumption per cubic meter (kWh/m³) and the initial moisture content (% dry basis) for 115 drying runs. The dashed line represents the estimated total energy use per cubic meter to dry the timber to 10% moisture content. (Adapted from Burnett, M.D. et al., *Drying Technol.*, 33, 455–465, 2015. With permission.)

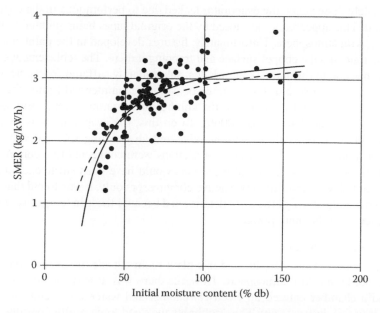

FIGURE 8.21 Specific moisture extraction rate (kg/kWh) for 115 complete runs drying to 20% moisture content (dry basis) as a function of the initial moisture content. The solid line shows the correlation obtained from Figure 8.20 and the dashed line shows the estimated SMER had the timber been dried to 10% moisture content (db). (Adapted from Burnett, M.D. et al., *Drying Technol.*, 33, 455–465, 2015. With permission.)

8.6.3.2 Thermal Control

The primary thermal control parameter for the kiln was the wet-bulb temperature. This was generally satisfactory, except that the wet-bulb wick dried out from time-to-time due to the failing water supply. As a consequence, the indicated wet-bulb temperature rose, so the actual wet-bulb temperature of the kiln was controlled at a lower level than the set point, reducing the drying rate and efficiency.

Another wet-bulb problem was caused by the airflow over the wet-bulb sensor stopping briefly when the airflow reversed. This resulted in the indicated wet-bulb temperature rising by 0.5°C–1.0°C, prompting the controller to increase the rate of heat rejection by subcooling and causing a rapid decrease in the actual wet bulb and saturated suction temperatures, accompanied by a temporary increase in the drying rate. This change reversed when the actual wet-bulb temperature had fallen by about 1°C. As a result, the drying rate typically exhibited short-term fluctuations of 10%–20% each time the airflow reversed. These instabilities had an adverse effect on the capacity and efficiency of the dryer.

This difficulty could have been avoided by disabling the subcooling controller when the fans were stopped and by using proportional-integral control for the water flow to the subcooler. An alternative approach would be to control the kiln temperature indirectly using the saturated suction temperature of the dehumidifier as the control parameter. This would have provided a more reliable control parameter than the wet-bulb temperature, since it would not depend on a continuous supply of water. But the heat rejection controller would nevertheless need to be disabled during airflow reversal.

On the whole, refrigerant subcooling was effective for discharging surplus heat from the kiln. Other possibilities include controlled venting or the use of an auxiliary finned-tube moisture condenser in parallel with the refrigerant evaporator, with an external supply of cooling water. Both options would provide efficient additional drying capacity as surplus heat is discharged. Compared with cooling water-driven refrigerant subcooling, as used at Whangarei, these methods for heat rejection would not affect the heat pump evaporating temperature directly. This would make it easier to maintain the dehumidifier in its most efficient operating state.

8.6.3.3 Evaporator

After approximately three years, the evaporator leaked due to perforations in the copper pipe-work and was replaced. This appeared to be caused by the original pipes being painted to prevent corrosion in the acidic kiln atmosphere. Unfortunately, fissures developed in the paint coating, allowing moisture to penetrate into the copper surface and trapping it there. The replacement evaporator was not painted and has not, to the author's knowledge, subsequently suffered from corrosion.

To keep the suction pressure of the compressor within recommended limits, the area of active evaporator surface was varied in step with the number of operating compressors as the capacity was changed (Scharpf and Carrington 2006b). To do this, the evaporator was divided into four sections, each with a thermostatic expansion valve and an isolating solenoid valve in the liquid feed line. The suction outlets of the evaporator sections were connected to a common compressor suction header. In practice, two evaporator sections would have been sufficient, since the system occasionally operated with fewer than two active compressors only. It was found that when just one compressor operated, the drying efficiency deteriorated because the kiln temperature fell, due to the heat loss rate exceeding the input power.

8.6.3.4 Electrical Equipment

The power relays and controls were located in a plant room outside the drying chamber to protect them from the hot humid kiln atmosphere. However there was at least one occasion when moist air from the kiln chamber entered an electrical cabinet and water vapor condensed on a power relay, causing it to fail destructively. This highlights the need to carefully consider the placement of electrical cables that penetrate the kiln wall. There should be a ventilated gap between the outer insulated wall of the kiln and the electrical cabinet which should also be ventilated to reduce the risk of moisture condensing on electrical equipment. The possibility of moisture moving along a power cable between the conductor and the insulation should also be considered.

8.6.3.5 Compressors

The use of hermetic scroll compressors located inside the kiln chamber was satisfactory, as was the practice of pumping the evaporator down each time the dehumidifier stopped. The scrolls coped well with variations in the saturated suction temperature without overheating or motor overload. The use of evaporating temperatures of 25°C, and occasionally up to 30°C, with HFC-134a did not appear to impact significantly on the compressor durability. One compressor failed during year nine after an operating time of approximately 50,000 h and a second failed shortly afterwards. The causes are unknown.

Generally the compressor oil temperature was close to the saturated discharge temperature. Table 8.4 and Figure 8.14 also show that the compressor discharge and oil temperatures were not especially high compared to those often found in air-conditioning applications. This suggests that the compressors were not particularly stressed, so it is possible that higher operating temperatures could have been used without affecting the compressor durability. Potentially this would enable the system to bridge the gap between the dehumidifier and the conventional Radiata pine drying regimes, although some of the low temperature product quality elements could be lost.

8.6.3.6 Airflow

The airflow configuration, with the evaporator and condenser in parallel airstreams, and periodic airflow reversal, caused few problems although one fan motor failed during the first three years for reasons unknown.

The baffle arrangement used to restrict the airflow bypassing the timber stack is a significant concern because its effectiveness varied, depending on the efforts of the operator when the kiln was loaded. This could be seen by the changes in the dry-bulb temperature when the airflow was reversed, discussed in Section 8.6.1.1. Under normal operating practice, the fraction of circulating air bypassing the timber stack has been estimated by kiln modeling (Sun et al, 2000) to be 30%–50%. This means that the relative humidity of air entering the dehumidifier was lower than it would have been, causing a loss of efficiency and capacity. The presence of a strong cyclic signature in the dry-bulb temperature accompanying airflow reversal, as in the lower panel of Figure 8.12, is a potentially useful indicator of inadequate bypass control.

Another airflow difficulty arose from the evaporator location adjacent to the timber stack (Figures 8.2 and 8.5). Because the stack was closer to the evaporator than expected, the cold evaporator airstream did not mix well with the warm condenser air when the air circulation was clockwise (Figure 8.5), creating a cooler zone where the timber did not dry as fast as elsewhere. This would have been avoided by redirecting the evaporator airstream to ensure that it mixed better with the condenser airflow during clockwise air circulation.

8.6.4 FUTURE OPPORTUNITIES

8.6.4.1 Batch Dryers

For small operations with a throughput of 3000 m³ per annum or less it is probably simplest to use a batch-mode dryer. Table 8.5 shows that the multicycle average MER is 71% of the peak short-term value shown in Table 8.4 and the average SMER is 67% of the corresponding short term value. This suggests that there are significant opportunities for increasing the efficiency and drying capacity of a batch-mode dryer by ensuring the average performance is closer to the peak value. These possibilities include improved temperature control, reduced heat losses, better use of liquid subcooling and better control of the airstream bypassing the timber stack. Other opportunities for improving dehumidifier performance include optimizing the capacity of the heat exchangers and airflow rate, use of more suitable compressors, and alternative refrigerants.

Air leaks represented the largest uncontrolled heat loss, being responsible on an average for 26% of losses (Table 8.6). The efficiency with which the leaks used the surplus heat was approximately 1 kg/kWh, which is comparable with the drying efficiency for heat rejection by subcooling. Thus, the main disadvantage of the air leaks is that they were uncontrolled. This had an adverse effect on the performance of the system when only one or two compressors were used, because the tempera-ture fell when the heat loss rate exceeded the power input (Carrington et al. 2000a).

Electricity used for resistive preheating at the start of each cycle represented 20% of the average energy input during a drying cycle (Table 8.6). This was a relatively inefficient use of electricity but it is difficult to avoid in a batch-mode dryer.

Another opportunity for increasing the capacity and efficiency is to subcool the liquid refrigerant using the cooled air leaving the evaporator (Bannister et al. 1995). In a reversible airflow configura-tion this could be achieved in several ways, such as by actively managing the evaporator airflow. While this would be a complication, the increase in drying capacity and energy efficiency is likely to be significant, around 20%.

If the product requires reconditioning, the use of high-pressure water sprays (Pearson and Haslett 2002) is a promising option.

8.6.4.2 Progressive Dryers

For an operation drying more than 3000 m^3 per annum, a progressive dryer with four or more stages would provide further opportunities for increasing the drying speed and energy efficiency, in addi-tion to those listed for batch dryers. In this configuration, the kiln chamber would be a tunnel with doors at each end for timber entry and exit on track-mounted trollies. There would be baffles or curtains between adjacent trollies, which would move progressively along the tunnel, perpendicular to the primary direction of air circulation. The trollies would move one stage along the drier when dried timber exits, making space for a new load to enter the first stage.

Once operating, the first stage could be heated using surplus heat from the other stages to reduce the power demand of the kiln and peak-power costs. Each stage would have a dehumidifier module similar to a batch-mode dryer, optimized for the stage temperature and relative humidity.

Although the main circulation of air would be transverse to the tunnel axis, some should also circulate axially in the direction of movement of the trollies when fresh timber is being heated. Heat rejection would take place at each stage as required, based on individual temperature settings. If reconditioning is required, high-pressure water sprays (Pearson and Haslett 2002) are an option, probably in a chamber separate from the dryer tunnel.

It is feasible that such a progressive dryer could achieve an average operating SMER of more than 4 kg/kWh using electricity alone. For drying Radiata pine from 100% to 10% moisture content (dry basis), an average power consumption significantly less than 100 kWh/m^3 is feasible, with a drying time of less than 110 h before reconditioning. Compared with a batch-mode dryer, the large peak in the power demand that occurs when fresh timber is loaded into the dryer would be avoided, which is likely to reduce electricity costs. The potential for better performance would need to be weighed against the risk of increased capital costs for the dryer and the possibility of higher timber handling and dryer management costs.

8.7 INDUSTRY ACCEPTANCE

Notwithstanding the positive features of heat pump timber dryers (Section 8.3.3), combustion-heated dryers are strongly preferred by the timber processing sector in New Zealand (Section 8.2.3). In this section, the energy cultures framework (Stephenson et al. 2010) is used to examine the reasons for this preference and to consider opportunities for usefully expanding the role of heat pump timber dryers in the future.[*]

---

[*] In this section, extracts are used from the paper, Socio-technical barriers to the use of energy-efficient timber drying technology in New Zealand, by Bell, M. et al., *J. Energy Policy*, 67, 744–755, 2014. Used with permission of the publisher, Elsevier BV.

8.7.1 Energy Cultures

The energy cultures framework is a theoretical model for decisions affecting energy demand, derived, in part, from the sociotechnical systems literature (Rotmans et al. 2001; Geels 2004). The framework distinguishes the internal drivers for energy decisions, over which an organization has agency, from the external ones, such as commercial pressures and regulatory intervention that are beyond its control. The framework identifies three internal drivers: energy-using activities (practices), physical infrastructure (material culture), and mental models of what is normal (norms). The framework is illustrated symbolically in Figure 8.22. A feature of the framework is that energy behaviors can be self-sustaining when the internal drivers are coupled, thus creating a durable energy culture. The internal interactions are also affected by external influences that may reinforce an established energy culture or perhaps change it.

8.7.2 Industry Survey

The data that inform this analysis were obtained mainly by Bell et al. (2014) from interviews with the managers of 20 New Zealand timber processing firms in 2009. The interviewer had no previous association with the timber industry and had a strong background in social science research. The interviews were semistructured and used open-ended questions to investigate the activities that could influence a firm's energy decisions. They were recorded and later transcribed verbatim for interpretive analysis. Relevant numerical data and attributes were set up in a data matrix for framework analysis (Ritchie and Lewis 2003) and the text was analyzed to identify relevant themes.

For analysis, the firms were clustered by their drying methods: 10 with combustion-heated dryers, 5 with heat pump dryers, and 3 with both technologies, although their heat pump dryers were not used intensively. Two firms had no dryers. Specific characteristics of each group were compared using indicator scales developed from the data matrix to provide a numerical measure of differences between them.

Because the cost information was incomplete, some firms were excluded from the assessment of operating costs and the values presented here are the averages for the remaining firms.

8.7.2.1 Business Issues

The interviews took place at a time when the timber processing industry in New Zealand was in difficulty (Section 8.2.1). Some firms had acute problems, those selling into commodity markets especially. Their response was to trim processing costs and improve control of their operations.

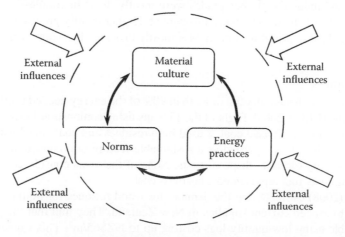

FIGURE 8.22 The energy cultures framework showing coupling between the material culture, energy practices, and norms supporting a durable energy culture. (Adapted and reprinted from Stephenson, J. et al., *Energy Policy*, 38, 6120–6129, 2010. With permission.)

A measure of their distress is that 3 of the 10 firms with combustion-heated dryers, and 1 of the 5 with heat pump dryers, closed down within 5 years of the interviews.

For many firms, continuing changes in the global market was a significant external influence, an instability that demanded operating flexibility. Variations in the exchange rate of the New Zealand currency exacerbated these problems. External influences on energy use were also embedded in the movement of the market. This had further implications on the need for flexibility in these firms and influenced their drying technology preferences.

Some firms had purchased second-hand equipment to reduce their costs, but this often created other problems. Older boilers had higher emissions of particulates and the efficiency of a second-hand sawmill was usually lower than newer equipment. Some of those with heat pump dryers had reduced costs by building their own drying chambers, which carried the risk of poor insulation and air-sealing and downstream problems.

The interviews provided data from which the capacity of firms to set their product prices was assessed. On average, there was little difference between the firms with combustion-heated dryers and those with heat pump dryers in reference to their capacity to obtain profitable returns. By contrast, the group of firms using both types of kiln was better able to secure attractive prices. This group engaged in niche markets more effectively and was also the most profitable, although its operating costs were higher than the others (Section 8.7.2.4).

Indicators of the profitability of firms were also obtained from the interviews, such as their business confidence, frustrations, new investment plans, and their focus on cost-cutting. Based on these responses, there was no real difference in the average profitability of the combustion-heated and heat pump dryer groups, while the three firms using both technologies were significantly better off. This suggests that the success of the mixed-technology group was determined by their capacity as price-setters, rather than their choice of drying technology. Nevertheless, these firms may have been assisted by the extra flexibility they achieved by having multiple drying methods.

8.7.2.2 Processing Issues

Eighteen of the 20 firms interviewed had sawmilling operations. Some sold the sawmill residues, representing typically 50% of the log, at NZ$20–30/tonne (wet). This was much less than the original log cost of around NZ$113/m^3 (MPI 2015b), bearing in mind that the wet log density is approximately 1 tonne/m^3.

Twelve of the firms with drying facilities dried only Radiata pine, four processed both Radiata pine and other species, and two focused on specialty timbers, such as imported species and swamp kauri. Structural and industrial timber grades were mostly dried in combustion-heated dryers at temperatures up to 120°C to achieve rapid throughput. Higher quality grades were normally dried at 70°C–90°C. The heat required for drying was mostly provided by steam or hot water heated by fuel combustion.

8.7.2.3 Fuel Choices

Typically for combustion-heated dryers, more than 90% of the energy needed for timber drying was used for kiln heating (EECA 2005). Eight of the 13 firms using combustion-heated dryers used sawmill residues as heating fuel, 2 used coal, 2 used both residues and coal, and 1 used gas. Generally the ability to use residues was regarded as a sustainable feature of combustion-heated dryers and it avoided the possibility of waste disposal costs. None of the firms considered the potential sales value of residues used as fuel represented a business cost.

Hall and Jack (2008) estimate that the demand for wood residues for liquid biofuel production has the potential to exceed current supplies in New Zealand. They find that the process would be economically viable using low-quality logs costing up to NZ$65/m^3. This suggests that if biofuel production from wood residues becomes established, the opportunity cost of wood residues is likely to increase beyond its present level of NZ$20–30/tonne, which could dispel the current perception that residues provide cost-free fuel for combustion-heated dryers.

8.7.2.4 Firm Scale

The production rate and the number of employees of the firms with combustion-heated dryers were much larger than those of the heat pump firms (Table 8.7). Nevertheless, the average operating cost per cubic meter of sawn timber was not significantly different. Comparison of the operating costs for firms with combustion-heated and heat pump dryers with the average export value for sawn timber, NZ$397/m^3 in 2009 (MPI 2015b), suggests that profit margins were in the order of NZ$60/m^3.

The average production rate of the heat pump firms is similar to that shown in Table 8.1 from an earlier survey of firms using heat pumps to dry Radiata pine, and their electricity costs are also similar. The scale of the firms using both types of kiln was between that of the other two groups, but their average operating cost per cubic meter of sawn timber was about 40% higher than the others.

8.7.2.5 Drying Costs

None of the participants indicated they had a detailed understanding of the factors affecting the cost of drying, such as capital amortization, maintenance, labor, electricity, and other fuel. This suggests they were not well equipped to make decisions that impacted their costs, opportunities, and risks. To illustrate these issues, the cost of drying is estimated here using data supplied in the interviews, other published data, and background information from the Whangarei dryer.

There is a significant difference in the initial costs, and in the production rates, of combustion-heated and heat pump dryers at the entry level. The capital cost of an entry-level heat pump dryer, that would dry some 3000 m^3 per annum (pa) is approximately NZ$400k. In comparison, an entry-level combustion-heated dryer would have a larger drying capacity, around 13,000 m^3 pa. Its cost, including the furnace and facilities for wood residue drying and storage, is in the order of NZ$2M.

The cost of drying is the income required (NZ$/m^3) to just cover the capital, interest, and operating costs over the financial life of the dryer, with no profit. Capital and interest are amortized in equal annual payments over this time.

For this analysis, the energy requirement for a combustion-heated dryer is based on data published by EECA (2005) and that for a heat pump dryer is based on Figure 8.20. Other assumptions include: financial life of the equipment 10 years; interest on borrowed funds 10% pa; maintenance costs 3% of capital pa; electricity cost in 2009, NZ11.7¢/kWh excl. gst (MBIE 2013); value of wood residue for fuel NZ$20/tonne; wood residue moisture content 100% dry basis; labor NZ$50,000 pa per capita; final sawn-wood moisture content 10% dry basis; kiln utilization 90%; total drying cycle time for 100% initial moisture content Radiata pine, including reconditioning

---

TABLE 8.7
Scale Data for 18 Timber Processing Firms

Technology Group	Combustion-Heated Dryers	Both Types of Kiln	Heat Pump Kilns
Number of firms in category	10	3	5
Average age of the business (year)	32	41	36
Average volume sawn timber per month (m^3)	4193	1253	256
Average number of employees	61	33	8
Average volume sawn timber/month-employee (m^3)	69	38	32
Operating cost/volume sawn timber (NZ$/m^3)	339	475	344
Purchased-energy-cost/volume sawn timber (NZ$/m^3)	10	16	20

Source: Bell, M. et al., *Energy Policy*, 67, 744–755, 2014. With permission.

Note: The operating cost (including the log cost) and the energy cost have been estimated from interview responses. Purchased energy excludes the value of own timber residues; costs exclude goods and services tax.

and reloading, 35 h for an accelerated conventional temperature combustion-heated dryer; for a heat pump dryer 150 h. The heat pump drying time is based on Figure 8.19 plus 20 h for drying from 20% to 10% moisture content and 10 h for reconditioning and reloading. This time is probably shorter than that of other heat pump dryers on which it is likely that the companies that were surveyed based their decisions.

Under these assumptions, the average cost of drying is essentially the same for both combustion-heated and heat pump kilns, NZ\$47/m³, although there are differences in the cost makeup, shown in Table 8.8. The estimated purchased energy costs are some 20% smaller than those reported in the interviews (Table 8.7), which is reasonable for new installations.

The average drying cost in Table 8.8 disguises considerable variations in practice. For instance, the energy use per cubic meter of timber varies between sites by typically ±30% (EECA 2005), in part, due to variations in the timber and the drying schedule. If a firm were to dry high-quality timber more slowly (say in 80 h) in a combustion-heated dryer, the cost of drying would go up to NZ\$87/m³, while the cost of drying structural timber at a higher temperature in 24 h would be just NZ\$37/m³. The cost of drying at the same temperature as a heat pump dryer in a combustion-heated one would be NZ\$152/m³, a substantial increase. In view of the slim profit margin noted in Section 8.7.2.4, firms were strongly incentivized to dry at higher temperatures. Drying at lower temperatures could force them into deficit, unless they obtained a significant price premium for quality improvements.

It was evident from the interviews that managers tend to focus on either their cash operating costs or their capital costs, not the combined cost of production. Those under stress were especially concerned about current costs, such as electricity costs. It was generally accepted that the cost of drying was lowest for combustion-heated dryers that used wood residues as fuel, although the opportunity cost of residues, which was ignored, was close to the cost of electricity (Table 8.8). The managers of firms using heat pump dryers were especially concerned about electricity costs, which were twice those for firms with combustion-heated dryers (Tables 8.7 and 8.8).

Some firms were attracted to lower temperature drying because it had the potential to yield a higher quality product (Section 8.3.3) and realize a higher price. On the other hand, lower temperature drying with combustion-heated dryers carried significant costs and the premium was accessible only to firms with appropriate marketing strengths. Thus, the pursuit of higher profits by drying at lower temperatures was a risky business strategy for firms with combustion-heated dryers.

8.7.2.6 Technology Concerns

Of the 10 firms using only combustion-heated dryers, 7 expressed concerns about their particulate emissions, especially those with older boilers. Because emissions are regulated, they represent a liability that could force firms to replace boilers at considerable cost. Some tried to minimize their emissions during testing by purposefully increasing their heat demand. When

TABLE 8.8

Indicative Drying Costs for Combustion-Heated and Heat Pump Dryers in NZ\$ per Cubic-Meter of Timber, Excluding Goods and Services Tax

Cost Category	Combustion-Heated Dryer (NZ\$/m³)	Heat Pump Kiln (NZ\$/m³)
Capital, interest, labor and maintenance	32	31
Electricity	8	16
Wood residues	7	0
Total	47	47

Source: Adapted from Bell, M. et al., *Energy Policy*, 67, 744–755, 2014. With permission.

asked, none of these firms said they would like to change to heat pump dryers to avoid emission problems; they were more apprehensive about the risk of unsatisfactory heat pump performance. None said they were worried about the product quality achieved in combustion-heated dryers and none felt they were trapped with their current drying technology.

By contrast, users of heat pump dryers had a range of concerns, especially electricity use and drying speed. Four of the eight firms with at least some heat pump dryers indicated that they would like to change to combustion-heated dryers. Three of the four were struggling financially. They produced mainly structural timber products, were price-takers, and had chosen heat pump dryers only because they were affordable. It appears that these firms were not well adapted to the limitations of their heat pump dryers and did not have the capacity to obtain commercial benefits from drying at lower temperatures.

8.7.2.7 Summary Indicators

Table 8.9 shows average business indicators and concerns of the three groups of firms, developed from the interview data. For the business indicators (top four rows), the differences between firms with combustion-heated and heat pump dryers are relatively small. By contrast, the two indicators of technology concerns (bottom two rows) show clear differences.

The firms with both types of dryers stand out as having higher marketing strengths and business success, while also being less concerned about cost control. They did not have high confidence in their drying technology and had only minor concerns about emissions.

8.7.3 SYNTHESIS

In this section, the energy cultures framework is used to analyze the success factors of the firms surveyed, their choice of drying technology, and the influence of the wider sociotechnical regime.

8.7.3.1 Success Factors

The firms that were price-takers in the commodity markets responded quickly to varying market conditions and minimized their operating costs by drying at higher temperatures with combustion-heated dryers, if they were big enough to afford them. It appears that these firms followed the incentives to dry at higher temperatures, discussed in Section 8.7.2.5, and were unable to access more profitable markets. The four firms surveyed that subsequently failed were price-takers.

The price-setters had well-developed consumer market relationships and were more profitable. This group included 4 of the 10 using combustion-heated dryers, 3 of the 5 with heat pump kilns, and all

---

TABLE 8.9

Average Business Indicators and Concerns of the 18 Timber Firms Surveyed with Drying Facilities

Indicator	Scale		Combustion-Heated Dryers	Users of Both Types	Heat Pump Dryers
	1	5			
Price setter or taker	Taker	Setter	2.0	3.3	2.6
Commercial success	Failing	Profitable	3.7	4.3	3.8
Concern for cost control	Low	High	3.2	2.5	3.6
Does the firm sell primarily into commodity markets?	Yes	No	2.8	3.8	3.0
Confidence in their drying technology	Low	High	4.9	3.0	2.7
Concerns about emissions	Low	High	2.8	1.3	1.0

3 with both types of dryer. This last group was less concerned about cost control; it engaged success-fully with profitable niche markets and was better able to carry higher processing costs than the others.

Overall, the evidence indicates that the most important business success factor was a timber firm's capacity to establish strong relationships with consumers in profitable markets. Its choice of drying technology was not a significant success factor.

8.7.3.2 Dryer Preferences

Combustion-heated timber drying was preferred by the larger firms, and was an aspiration of smaller firms wishing to expand their business. They were seen as the industry norm; firms were satisfied with them and had confidence in them. Heat pump dryers were cheaper initially, so were attractive for small firms. One might expect that this opening into the timber processing sector would enable heat pump dryers to succeed in the industry, since suppliers needed only to satisfy their first custom-ers. And there was indeed a group of satisfied heat pump kiln operators, mostly the more successful heat pump users. But others were dissatisfied. For example, the Whangarei firm that acquired the Delta S dehumidifier was successful but it subsequently installed a combustion-heated dryer to expand its drying capacity. Since the firm was drying for pressure preservative treatment, quality was not a key issue, so a heat pump dryer offered no particular advantages. The fact that Delta S did not have the capacity to provide the firm with strong continuing support may have been a factor in its decision.

On balance, did the firms that changed from heat pump to combustion-heated dryers make good business decisions? From their perspective the main benefits were: (1) shorter drying times; (2) ability to use wood residues for fuel; (3) lower electricity costs; (4) more control over drying speed; and (5) better technical and research support. On the other hand, the cost of slower drying with a heat pump dryer is a lot lower than low temperature drying with a combustion-heated kiln (Section 8.7.2.5). In addition, heat pump dryers are scalable; at least one firm in New Zealand uses a cluster of heat pump timber dryers to achieve capacity (Perré and Keey 2006). And the use of wood resi-dues in combustion-heated dryers provides little cost saving when the opportunity cost of residues is included (Table 8.8). Combustion-heated dryers also provide no operating cost or profitability advan-tages (Sections 8.7.2.4 and 8.7.2.5); they are not a business success factor (Section 8.7.3.1); they cap the timber quality (Section 8.3.3); they incentivize firms to dry at higher temperatures (Section 8.7.2.5); and firms must deal with combustion emissions and source heating fuel (Sections 8.7.2.3 and 8.7.2.6).

Thus, a decision to move from a heat pump to a combustion-heated dryer carried mixed benefits, suggesting that the preference for combustion-heated dryers was supported by other factors.

8.7.3.3 Sociotechnical Regime

Naturally, the suppliers of combustion-heated dryers promoted their technology in order to maintain its role in the timber processing industry. But the sustaining influences were more pervasive than this. The larger timber companies confidently endorsed combustion-heated dryers and also supported drying standards and timber drying research, so the dominance of combustion-heated dryers was embedded in the standards and in the sector research agenda. These firms were comfortable with the links between the combustion-heated dryer suppliers and the timber research providers. The prefer-ence for combustion-heated kilns was also endorsed by the organizations that provided timber firms with technical support, while their disadvantages were accepted. By contrast, heat pump dryers were used primarily by the smaller timber firms and were seen as being suitable only for such firms.

Evidently, the technology decisions of the timber processing firms were influenced by the wider sociotechnical regime (Rotmans et al. 2001; Geels 2004). From the energy cultures perspective, there was a culture of timber drying that centered on the use of combustion-heated dryers. Norms, technologies, and practices generated a sustaining culture, supported by the feedback processes of the regime. For individual firms (Figure 8.23), their commitment to combustion-heated dryers was reinforced by managers accepting the technology as the sector norm and that fast high-temperature

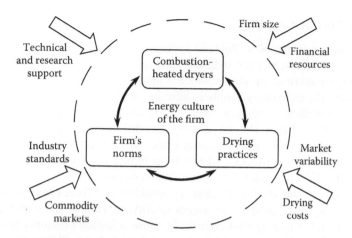

FIGURE 8.23 The self-sustaining energy culture of firms using combustion-heated dryers and key external drivers. (Adapted and reprinted from Bell, M. et al., *Energy Policy*, 67, 744–755, 2014. With permission.)

drying is required for economic viability. This was underpinned by their expectations for processing flexibility and the use of sawmill residues for fuel. Other drivers included firm size, access to financial resources, market variability, the availability of strong technical and research support, and a focus on electricity cost rather than the total cost of drying.

This set of influences sat within broader industry practices that reinforced acceptance of combustion-heated dryers across the industry (Figure 8.24). The culture that normalized combustion-heated drying resided at the level of the timber industry, its supporting organizations, and the technology suppliers. At this level, the predominance of combustion-heated dryers, the industry norms, and good technical and research support created an encompassing energy culture that excluded alternative technologies. Questions about the trade-off between drying temperature, product quality, and access to more profitable markets were not considered.

The long lifetime and replacement schedule of drying equipment also contributed to technology inertia. Because the industry did not grow significantly after the year 2000 (MPI 2015a), new drying investments were made only as replacements to existing capacity, or as small incursions on the fringe when specific new opportunities arose. Combining this with an aversion to capital spending in a tight market, these additionally installed capital considerations contributed to drying technology stasis.

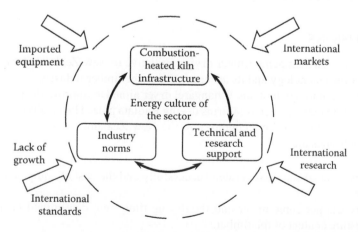

FIGURE 8.24 The energy culture of the timber drying sector, showing the external influences and the internal feedbacks that reinforced the use of combustion-heated dryers. (Adapted and reprinted from Bell, M. et al., *Energy Policy*, 67, 744–755, 2014. With permission.)

8.7.4 CAN THE CULTURE CHANGE?

The major stakeholders in the New Zealand timber industry—the larger processing firms, the drying technology providers, and the industry research organizations—are strongly committed to the current energy culture of the sector, which is dominated by combustion-heated drying. The processing firms and the technology suppliers have invested heavily in combustion-heated drying and the research sector relies on industry support for funding. As a result, none of them is in a position to initiate or promote change. But the industry is in difficulty. Many sawmill firms have closed down and the sector has not grown significantly since 2000, despite the timber harvest increasing by 66% by 2014. The sector also faces threats from tighter emissions restrictions, higher value wood residues, and a difficult market for commodity sawn timber.

This analysis suggests that the timber industry would benefit from engaging with higher value markets and producing more products of higher quality. In view of the opportunities they provide for increasing timber quality cost-effectively, these selection pressures favor greater use of heat pump dryers. This raises the question, how could this happen? The experiences with the Whangarei dryer described in this chapter show that the current limitations of heat pump timber dryers are amenable to development. But technology change driven by incremental market pressures and technology advancements is likely to be slow. The heat pump suppliers would need considerable resources and commercial capacity to engineer a culture shift.

To succeed faster, it would be better if the heat pump suppliers were to collaborate with other stakeholders. For example, a large new processing firm with a mature and independent approach may enter the industry with a view to using heat pump drying to further its own business interests. Other possibilities include an electricity supplier, or a demand aggregator, engaging with the technology suppliers with the aim of gaining access to drying electrical base-load suitable for demand management. Larger companies such as these would have the capacity to advance heat pump timber drying as a mainstream technology. It is also possible that one of the main existing stakeholders, or a consortium of them, could support change, but this is less likely in view of their current commitment to combustion-heated dryers.

The government is another potentially influential stakeholder since it has interests in raising the productivity of the timber processing sector, reducing its atmospheric emissions, and ensuring it manages its current difficulties and emerging risks prudently. If the government decides to take action, it could influence the energy culture of the industry by offering targeted R&D assistance, supporting the marketing capacity of the timber firms, placing an appropriate price on carbon, supporting the biofuels industry, and tightening emissions requirements.

8.8 CONCLUSIONS

This chapter examines a heat pump timber drying initiative in New Zealand from two viewpoints: the application of the technology and its acceptability to the timber industry.

Considering the technology first, the Whangarei dryer aimed to improve the performance of heat pump timber dryers to meet the needs of firms drying Radiata pine. The results of this study provide insights into opportunities for the technology in the New Zealand softwood processing industry. The primary conclusions are:

- The airflow reversal scheme operated successfully and the cost of equipment failures was reasonable.
- The energy use per cubic meter and the drying time were determined primarily by the initial moisture content of the timber.
- The cost of drying was essentially the same as a combustion-heated dryer using an advanced conventional temperature schedule.

- Opportunities for improving the drying speed and efficiency of the dryer include: better control of air bypassing the timber stack, better control of heat-rejection, the use of a progressive configuration, use of liquid subcooling, and further optimizing the heat exchangers and airflow rates.
- With the improvements proposed, a heat pump dryer could dry 50 mm Radiata pine from 100% to 10% moisture content using less than 100 kWh/m³ in less than 110 h.

At the industry level, the evidence indicates that the commercial success of New Zealand timber processing firms currently depends more on their marketing capacity than on their choice of drying technology. At present, however, the industry is in difficulty. It is not growing and it is facing downward pressure on commodity timber prices. There is also increasing demand for sawmill residues, there is a possibility of tighter emissions standards, and there is increasing global demand for higher energy efficiency. In this situation, the dominance of combustion-heated dryers raises the risk the industry will not recover from its present difficulties, and will be technologically stranded while locked into low-value markets.

Of course there are good reasons for the industry to retain combustion-heated dryers in its technology arsenal. But this is a mature technology, whereas heat pump drying is at an earlier stage of development with a clear pathway toward higher productivity and lower costs. The technology also carries the prospect of continuing learning bonuses as industrial experience expands. Thus, it is not desirable for timber firms to remain committed to their traditional drying technology, rather than engaging with another that offers higher productivity and improved resilience.

The existence of smaller firms supplying and using heat pump dryers indicates that the culture of combustion-heated dryers is being contested, although they are unlikely to achieve rapid change by themselves. Change will occur faster if they collaborate with larger independent processing firms intent on stimulating the New Zealand softwood sector for their own business interests. Alternatively, an electricity supplier or a demand aggregator might engage with the industry with a view to create new opportunities in the electricity market. It is less likely that an existing stakeholder, or stakeholder consortium, will promote change. The government could improve the prospects of the sawn wood sector by supporting industry R&D and helping it to build its marketing skills. Industry change would be further promoted by pricing carbon emissions appropriately, encouraging the market for wood residues, and setting tighter emissions requirements.

In a nutshell, it appears that heat pump softwood drying is already a viable technology that offers new opportunities for the ailing New Zealand wood-processing sector. It also has the potential for ongoing development. Whether or not the suggestions put forward in this chapter are useful remains to be seen.

ACKNOWLEDGMENTS

The author is grateful to Dr. E.W. Scharpf, Dr. J.R. Stephenson, and Professor Z.F. Sun, who made important contributions to the work presented in this chapter. Dr. Scharpf kindly read a draft and made a number of valuable suggestions. He also thanks many collaborators for their support: Mr. J. Anderson, Dr. P. Baines, Dr. P. Bannister, Dr. M. Bell, Mr. M. Burnett, Dr. G. Chen, Dr. I. Cox-Smith, Mr. D. Croft, Dr. C. Davis, Mr. A. Firth, Dr. R. Ford, Mr. A. Haslett, Professor R. Lawson, Professor S. Pang, Dr. I. Simpson, Mrs. Q. Sun, and Dr. M. Van der Pal. The author is grateful to the former New Zealand Foundation for Research Science and Technology and the University of Otago Research Committee for research funding. The cooperation of New Zealand timber processing firms, and the support of the Delta S Technologies shareholders is gratefully acknowledged.

REFERENCES

Alexiadis, P. 2003. Kiln drying problems and issues in Canada: Benchmarks and comparisons to Europe. MSc thesis, University of British Columbia, Vancouver, British Columbia, Canada.

Ananias, R.A., J. Ulloa, D.M. Elustondo, C. Salinas, P. Rebolledo, and C. Fuentes. 2012. Energy consumption in industrial drying of radiata pine. *Drying Technology*, 30:774–779.

Baines, P.G. and C.G. Carrington. 1988. Analysis of Rankine cycle heat pump driers. *International Journal of Energy Research*, 12:495–510.

Bannister, P., G. Carrington, and G. Chen. 2002. Heat pump dehumidifier drying technology—Status, potential and prospects. In *Proceedings of the 7th International Energy Agency Conference: Heat pumps – Better by Nature,* Beijing, China, May 19–22. IEA Heat Pump Centre, c/- SP Technical Research Institute of Sweden, Borås, Sweden, pp. 219–230.

Bannister, P., G. Carrington, and Q. Liu. 1995. Influence of enhancing features on dehumidifier performance: Laboratory measurements. *International Journal of Energy Research*, 19:397–406.

Bell, M., G. Carrington, R. Lawson, and J. Stephenson. 2014. Socio-technical barriers to the use of energy-efficient timber drying technology in New Zealand. *Energy Policy*, 67:744–755.

Burnett, M.D., C.G. Carrington, E.W. Scharpf, and Z.F. Sun. 2015. Operational performance of a batch-mode dehumidifier timber dryer. *Drying Technology*, 33:455–465.

Carrington, C.G. and P. Bannister. 1996. An empirical model for a heat pump dehumidifier drier. *International Journal of Energy Research*, 20:853–869.

Carrington, C.G., P. Bannister, and Q. Liu. 1995. Performance analysis of a dehumidifier using HFC134a. *International Journal of Refrigeration*, 18:477–485.

Carrington, C.G., P. Bannister, and Q. Liu. 1996. Performance of a scroll compressor with R134a at medium temperature heat pump conditions. *International Journal of Energy Research*, 20:733–743.

Carrington, G., Z.F. Sun, and P. Bannister. 2000a. Dehumidifier batch drying—Effect of heat-losses and air leakage. *International Journal of Energy Research*, 24:205–214.

Carrington, G., Z.F. Sun, P. Bannister, and G. Chen. 2002a. Opportunities for dehumidifier dryers in the timber industry. In *Proceedings of the IRHACE Technical Conference*, Christchurch, New Zealand, April 26. Institute of Refrigeration, Heating, and Air Conditioning Engineers of New Zealand, Auckland, New Zealand, pp. 64–69.

Carrington, C.G., Z.F. Sun, Q. Sun, P. Bannister, and G. Chen. 2000b. Optimizing efficiency and productivity of a dehumidifier batch dryer. Part 1—Capacity and airflow. *International Journal of Energy Research*, 24:187–204.

Carrington, C.G., Z.F. Sun, Q. Sun, P. Bannister, and G. Chen. 1999. Dehumidifier dryers for hard-to-dry timbers. In *Refrigeration into the Third Millennium, Proceedings of the 20th International Congress of Refrigeration*, Sydney, New South Wales, Australia, September 19–24. International Institute of Refrigeration, Paris, France. Paper 548, pp. 2718–2725.

Carrington, C.G., C.M. Wells, Z.F. Sun, and G. Chen. 2002b. Use of dynamic modelling for the design of batch-mode dehumidifier driers. *Drying Technology*, 20:1645–1657.

Ceylan, İ., M. Aktaş, and H. Doğan. 2007. Energy and exergy analysis of timber dryer assisted heat pump. *Applied Thermal Engineering*, 27:216–222.

Chen, G., P. Bannister, C.G. Carrington, P. Ten Velde, and F.C. Burger. 2002. Design and application of a dehumidifier dryer for drying pine cones and pollen catkins. *Drying Technology*, 20:1633–1643.

Cooper, G. 2003. Methods of reducing the consumption of energy on wood drying kilns. Client report number 211-915 prepared for the Building Research Establishment, Construction Division, Watford, UK.

Cox-Smith, I.R., P.G. Baines, and C.G. Carrington. 1987. Experiences with heat pump timber driers. In *Proceedings of the XVIIth International Congress of Refrigeration*, Vienna, Austria, August 24–29. International Institute of Refrigeration, Paris, France, Vol. V, pp. 207–212.

Cox-Smith, I.R. and C.G. Carrington. 1991. Performance of commercially operated heat pump dehumidifiers for timber drying. In *Proceedings of the XVIII International Congress of Refrigeration*, Montréal, Quebec, Canada, August 10–17. International Institute of Refrigeration, Paris, France, pp. 1587–1591.

EECA. 2005. Summary of 9 Sawmill Energy Audits. Energy Efficiency and Conservation Authority, Wellington, New Zealand.

Geels, F.W. 2004. From sectoral systems of innovation to socio-technical systems: Insights about dynamics and change from sociology and institutional theory. *Research Policy*, 33(6–7):897–920.

Geeraert, B. 1976. Air drying by heat pumps with special reference to timber drying. In *Heat Pumps and Their Contribution to Energy Conservation*, E. Camatini and T. Kester (eds.). NATO Advanced Study Institute, Series E, Applied Sciences, No. 15. Noordhoff, Leyden, the Netherlands, pp. 219–246.

Hall, P. and M. Jack. 2008. Bioenergy options for New Zealand: Pathways analysis. Scion Energy Group Report, Scion, Rotorua, New Zealand.

Haslett, T. and M. Dakin. 2001. Assessment of potential for dehumidifier drying to reduce internal checking, collapse and kiln brown stain in radiata pine sawn lumber. Research Report, Forest Research, Rotorua, New Zealand.

Kreber, B. and A.N. Haslett. 1997. A study of some factors promoting kiln brown stain formation in radiata pine. *Holz als Roh- und Werkstoff* 55(2–4):215–220.

Laytner, F. and C.G. Carrington. 1994. Dehumidification drying revisited. *NZ Forest Industries* 25(11):31–32.

MBIE. 2012. Greenhouse gas emissions report, greenhouse gas emissions tables. Ministry of Business, Innovation and Employment, Wellington, New Zealand.

MBIE. 2013. Historic electricity prices. Ministry of Business, Innovation and Employment, Wellington, New Zealand.

McCurdy, M., S. Pang, and R. Keey. 2004. Drying schedule optimization for producing bright colored softwood timber. In *Drying 2004—Proceedings of the 14th International Drying Symposium*, São Paulo, Brazil, August 22–25. Vol. B, pp. 1360–1368.

McDonald, A.G., P.H. Dare, J.S. Gifford, D. Steward, and S. Riley. 2002. Assessment of air emissions from industrial kiln drying of *Pinus radiata* wood. *Holz als Roh- und Werkstoff* 60(3):181–190.

Ministry of Economic Development. 2012. New Zealand energy data file: 2011 calendar year edition. Ministry of Economic Development, Wellington, New Zealand.

Minea, V. 2010. Improvements in high temperature drying heat pumps. In *Proceedings of the 13th International Refrigeration and Air Conditioning Conference at Purdue*, West Lafayette, IN, July 12–15, Paper 2372.

Minea, V. 2013. Drying heat pumps—Part I: System integration. *International Journal of Refrigeration* 36:643–658.

Morrison, T. 2014. Stars align for NZ foresters as 'wall of wood' comes on stream, prices reach record highs. In *The National Business Review*, N. Gibson (ed.), Auckland, New Zealand, February 12.

MPI. 2015a. Production of rough sawn timber in New Zealand, 1970–2014. Estimated roundwood removals from New Zealand forests, year ended 31 March, 1951 to 2014. Ministry of Primary Industries, Wellington, New Zealand.

MPI. 2015b. Exports of Forestry Products from New Zealand years ended 30 June 1981 to 2014. Ministry of Primary Industries, Wellington, New Zealand.

NBR. 2008. Big production loss seen on timber horizon. In *The National Business Review*, Auckland, New Zealand, June 6.

Pang, S., H.F. Ngu, P. Davison, and J. Slovak. 2006. New instrument for measurement of kiln emissions in drying of softwood timber and result analysis. In *Drying 2006—Proceedings of the 15th International Drying Symposium*, Budapest, Hungary, August 20–23. Vol. B, pp. 802–807.

Pearson, H. and A.N. Haslett. 2002. Evaluation of effectiveness of water spray stress relief after dehumidification drying. Research Report, Forest Research, Rotorua, New Zealand.

Perré, P. and R. Keey. 2006. Drying of wood: Principles and practices. In *Handbook of Industrial Drying*, Arun S. Mujumdar (ed.), 3rd ed. CRC Press/Taylor & Francis Group, Boca Raton, FL, pp. 821–872.

Ritchie, J. and J. Lewis. 2003. *Qualitative Research Practice: A Guide for Social Science Students and Researchers*. London, UK: Sage.

Rotmans, J., R. Kemp, and M. van Asselt. 2001. More evolution than revolution: Transition management in public policy. *Foresight* 3(1):15–31.

Scharpf, E.W. and C.G. Carrington. 2006a. Heat pump drier with improved control. New Zealand patent #524470.

Scharpf, E.W. and C.G. Carrington. 2006b. Heat pump drier with improved efficiency. New Zealand patent #524469.

Scharpf, E.W., C.G. Carrington, and Z.F. Sun. 2006. Dehumidifier drier with reversible airflow. New Zealand patent #524471.

Simpson, I. and N. Haque. 2007. Effect of airflow reversal on drying quality in dehumidifier dryers, Progress report 180507. Scion, New Zealand.

Simpson, I.G. 2004. Comparison of a closed cycle dehumidifier with conventional dehumidifiers. Project No. A15216. New Zealand Forest Research Institute Ltd (Scion), Rotorua, New Zealand.

Stephenson, J., B. Barton, C.G. Carrington, D. Gnoth, R. Lawson, and P. Thorsnes. 2010. Energy cultures: A framework for understanding energy behaviours. *Energy Policy* 38:6120–6129.

Sun, Z.F. and C.G. Carrington. 1999a. Dynamic modelling of a dehumidifier wood drying kiln. *Drying Technology* 17:731–743.

Sun, Z.F. and C.G. Carrington. 1999b. Effect of stack configuration on wood drying processes. In *Proceedings of the Sixth IUFRO International Wood Drying Conference*, Stellenbosch, South Africa, January 25–28. pp. 89–98.

Sun, Z.F., C.G. Carrington, J.A. Anderson, and Q. Sun. 2004. Air flow patterns in dehumidifier wood drying kilns. *Transactions of the Institution of Chemical Engineers, Chemical Engineering Research and Design* 82(A10):1344–1352.

Sun, Z.F., C.G. Carrington, and P. Bannister. 1999. Validation of a dynamic model of a dehumidifier wood drying kiln. *Drying Technology* 17:731–743.

Sun, Z.F., C.G. Carrington, and P. Bannister. 2000. Dynamic modelling of the wood stack in a wood drying kiln. *Transactions of the Institution of Chemical Engineers* 78A:107–117.

Sun, Z.F., C.G. Carrington, C.P. Davis, Q. Sun, and S. Pang. 2003. Drying radiata pine timber under dehumidifier conditions: Comparison of modelled results with experimental results. In *Proceedings of the 8th International IUFRO Wood Drying Conference*, Brasov, Romania, August 24–29. Transylvania University of Brasov, Brasov, Romania. pp. 39–44.

Van der Pal, M., E. Scharpf, Z. Sun, and G. Carrington. 2005. Performance of an industrial dehumidifier timber dryer. In *Proceedings of the 3rd Nordic Drying Conference*, Karlstad, Sweden, June 15–17. University of Karlstad, Karlstad, Sweden. Session S-V, paper P-32.

Index

Note: Page numbers followed by f and t refer to figures and tables, respectively.

Printed and bound by CPI Group (UK) Ltd, Croydon, CR0 4YY

01/11/2024

01782601-0003